EARTH SCIENCES IN THE 21ST CENTURY

RUTILE

PROPERTIES, SYNTHESIS AND APPLICATIONS

EARTH SCIENCES IN THE 21ST CENTURY

Additional books in this series can be found on Nova's website under the Series tab.

Additional E-books in this series can be found on Nova's website under the E-book tab.

EARTH SCIENCES IN THE 21ST CENTURY

RUTILE

PROPERTIES, SYNTHESIS AND APPLICATIONS

IT-MENG (JIM) LOW
EDITOR

Nova Science Publishers, Inc.
New York

For permission to use material from this book please contact us:
Telephone 631-231-7269; Fax 631-231-8175
Web Site: http://www.novapublishers.com

Library of Congress Cataloging-in-Publication Data

Low, It-Meng.
Rutile : properties, synthesis and applications / Jim Low.
p. cm.
Includes index.
ISBN 978-1-61942-233-9 (hbk.)
1. Rutile. 2. Titanium--Metallurgy. I. Title.
TN799.T5.L69 2011
669'.7322--dc23

2011045792

Published by Nova Science Publishers, Inc. † New York

CONTENTS

PREFACE

This book deals with the synthesis, properties and applications of titanium dioxide (TiO_2) which is a naturally occurring oxide of the element titanium. In nature these oxides are found in well-known minerals such as Rutile, Anatase and Brookite. However, it is most commonly extracted from titanium tetrachloride by carbon reduction and re-oxidization. Alternatively, it may be processed from another oxide called ilmenite, which is subjected to reduction with sulphuric acid to achieve pure titanium dioxide.

The most common form is rutile which is also the equilibrium phase at all temperatures. At elevated temperature, both metastable anatase and brookite phases convert to rutile. Six coordinated titanium ions are found in rutile, anatase and brookite where they all have the same composition but with different crystal structures. TiO_2 is a multifaceted compound and it has been studied for its interesting qualities, such as electronic, optical, magnetic, photo-catalytic and electrochemical properties where it has been very widely used in pigments, catalysts and sensors.

There are a number of industrial applications for this mineral. In particular, it has very high refraction properties and is often included in many cosmetic preparations to reflect light away from the human skin. It is also a major component of sun screen to minimise the harmful effects of ultraviolet (UV) rays from the sun which is the main cause of skin cancer. As a pigment, titanium dioxide is used to enhance the white color of certain foods, such as dairy products and candy. It also lends brightness to toothpaste and some medications. However, it is also used as a food additive and flavor enhancer in a variety of non-white foods, including dried vegetables, nuts, seeds, soups, and mustard, as well as beer and wine. Since titanium dioxide reflects light so well, it is ideal for use as a protective coating for many products, such as automobile parts and optical mirrors.

With the advent of nanotechnology, nanostructured TiO_2 has been used as photocatalysts for fascinating applications such as solar-hydrogen production, waste water treatment, gas sensors and dye-sensitized solar cells. In particular, there is great interest in using TiO_2 nanotubes or nanofibres because of its high photoactivity, photodurability, chemical and biological inertness, mechanical robustness and low cost. Since photocatalytic reactions mainly take place on the surface of the catalyst, a high surface-to-volume ratio is of great significance for increasing the decomposition rate. Very high photocatalytic activity has already been demonstrated for nanostructured TiO_2 with various morphologies such as nano-powders, nano-rods, nano-wires, nano-fibers, nano-belts, nano-tubes, thin films, and porous nanostructures. Although a lot of progress has been achieved in these forms of TiO_2, the poor recuperability and reutilization limitation for nano-powders and processing difficulty for nano-fibres or nano-tubes are still challenges for their commercial applications.

Hitherto, various processes of materials synthesis have been developed for nanostructured TiO_2 and these include the sol-gel method, micelles and inverse micelles method, hydrothermal method, and the solvothermal method. More recently, direct oxidation of anodized titanium foils to grow a thin surface coating of nano-structured TiO_2 has become quite popular. Using indirect oxidation methods, nanorod arrays have been prepared on the surface of Ti substrates by oxidizing titanium with $(CH_3)_2CO$, H_2O_2 solution, and KOH. However, few literature sources have reported the growth of aligned TiO_2 nanostructures on the substrate by electrochemical anodization of Ti metal and subsequent annealing in air. The growth of aligned TiO_2 nanostructures on the substrates may have potential for photoelectrochemical and optoelectronic applications.

However, due to its wide band gap energy, TiO_2 is active only under near-ultraviolet irradiation. Therefore, numerous researches have been carried out over the last 20 years to develop modified TiO_2 so that they are active under visible light irradiation (λ > 400 nm). One of the most studied methods is by doping the TiO_2 materials with metal ions (iron, nickel, vanadium, and chromium) or nonmetallic elements (nitrogen, sulphur and carbon). For instance, Cr or N-doping of TiO_2 can result in visible light response by virtue of a narrowing in the band gap due to the mixing of p states of nitrogen with O 2p states. Among these doping methods, doping with transitional metals is one of the most efficient methods. X-ray photoelectron spectroscopy (XPS) results showed that the doped samples contained N, C, B and F elements and the doped TiO_2 showed the shift in the band gap transition down to 2.98 eV. The photocatalytic activity of the doped TiO_2 was 1.61 times better than undoped TiO_2. Researches have also been carried out to dope bismuth into TiO_2 to enhanced the photocatalytic activities in these systems. It has been reported that the metal/TiO_2 nanostructures enhance the efficiency of photocatalysis in water-splitting and dye-sensitized solar cell (DSSC) DSSCs. Metals compounded on semiconductor materials increase charge-collection efficiency due to a much slower electron-hole recombination, giving rise to longer electron lifetime, which will result in an increasing interfacial electron-transfer process.

Hitherto, the doping of TiO_2 has been widely synthesized using the sol-gel method. However, ion-implantation has now emerged as an alternative but effective method to improve the separation of the photogenerated electron–hole pairs or to extend the wavelength range of the TiO_2 photoresponses into the visible region. XPS measurements revealed that the implanted nitrogen species were mainly interstitial ones. The nitrogen concentration was increased with increasing ion flux which could be controlled by adjusting the gas flow rate of the ion source, resulting in improved visible-light photocatalytic activities. Higher visible-light photocatalytic efficiency was achieved with higher implanted nitrogen concentration.

Nine peer-reviewed Chapters are presented in this book. Each chapter has been written by a leading researcher of international recognition on TiO_2. This book is concerned with the synthesis, characterisation, microstructure, properties, modelling and potential applications of TiO_2 phases such as anatase and ruitle. The synopsis of each Chapter is as follows:

Chapter 1: This chapter provides an overview of the photocatalytic activity of titanium dioxide (TiO_2). It is shown that TiO_2 exists in a large number of polymorphs that not only are the abundant rutile, anatase and brookite phases, but also the high-pressure columbite, baddeleyite and cotunnite (*c*-TiO_2) structures. Amongst such polymorphs, rutile is the thermodynamically stable one: anatase and brookite – both metastable phases – transform into rutile – stable phase – upon heating. The photocatalytic activity of TiO_2 can be enhanced by doping with certain elements (e.g. N, S, Cr. Bi) to reduce its bandgap for response in visible-light.

Chapter 2: This chapter reviews and discusses the stability of the different phases of TiO_2, the transformation of the metastable forms to rutile, and the methods through which this transformation can be enhanced or impeded. With the exception of various high-pressure phases, rutile is the equilibrium phase of titanium dioxide at all temperatures and is generally the preferred phase for pigmentary applications. As a result of their more rapid formation kinetics, the metastable phases anatase and, to a lesser extent, brookite, which exhibit photocatalytic activity superior to that of rutile, are the first phases formed during many production routes. At low temperatures, the metastable phases are energetically stabilised and transform to rutile through thermal treatment. The onset temperature of the anatase (or brookite) to rutile phase transformation and the rate at which it proceeds can be affected significantly by dopants, firing atmosphere, microstructure, sample morphology, and the presence of impurities in the material.

Chapter 3: This chapter discusses the various rutile nanostructures of pure and Fe^{3+}-doped titania which were synthesized by co-precipitation in acidic aqueous solution. The synthesized TiO_2 was characterized by structural (XRD), morphological (TEM and HRTEM), optical (UV/vis absorption, reflection, and photoluminescence), and analytical (XPS and ICP-OES) techniques. Also, photocatalytic activity measurements were performed, using photodegradation of specific herbicides as model systems. Applied synthetic procedure resulted in formation of flower like agglomerates of 200-400 nm in diameter. Agglomerates represent assembly of nanorods consisted of "chains" of spherical particles, 5-7 nm in diameter, most likely interconnected through the so-called oriented attachment or grain-rotation induced-grain coalescence (GRIGC) process. The presence of dopant iron ions didn't distract morphology of rutile nanoparticles, nor agglomerates, but significantly changed optical properties of titania compared to pure material. The photoluminescence of pure and Fe-doped rutile nanoparticles is characterized with several well-resolved peaks extending in the visible spectral region. Photon energy up-conversion from rutile TiO_2 nanoparticles was also observed at low excitation intensities. The energy of up-converted photoluminescence spans the range of emission of normal photoluminescence. The explanation of photon energy up-conversion involves mid-gap energy levels originating from oxygen vacancies.

Chapter 4: This Chapter reports the use of sol spray hydrolysis to synthesize hollow micro-spherical anatase and rutile particles with mesoporosity using $TiCl_4$ as sole precursor. The wall of the hollow micro-sphere is constituted of nanoparticles and mesoporosity. Its formation mechanism can be attributed to the spray dry process and the property of $TiCl_4$ aqueous sol. The optical property of hollow sphere mesoporous titania is better than that of nano-anatase and nano-rutile in aqueous suspensions, and related to its crystal phase, specific surface area, crystal size, porosity size and crystallinity of titania, and these characteristics are changing as the calcined temperature increasing. The merit of this approach is template free and without any additives, organic or inorganic. The sol-spray hydrolysis approach can be used to prepare other materials with the same structure at macroscale.

Chapter 5: The comparative analysis of experimental data over titanium dioxide powders prepared by various ways under ambient air and investigated in conditions close to those of the real troposphere is discussed in this chapter. The results of TiO_2 prepared by high-temperature heating of anatase, produced by burning of titanium micro particles and grinding of rutile crystal are used for comparison. It is shown that Earth's atmosphere ecology can be keep in safe level in the result of photocatalytic and photoadsorption processes proceeding on aerosol particle surface under troposphere solar irradiation. In addition, it is possible to clean the atmosphere near a pollution source by the production of titania aerosol in direct place of a contamination. The concept of oxidative photocatalysis as nature phenomenon is also developed which suggests that photocatalytic reactions proceed only under light quanta from the fundamental absorption band of photocatalyst; oxygen photoadsorption is not a limiting step of photocatalytic oxidation reaction; photocatalytic activity is only inherent in metal oxides, the top level of whose valence band is formed by 2p - orbital of oxygen; during of photocatalytic oxidation reaction the oxidized gas-phase molecule interacts with surface lattice oxygen at the moment of its discharge by a free mobile hole. The quantum efficiency of photocatalytic reaction mainly depends upon the bond energy of surface lattice oxygen; the quantum yield is also determined by physicochemical characteristics of photocatalysts. In agreement with the proposed mechanism the photocatalytic oxidation reaction does not require an active surface center.

Chapter 6: This chapter reports the preparation, morphology and catalytic activity of rutile as superacid catalysts. In recent years, heterogeneous catalysts have begun to replace highly corrosive, hazardous and polluting homogenous catalysts. The replacement of liquid acids by solid superacids as catalysts is expected to ease their separation from reacting mixture, allowing continuous operation as well as regeneration and neutralization of the catalyst and lowering the cost of process installation and maintenance. Metal oxides such as SnO_2, TiO_2, VO_2 and ZrO_2 have been used as a precursor of solid superacids preparation. Among them, TiO_2 and TiO_2-based materials are of importance for sensors, pigments and photocatalysis. These materials have been shown to exhibit higher acidity than concentrated H_2SO_4 and their acidity and morphology can vary depending on the synthetic methods. They can also be used for several acidic reactions e.g. alkylation, isomerization, estrification, cracking and condensation.

Chapter 7: This chapter reviews the essential role of surfaces in physico-chemical processes and identify the structure of the surface layer within the thickness of several nanometers. One of the effective physico-chemical techniques to evaluate the electronic state of the solid materials surface is the Electronic Spectroscopy of Diffusion Reflection (ESDR) in the UV-Vis-spectrum. On the example of titanium dioxide (rutile) of various origins and two-phase pyrogenic titanium dioxide P25 (Degussa) was demonstrated the possibility of mathematical processing of ESDR-spectra to determine the coordination state of polyhedron with different spatial environment of titanium atoms. The results of mathematical processing of experimental spectra are compared with other methods of ESDR-spectra presentations (definition of the absorption edge by the method of tangents, the representation in coordinates of Gurevich-Kubelka-Munk (F_R), $(\alpha h\nu)^m \propto (h\nu - E_g)$, $(F_R\, h\nu)^m \propto (h\nu - E_g)$) and hardware differentiation of the spectra. It is shown that the technological prehistory and chemical modification of surface of industrial dispersed TiO_2 having a significant impact on the coordination state of titanium atoms on the titanium dioxide surface.

Chapter 8: This chapter reviews the theoretical investigation of transition-metal (TM) doped rutile TiO_2 system (TM = Mn, Fe, Co, Ni, and Cu) for different concentrations and distributions of these impurities. The effect of oxygen vacancies on the magnetic moments and the magnetic alignment for pure and doped TiO_2 is studied as a function of vacancy concentration and distribution. The experimental work done by the authors using X-ray absorption, magnetometric techniques and Mössbauer spectroscopy is also reviewed. The combined theoretical and experimental studies have enabled them to characterize the local structure around the TM impurities in rutile TiO_2 unambiguously and support the conclusion that oxygen vacancies are fundamental for the ferromagnetic response of these systems, explaining the diversity of magnetic moments observed experimentally for samples grown in different conditions.

Chapter 9: A different method for the production of DSA®-type electrodes that contain different amounts of RuO_2, TiO_2, and Ta_2O_5, and to describe their morphology and physical-chemical characterization is described in this work. The performance of these electrodes as anode for organic compounds oxidation is presented. On the basis of the data obtained by the authors, this chapter focuses on recent their results pertaining to the preparation of ruthenium-based electrodes for application in the electrochemical degradation of toxic organic compounds. Thermally prepared ruthenium, titanium, and tantalum metallic oxides are natural candidates when properties such as high corrosion resistance and physical-chemical stability under high positive potentials are required. DSA®-type electrodes are frequently prepared by either thermal decomposition of the precursor metal salts or thermal decomposition of polymeric precursors.

Chapter 10: This chapter provides an overview of the origin, mineral chemistry and processes controlling the trace element composition of natural multi-doped hydrothermal rutile. The performance of the natural-multi-doped rutile in photocatalytic oxidation cells (POC), microbial fuel cells (MFC) and other applications has not been entirely evaluated, even when it could be a promising alternative considering the broad spectrum of different dopants present in this mineral phase combined with its abundance in nature and low recovery costs. Natural multi-doped rutile can potentially be recovered, using existing low-cost technologies, as a by-product from large-scale mining operations, such as those associated with giant world-class porphyry Cu-(Mo-Au) deposits. The economic importance of porphyry-type ore deposits as a source for natural rutile and its recovery as a by-product are also briefly discussed. Finally, an outline of the hydrothermal rutile applications as semiconductor highlights the necessity for further mineralogical studies on natural catalysts, a new developing research field in Earth Sciences.

RECOMMENDED READINGS

Asahi, R., *et al.*, Visible-Light Photocatalysis in Nitrogen-Doped Titanium Oxides, *Science* 293 (2001) 269-271.

Fujishima, A. and Honda, K., Electrochemical Photolysis of Water at a Semiconductor Electrode. *Nature*, 238 (1972) 37-38

Grätzel, M., Photoelectrochemical Cells. *Nature,* 414 (2001) 338-344.

Grimes, C.A., *et al.*, *Light, Water, Hydrogen: The Solar Generation of Hydrogen by Water Photoelectrolysis*. 2008, New York: Springer.

Hoffmann, M.R., S.T. Martin, W. Choi and D.W. Bahnemann, Environmental Applications of Semiconductor Photocatalysis, *Chem. Rev.* 95 (1995) 69-96.

Nowotny, J., *et al.*, Titanium Dioxide for Solar-Hydrogen I. Functional properties. *Int. J. Hydrogen Energy*, 32 (2007) 2609-2629

Taga, Y., Titanium Oxide Based Visible Light Photocatalysts: Materials Design and Applications, *Thin Solid Films*, 517 (2009) 3167-3172. www.wisegeek.com/what-is-titanium-dioxide.htm).

In: Rutile: Properties, Synthesis and Applications
Editor: It-Meng (Jim) Low

ISBN 978-1-61942-233-9

Chapter 1

TITANIUM DIOXIDE AND RUTILE: AN OVERVIEW AND ITS PHOTOCATALYTIC ACTIVITY

David Maria Tobaldi*

Ceramics and Glass Engineering Department, CICECO,
University of Aveiro, Aveiro, Portugal

ABSTRACT

Titanium dioxide (TiO_2) exists in a large number of polymorphs that not only are the abundant rutile, anatase and brookite phases, but also the high-pressure columbite, baddeleyite and cotunnite (*c*-TiO_2) structures. Amongst such polymorphs, rutile is the thermodynamically stable one: anatase and brookite – both metastable phases – transform into rutile – stable phase – upon heating.

Anatase (tetragonal, space group $I4_1/amd$) crystal structure is constituted by distorted TiO_6 octahedral sites sharing four corners; the structure of brookite (orthorhombic, space group *Pbca*, the structure with lower symmetry among the counter-polymorphs), is also made by distorted octahedral sites elongated along *c*-axis and each octahedron shares three edges with adjacent octahedra; similarly, the structure of rutile (tetragonal, space group $P4_2/mnm$) is made by chains of TiO_6 octahedra that share a vertex along *c*-axis. The anatase-to-rutile phase transformation (A → R) – exothermal and irreversible – follows a nucleation-growth mechanism.

Titanium dioxide is broadly used in every day life applications, such as paintings, ceramic pigment, electrochemical electrodes, solar cells or whitener agent for food. More recently, titanium dioxide has definitely been the most investigated material for photocatalytic applications. Though anatase is considered – by most of the scientific research – as the photocatalytically active phase among the TiO_2 polymorphs, recent research works state rutile as an excellent photocatalyst as well.

* E-mail address: david.tobaldi@ua.pt

Commentary

Titanium dioxide is widely used in every day life applications such as paintings, ceramic pigments, electrochemical electrodes, solar cells or whitener agent for food [1-4]. In more recent years, titanium dioxide has been the most investigated material for photocatalytic applications, not only for outdoor and indoor air purification [5-6] but also in the decontamination of polluted waters [7].

Titanium dioxide (TiO_2) exists in a large number of polymorphs – not only the well known rutile, anatase and brookite phases [8], but also the high-pressure columbite [9], baddeleyite [10] and cotunnite (c-TiO_2) structures [11]. Amongst such polymorphs, the most occurring and best known TiO_2 phases are: rutile, anatase and brookite. Rutile is the thermodynamically stable one: anatase and brookite – both metastable phases – transform into rutile – stable phase – upon heating. It is widely accepted that the phase transformation brookite → rutile occurs in the thermal range of 500-700 °C [12], whilst anatase transforms into rutile from 600 to 1100 °C [13].

Anatase (tetragonal, space group $I4_1/amd$) crystal structure is constituted by distorted TiO_6 octahedral sites sharing four corners [14]; the structure of brookite (orthorhombic, space group *Pbca*, the structure with lower symmetry among the counter-polymorphs), is also made by distorted octahedral sites elongated along *c*-axis and each octahedron shares three edges with adjacent octahedral [15]; similarly, the structure of rutile (tetragonal, space group $P4_2/mnm$) is made by chains of TiO_6 octahedra that share a vertex along *c*-axis [16]; see Table 1 and Figure 1 for data and the structure of the three major titania polymorphs.

The anatase-to-rutile (A → R) phase transformation – exothermal and irreversible – follows a nucleation-growth mechanism [18]. Upon heating, anatase crystallites increase their size till they reach a critical dimension (r_c) and the A → R transformation begins [19]. The polymorphic reaction occurs through a cooperative movement of Ti and O atoms after the breaking of anatase Ti–O bonds [20].

a) b) c)

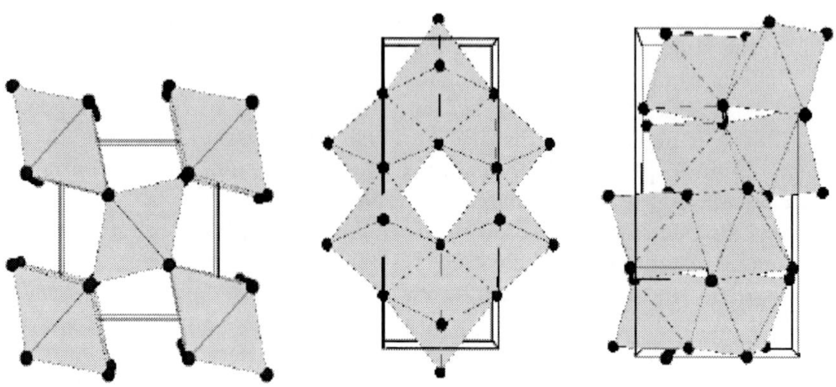

[a] Structure of rutile, along plane *ab*.
[b] Structure of anatase, along plane *ac*.
[c] Structure of brookite, along plane *ab*.

Figure 1. Structure of rutile, anatase and brookite. The crystal structures were obtained using the software DRAWxtl [17].

Table 1. Data for rutile, anatase and brookite

Structure	Space group	Density, g/cm^3	Unit cell parameters, Å	Reference
Rutile	$P4_2/mnm$	4.25	a=b=4.593 c=2.959	JCPDS 211276
Anatase	$I4_1/amd$	3.89	a=b=3.785 c=9.513	JCPDS 211272
Brookite	*Pbca*	4.12	a=5.455 b=9.181 c=5.142	JCPDS 291369

The temperature of the A → R phase transformation is affected by several factors, such as: crystallite dimension [21], grain size [22], atmosphere [23] or the presence of dopants [24]: in such a case the polymorphic transformation is accelerated by doping TiO_2 with an ion having valence lower than 4+, because of the formation of oxygen vacancies. On the other hand, if the dopant has valence higher than 4+, the phase transformation retards due to the reduction of the oxygen vacancies and forming interstitial Ti^{3+}, that repress the atomic transfer in the anatase structure [25].

Considering the optical properties, Ti^{4+} has no valence *d*-electron; therefore it does not have absorption in the visible region of the spectrum. Nonetheless, a blue colour can easily be induced in synthetic rutile mixing titanium oxidation states: Ti^{4+} can enter into intervalence charge transfer with Ti^{3+} resulting in an absorption in the red to near-infrared [26].

By an electronic point of view, titania is a *n*-type semiconductor. The energy band gap E_g of anatase is equal to 3.2 eV, whilst rutile one is equal to 3.0 eV [27]. This entails that, in order to be used as photocatalysts, both anatase and rutile are activated by UV-light. This is the major drawback of titania as a photocatalyst, as it exploits 3–5% of the solar energy in the photocatalytic reaction, being transparent for the most of solar radiation [28]. Among these two titania polymorphs, anatase is considered to be more photoactive than rutile because of its stronger reducing power and a better hole-trapping ability [29]. Actually, for sake of clarity, rutile can absorb radiation which is near to visible-light [30], but its photocatalytic activity is lower than anatase one due to the above-mentioned reasons.

Therefore, though the majority of international scientific papers designate anatase as the only titania polymorph that is photocatalytically active, [31-32], more recent research works state also rutile as an active photocatalyst, being itself a semiconductor [33-36]. Actually, the powder Degussa P25, considered as one of the more active photocatalysts available, is composed by ~85 wt% of anatase and ~15 wt% of rutile [37].

In order to have a considerable photocatalytic activity, rutile must possess a high specific surface area and crystallinity. Actually, the lowering of the photocatalytic activity due to the A → R phase transformation, is to be ascribed more to variations of the specific surface area and porosity than to variations on the crystal structure, which depend on the production techniques [34]. Thus, the road to follow, in order to obtain photocatalytically active rutile, should be the one to modulate its preparation techniques and preparative conditions, since rutile absorbs light in the UVA region (360-400 nm) in a more effective way than anatase. Several attempts were made on the synthesis of rutile to make it a more effective photocatalyst. Kim *et al.* [38], synthesised rutile nanopowders with acicular shape and high specific surface area, have a superior photocatalytic activity. Due to such morphology, the photo-excited electrons and holes can easily and effectively participate in the photo-reaction without a large recombination that is otherwise due if in presence of a rounded morphology. Also according to Kobayashi *et al.* [39] the morphology of nano-rutile plays a basic role in its photocatalytic activity. They prepared highly anisotropic whisker-like rutile nanoparticles and, owing to this morphology, rutile nanoparticles showed a high photocatalytic activity. The road of modulate rutile morphology in order to make it an effective photocatalyst was followed by Zhang *et al.* [40] who synthesised rod-like rutile nanoparticles grown along [001] direction. They obtained rutile nanocrystals that exhibit higher photocatalytic activity than the commercially available powder Degussa P25.

Other morphologies to be taken into account, in order to improve titania photocatalytic activity, are nanotubes and nanofibers. Actually, it is possible to tailor the morphology of titania by anodization of Ti-foils [41]. Fang *et al.* [42] synthesised titanium oxide nanotubes – by electrolytic anodization of titanium electrodes – that showed to posses superior photocatalytic activity. Chen *et al.* [43], fabricated an array of titania nanotubes by such methodology. Titania nanofibers too can be synthesised via the anodic oxidation process [44].

As above-mentioned, the major drawback of titania employed as a photocatalyst is that the photocatalytic reaction is promoted by UV-light. There have been several challenges in order to override such a drawback, basically, doping titania – and therefore rutile – with transition metal ions or non-metal atoms: the doping of titania with such atoms should produce states in the band gap of TiO_2 that absorb visible-light [45].

Amongst transition metal ions, the most used to improve the visible-light activity of titania are: W^{6+}, Nb^{5+}, Cr^{3+}, Cu^{2+} and Fe^{3+}.

Tungsten and niobium doping improve titania photocatalytic activity due to an increase of the surface acidity: an excess of positive charges is determined by W^{6+} and Nb^{5+} cations [46-47]. Chromium doping promotes the charge separation of the photo-generated couple e^--h^+ [48] as well as photo-reduction reactions [49]. Cu-doping improves titania porosity and specific surface area, though the A → R transition temperature is considerably lowered [50]. Also Fe^{3+}-doping makes the A → R transition temperature lowering [51], but Fe-TiO_2 photocatalytic activity is enhanced because of oxygen vacancies on titania surface. Oxygen vacancies promote both water adsorption and the formation of hydroxyl surface groups, therefore the photocatalytic activity [52].

Transition metal ions doping revealed somewhat unsuccessful because, if on one hand absorption in the visible region is achieved, on the other the photocatalytic activity is reduced since dopants behave as electron-hole recombination centers: this way, they are no longer available for the photocatalytic reaction [53-54].

Among non-metal atoms, nitrogen is the most extensively used, even if the origin of its superior photocatalytic activity under visible-light irradiation is still controversial. Asahi *et al.* [45] attributed the visible-light-extended photocatalytic activity to the band gap narrowing due to a mixing of N p states with the O $2p$ states in titania valence band. In contrast, Di Valentin *et al.* [55] basing on DFT calculations, established that titania visible-light absorption was due not as much to a band gap narrowing, but rather to the formation of N $2p$ states just above the top of O $2p$ band, with a different behaviour whether on anatase or rutile. Several techniques are employed in order to synthesise titania with such a stoichiometry: $TiO_{2-x}N_x$. The most adopted of them are: annealing of the precursor TiO_2 under NH_3 flow [56]; nitridation of TiO_2 nanoparticle colloidal solution [57]; PVD via magnetron sputtering in a mixture of N_2 and O_2 atmosphere, followed by annealing [58]. Batzill *et al.* [59], doped rutile and anatase single crystal surfaces by low energy nitrogen ion-implantation: the formation of localized midgap states is proven by UV-photoelectron spectroscopy (UPS) studies on N-implanted TiO_2 single crystals.

N doping is able to exert a band gap narrowing until ~2.55 eV, therefore a visible-light absorption for $\lambda \leq 486$ nm. Recently, Liu and Gao [60] have synthesised sulphur and nitrogen co-doped rutile, that revealed to be a visible-light activated photocatalyst more effective than the singles S-TiO_2 and N-TiO_2, not only because of a shifting of the absorption edge of TiO_2 toward longer wavelength, but also because of two distinct optical absorptions in the visible-light region.

Conclusion

Titanium dioxide, and rutile thereafter, is widely used in every day life applications. In more recent years, titanium dioxide has been the most investigated material for photocatalytic applications. Since rutile is the thermodynamically stable phase amongst the TiO_2 polymorphs, could be interesting its photocatalytic application when treated at high temperature.

A future development of rutile as a photocatalyst should be the doping with transition metal ions or non-metal atoms in order to be activated by visible-light radiation. An attractive application in such a way should be the production of photocatalytically active building materials such as ceramic tiles, with the photocatalyst powder – activated by visible-light – dispersed on the surface.

References

[1] Kavan, L., Rathouský, J., Grätzel, M., Shklover, V., and Zukal, A. (2001). Mesoporous thin film TiO_2 electrodes. *Microporous and Mesoporous Materials*, 44–45, 653-659.

[2] Dondi, M., Cruciani, G., Guarini, G., Matteucci, F., and Raimondo, M. (2006). The role of counter-ions (Mo, Nb, Sb, W) in Cr-, Mn-, Ni- and V-doped rutile ceramic pigments: Part 2. Colour and technological properties. *Ceramics International*, 32, 393-405.

[3] Grätzel, M. (2003). Dye-sensitized solar cells. *J. Photochem. Photobiol. C Photochem. Rev.*, 4, 145–153.

[4] Meacock, G., Taylor, K. D. A., Knowles, M. J., and Himonides, A (1997). The improved whitening of minced cod flesh using dispersed titanium dioxide. *J. Sci. Food Agr.*, 73, 221–225.

[5] Chen, J. and Poon, C. S. (2009). Photocatalytic construction and building materials: from fundamentals to applications. *Build. Environ.*, 44, 1899–1906.

[6] Puddu, V., Choi, H., Dionysiou, D. D., and Li Puma, G. (2010). TiO_2 photocatalyst for indoor air remediation: influence of crystallinity, crystal phase, and UV radiation intensity on trichloroethylene degradation. *Appl. Catal. B Environ.*, 94, 211-218.

[7] Malato, S., Fernández-Ibáñez, P., Maldonado, M. I., Blanco, J., and Gernjak, W. (2009). Decontamination and disinfection of water by solar photocatalysis: recent overview and trends. *Catal. Today*, 147, 1–59.

[8] Muscat, J., Swamy, V., and Harrison, N.M. (2002). First-principles calculations of the phase stability of TiO_2. *Phys. Rev. B*, 65, 244112.

[9] Haines, J. and Leger, J. M. (1993). X-ray diffraction study of TiO_2 up to 49 GPa. *Phys. B Condens. Matter.*, 192, 233–237.

[10] Sato, H., Endo, S., Sugiyama, M., Kikegawa, T., Shimomura, O., and Kusaba, K. (1991). Baddeleyite-type high-pressure phase of TiO_2. *Science*, 251, 786–788.

[11] Dubrovinsky, L. S., Dubrovinskaia, N. A., Swamy, V., Muscat, J., Harrison, N. M., Ahuja, R., Holm, B., and Johansson, B. (2001). Materials science: the hardest known oxide. *Nature*, 410, 653–654.

[12] Li, J. G., and T. Ishigaki, T. (2004). Brookite rutile phase transformation of TiO_2 studied with monodispersed particles. *Acta Mat.*, 52, 5143-5150.

[13] Ovenstone, J., and Yanagisawa, K. (1999). Effect of hydrothermal treatment of amorphous titania on the phase change from anatase to rutile during calcinations. *Chem. Mater.*, 11, 2770-2774.

[14] Burdett, J. K., Hughbanks, T., Miller, G. J., Richardson, Jr, J. W., and Smith, V (1997). Structural-electronic relationships in inorganic solids: powder neutron diffraction studies of the rutile and anatase polymorphs of titanium dioxide at 15 and 295 K. *J. Am. Chem. Soc.*, 109, 3639-3646.

[15] Pauling, L., (1929). The principles determining the structure of complex ionic crystals. *J. Am. Chem. Soc.*, 51, 1010-1026.

[16] Swope, R. J., Smyth, J. R. and Larson, A. C. (1995). H in rutile-type compounds: I. Single-crystal neutron and X-ray diffraction study of H in rutile. *Am. Mineral.*, 80, 448-453.

[17] Finger, L. W., Kroeker, M. and Toby, B. H. (2007). DRAWxtl, an open-source computer program to produce crystal structure drawings. *J. Appl. Crystallogr.*, 40, 188-192.

[18] Shannon, R. D. and Pask, J. A. (1965). Kinetics of the anatase-rutile transformation. *J. Am. Ceram. Soc.*, 48, 391-398.

[19] Kumar, K. N. P. (1995). Growth of rutile crystallites during the initial stage of anatase-to-rutile transformation in pure titania and in titania-alumina composites. *Scripta Metall. Mater.*, 32, 873-877.

[20] Shannon, R. D. and Pask, J. A. (1964). Topotaxy in the anatase–rutile transformation. *Am. Mineral.*, 49, 1707-1717.

[21] Gribb, A. A. and Banfield, J. F. (1997). Particle size effects on transformation kinetics and phase stability in nanocrystalline TiO_2. *Am. Mineral.*, 82, 717-728.

[22] Ding, X. Z., Liu, X. H. and He, Y. Z. (1996). Grain size dependence of anatase-to-rutile structural transformation in gel-derived nanocrystalline titania powders. *J. Mater. Sci. Lett.*, 15, 1789-1791.

[23] Gamboa, J. A. and Pasquevich, D. M. (1992). Effect of chlorine atmosphere on the anatase–rutile transformation. *J. Am. Ceram. Soc.*, 75, 2934-2938.

[24] Matteucci, F., Cruciani, G., Dondi, M., and Raimondo, M. (2006). The role of counter-ions (Mo, Nb, Sb, W) in Cr-, Mn-, Ni- and V-doped rutile ceramic pigments. Part.1 Crystal structure and phase transformation. *Ceram. Int.*, 32, 385-392.

[25] Okada, K., Yamamoto, N., Kameshima, Y., Yasumori, A., and MacKenzie, K. J. D. (2001). Effect of silica additive on the anatase-to-rutile phase transition. *J. Am. Ceram. Soc.*, 84 1591-1596.

[26] Rossman, G. R. (1988). Optical Spectroscopy. In F.C. Hawthorne (Ed.), *Spectroscopic method in mineralogy* (207-254), Chelsea, MI: Mineralogical Society of America.

[27] Fujishima, A., Zhang, A. and Tryck, D. A. (2008). TiO_2 photocatalysis and related surface phenomena. *Surf. Sci. Rep.*, 63, 515-582.

[28] Karvinen, S. and Lamminmäki, R. J. (2003). Preparation and characterization of mesoporous visible-light-active anatase. *Solid State Sci.*, 5, 1159–1166.

[29] Augustynski, J. (1993). The role of the surface intermediates in the photoelectrochemical behaviour of anatase and rutile TiO_2. *Electrochim. Acta*, 38, 43–46.

[30] Li Puma, G., Bono, A., Krishnaiah, D., and Collin, J. G. (2008). Preparation of titanium dioxide photocatalyst loaded onto activated carbon support using chemical vapor deposition: a review paper. *J. Hazard. Mater.*, 157, 209–219.

[31] Tanaka, K., Capule, M. F. V. and Hisanaga T. (1991). Effect of crystallinity of TiO_2 on its photocatalytic action. *Chem. Phys. Lett.*, 187, 73-76.

[32] Sclafani, A. and Herrmann J. M. (1996). Comparison of the photoelectronic and photocatalytic activities of various anatase and rutile forms of titania in pure liquid organic phases and in aqueous solutions. *J. Phys. Chem.-US*, 100, 13655-13661.

[33] Sopyan, I., Watanabe, W., Murasawa, S., Hashimoto, K., and Fujishima, A. (1996). Efficient TiO_2 powder and film photocatalysts with rutile crystal structure. *Chem. Lett.*, 25, 69-70.

[34] Bacsa, R. R. and J. Kiwi, J. (1998). Effect of rutile phase on the photocatalytic properties of nanocrystalline titania during the degradation of p-coumaric acid. *Appl. Catal. B-Environ.*, 16, 19-29.

[35] Yang, S. and Gao, L. (2006). Synthesis and characterization of porous single-crystalline titanium dioxide nanorods. *J. Am. Ceram. Soc.*, 89, 720-723.

[36] Zhang, S., Liu, C. Y., Liu, Y., Zhang, Z. Y., and Mao, L. J. (2009). Room temperature synthesis of nearly monodispersed rodlike rutile TiO_2 nanocrystals. *Mater. Lett.*, 63, 127-129.

[37] Tobaldi, D. M., Tucci, A., Sever Škapin, A., and Esposito, L. (2010). Effects of SiO_2 addition on TiO_2 crystal structure and photocatalytic activity. *J. Eur. Ceram. Soc.*, 30, 2481-2490.

[38] Kim, S. J., Lee, E. G., Park, S. D., Jeon, C. J., Cho, Y. H., Rhee, C. K., and Kim, W. W. (2001). Photocatalytic effects of rutile phase TiO_2 ultrafine powder with high specific surface area obtained by a homogeneous precipitation process at low temperature. *J. Sol.Gel Sci. Techn.*, 22, 63-74.

[39] Kobayashi, M., Petrykin, V., Kakihana, M., and Tomita, K. (2009). Hydrothermal synthesis and photocatalytic activity of whisker-like rutile-type titanium dioxide. *J. Am. Ceram. Soc.*, 92, S21-S26.

[40] Zhang, S., Liu, C. Y., Liu, Y., Zhang, Z. Y., and Mao, L. J. (2009). Room temperature synthesis of nearly monodisperse rodlike rutile TiO_2 nanocrystals. *Mater. Lett.*, 63, 127-129.

[41] Xue, C., Zhang, F., Chen, S., Yin, Y., and Lin, C. Tailoring the surface morphology of TiO_2 nanotube arrays connected with nanowires by anodization. *Mater. Sci. Semicond. Process*, in press, doi:10.1016/j.mssp.2011.02.005.

[42] Fang, D., Luo, Z., Huang, K., and Lagoudas, D.C. (2011). Effect of heat treatment on morphology, crystalline structure and photocatalysis properties of TiO_2 nanotubes on Ti substrate and freestanding membrane. *Appl. Surf. Sci.*, 257, 6451-6461.

[43] Chen, Y. H., Huang, K. C., Chen, J. G., Vittal, R., and Ho, K. C. Titanium flexible photoanode consisting of an array of TiO_2 nanotubes filled with a nanocomposite of TiO_2 and graphite for dye-sensitized solar cells. *Electrochim Acta,* in press, doi:10.1016/j.electacta.2011.02.003.

[44] Beh, K. P., Yam, F. K., Tneh, S. S., and Hassan, Z. (2011). Fabrication of titanium dioxide nanofibers via anodic oxidation. *Appl. Surf. Sci.*, 257, 4706-4708.

[45] Asahi, R., Morikawa, T., Ohwaki, T., and Taga, Y. (2001). Visible-light photocatalysis in nitrogen-doped titanium oxides. *Science*. 293 269–71.

[46] Papp, J., Soled, S., Dwight, K., and Wold, A. (1994). Surface acidity and photocatalytic activity of TiO_2, WO_3/TiO_2, and MoO_3/TiO_2 photocatalysts. *Chem. Mater.*, 6, 496-500.

[47] Hirano, M. and Matsushima, K. (2006). Photoactive and absorptive niobium-doped anatase (TiO_2) nanoparticles: influence of hydrothermal conditions on their morphology, structure and properties. *J. Am. Ceram. Soc.*, 89, 110-117.

[48] Venezia, A. M., Palmisano, L. and Schiavello, M. (1995): Structural changes of titanium oxide induced by chromium addition as determined by an X-Ray diffraction study. *J. Solid State Chem.*, 114, 364-368.

[49] Palmisano, L., Augugliaro, V., Sclafani, A, and Schiavello, M. (1988). Activity of chromium-ion-doped titania for the dinitrogen photoreduction to ammonia and for phenol photodegradation. *J. Phys. Chem.-US*, 92, 6710-6713.

[50] Colón, G., Maicu, M., Hidalgo, M. C., and Navío, J. A. (2006). Cu-doped TiO_2 systems with improved photocatalytic activity. *App. Catal. B-Environ.*, 67, 41-51.

[51] Zhu, J., Chen, F., Zhang, J., Chen, H., and Anpo, M. (2006): Fe^{3+}-TiO_2 photocatalysts prepared by combining sol-gel method with hydrothermal treatment and their characterization. *J. Photoch. Photobio. A*, 180, 196-204.

[52] Hirano, M., Joji, T., and Inagaki, M. (2004). Direct formation of iron(III)-doped titanium oxide (anatase) by thermal hydrolysis and its structural property. *J. Am. Ceram. Soc.*, 87, 35-41.

[53] W. Choi, A. Termin and M.R. Hoffmann, The role of metal ion dopants in quantum-sized TiO_2: correlation between photoreactivity and charge carrier recombination dynamics. J. Phys. Chem., 98, 13669–13679, 1994.

[54] N. Serpone, D. Lawless, J. Disdier, and J. M. Herrmann, Spectroscopic, photoconductivity, and photocatalytic studies of TiO_2 colloids: naked and with the lattice doped with Cr^{3+}, Fe^{3+}, and V^{5+} cations. Langmuir, 10, 643–652, 1994.

[55] Di Valentin, C., Pacchioni, G. and Selloni, A. (2004). Origin of the different photoactivity of N-doped anatase and rutile TiO_2. *Phys. Rev. B*, 70, 085116.

[56] Irie, H., Watanabe, Y. and Hashimoto, K. (2003). Nitrogen-concentration dependence on photocatalytic activity of $TiO_{2-x}N_x$ powders. *J. Phys. Chem. B*, 107, 5483-5486.

[57] Gole, J. L., Stout, J. D., Burda, C., Lou, Y., and Chen X. (2004). Highly efficient formation of visible light tunable $TiO_{2-x}N_x$ photocatalysts and their transformation at the nanoscale. *J. Phys. Chem. B*, 108, 1230-1240.

[58] Li, Q. and Shang, J. K. (2008). Heavily nitrogen-doped dual-phase titanium oxide thin films by reactive sputtering and rapid thermal annealing. *J. Am. Ceram. Soc.*, 91, 3167-3172.

[59] Batzill, M., Morales, E. H. and Diebold, U. (2006). The influence of N-doping on the defect formation and surface properties of TiO_2 rutile and anatase. *Phys. Rev. Lett.*, **96**, 026103.

[60] Liu, H. and Gao, L. (2004). (Sulfur, Nitrogen)-codoped rutile-titanium dioxide as a visible-light-activated photocatalyst. *J. Am. Ceram. Soc.*, 87, 1582-1584.

In: Rutile: Properties, Synthesis and Applications
Editor: It-Meng (Jim) Low
ISBN 978-1-61942-233-9

Chapter 2

PHASE STABILITY AND TRANSFORMATIONS IN TiO_2

Dorian A. H. Hanaor, Pramod Koshy and Charles C. Sorrell

School of Materials Science and Engineering
University of New South Wales
Sydney, NSW 2052Australia

ABSTRACT

Titanium dioxide, a material primarily used as a white pigment, is attracting increasing attention for use as a semiconductor photocatalyst. With the exception of various high-pressure phases, rutile is the equilibrium phase of titanium dioxide at all temperatures and is generally the preferred phase for pigmentary applications. As a result of their more rapid formation kinetics, the metastable phases anatase and, to a lesser extent, brookite, which exhibit photocatalytic activity superior to that of rutile, are the first phases formed during many production routes. At low temperatures, the metastable phases are energetically stabilised and transform to rutile through thermal treatment. The onset temperature of the anatase (or brookite) to rutile phase transformation and the rate at which it proceeds can be affected significantly by dopants, firing atmosphere, microstructure, sample morphology, and the presence of impurities in the material. The present work discusses the stability of the different phases of TiO_2, the transformation of the metastable forms to rutile, and the methods through which this transformation can be enhanced or impeded.

1. OVERVIEW

Titanium dioxide (TiO_2) is used primarily as a white pigment. However, in recent decades, it has been attracting increasing attention as a photocatalyst, particularly in environmentally beneficial applications, such as water treatment and hydrogen production. Rutile is accepted widely to be the only equilibrium phase of TiO_2, although the metastable anatase phase is formed frequently as a product of TiO_2 synthesis processes. Another metastable phase, brookite, occurs often as a by-product of the synthesis of rutile or anatase [1, 2]. However, its natural scarcity and difficulties in synthesising it have resulted in its having very limited number of applications. The metastable phases transform to rutile when treated at elevated temperatures.

Additionally, five high-pressure phases of TiO_2 have been reported [3-5]. These phases exhibit orthorhombic, monoclinic, and cubic structures and form from anatase or rutile at high pressures [5].

The most notable of these is the α-PbO_2 structured TiO_2 (II) phase [6], which has been observed in pressure-shocked TiO_2 found in geological formations and meteor craters [7]. This phase is of particular importance since it has been suggested to exhibit an equilibrium phase boundary with rutile at high pressures.

In order to increase and optimise the use of TiO_2 for different applications, it important to understand the nature of the different phases, their properties, stability regimes, and factors affecting their transformations, which are discussed in the present work.

2. Properties of Anatase, Brookite and Rutile

Owing to the differences in crystallography and Ti-O bond lengths, anatase, rutile, and brookite exhibit differences in properties; these are summarised in Table 1.

The crystal structures of anatase and rutile are comprised of TiO_6 octahedra, with a Ti ion in the centre and an oxygen ion at each corner. These octahedra share four edges in anatase and two edges in rutile, as shown in Figure 1. Brookite has a more complex structure comprised of distorted TiO_6 octahedra sharing three edges, with varying Ti-Ti distances [8, 13].

Table 1. Properties of Anatase, Brookite, and Rutile Polymorphs of TiO_2 [8-12]

Parameter		Rutile	Anatase	Brookite
Crystallographic Data		Tetragonal, Z=2	Tetragonal, Z=4	Orthorhombic, Z=8
Lattice Parameters (nm)		*a*=0.459 *c*=0.296	*a*=0.378 *c*=0.951	*a*=0.918 *b*=0.545 *c*=0.524
Hardness (Mohs)		~7	~6	~6
Density ($kg{\cdot}m^{-3}$)		4200	3800	4100
Refractive Index (500 nm)		3	2.5	2.6
Band Gap	(eV)	~3.05	~3.2	3.1 - 3.4
	nm	~407	~387	365 - 400

a) b)

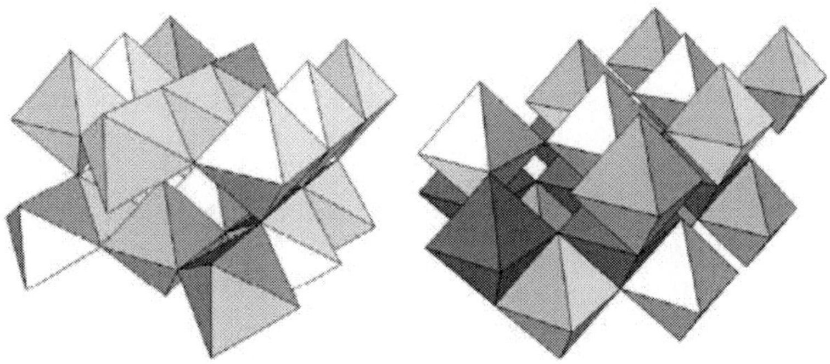

Figure 1. Arrangement of TiO_6 Octahedra in (a) Rutile and (b) Anatase.

3. Applications of TiO_2 Phases

Due to its very high refractive index, which is the highest of all oxides, TiO_2 is used extensively as a white pigment in paints, plastics, textiles, paper, and food. Pigments account for more than 95 wt% of TiO_2 usage and so the production of pigmentary titania is the main driver for the mining of titanium-bearing minerals [14]. For these applications, TiO_2 is present generally in the form of rutile or anatase. The metastable anatase phase is considered advantageous for use in the paper and textile industries owing to its lower costs, greater uniformity in particle size distribution, and lower abrasiveness in comparison with rutile pigments [15]. However, for use in paints, rutile is preferred owing to its higher refractive index and the tendency of anatase to undergo *chalking*, which is caused by photo-oxidation of the paint binder. Overall, rutile accounts for the majority of industrially produced titania pigments [14, 16, 17]. Brookite is not produced on an industrial scale and so is not used as a pigment.

In addition to its use for pigments, rutile also finds mainstream usage in low-purity form as a flux added to welding-rod coatings [15, 18]. While TiO_2 is used predominantly as a white pigment, new cutting-edge applications for titania as a semiconductor photocatalyst are attracting increasing attention. The photocatalytic effect in TiO_2 results from photo-generated electron-hole pairs, known as *excitons*, which facilitate reactions with adsorbed species on the catalyst surface. Hence, titania photocatalysts can be used for environmental purposes, including the photocatalysed destruction of pollutants for water purification [19-22] and the production of hydrogen by water splitting [23-26]. Further uses of titania photocatalysts include self-cleaning [27, 28] and anti-microbial coatings [29, 30].

For semiconductor photocatalysis, anatase is preferred generally over rutile owing to the higher levels of adsorbed species possible on anatase surfaces [31-33]. However, increasing numbers of publications report that a *mixed* titania phase composition exhibits superior photocatalytic performance owing to improved charge carrier separation relative to single-phase anatase [34, 35]. This is reported to occur through the trapping of conduction band electrons in the rutile phase while the holes remain in the anatase phase [34, 36]. It also has been reported that titania photocatalysts consisting of equal portions of anatase, rutile, and brookite exhibit enhanced photocatalytic activity in comparison to biphasic material [37]. In contrast, the formation of mixed-phase TiO_2 is undesirable in the production of pigments owing to the aforementioned phenomenon of chalking and the divergent particle sizes that are seen commonly in mixed-phase materials [16].

The brookite form of TiO_2 has been studied for use as a photocatalyst and is reported to exhibit good photocatalytic activity [38, 39]. However, this phase is unlikely to provide any significant advantages over anatase or mixed anatase-rutile photocatalysts as a result of its larger band gap and difficulties in its production.

Other emerging applications for anatase and rutile include gas sensors [40], super-hydrophilic coatings [41], porous gas- and liquid-separation membranes [42], and high-permittivity dielectrics [43]. The phase composition has been observed to have a significant impact on the performance of TiO_2 in these applications.

4. Synthesis Routes for TiO_2 Phases

Commercially, titania is synthesised from naturally occurring iron-containing rutile ($Ti_{1-x}Fe_xO_2$), ilmenite ($FeTiO_3$), leucoxene (a weathering product of ilmenite), and from ilmenite slags formed as by-products of iron production. These materials are processed by acid leaching (chloride process or sulphate process) in order to produce salts as precursors to calcining to form TiO_2. These industrial processes can be designed to yield a particular phase of titania by using TiO_2 seed grains, the addition of dopants, and variation of the pH [9, 14, 44].

The synthesis of titania for research purposes using organometallic precursors usually yields anatase as the initial crystalline product. This is the result of the more open crystal structure of anatase, where more positions are available for TiO_6 octahedral units to construct the crystal lattice [45-47]. The anatase structure exhibits closer structural similarity to amorphous TiO_2 than rutile and so anatase is the first phase to crystallise during the heat treatment of amorphous TiO_2.

Crystalline titania can be precipitated at or around room temperature using solvo-thermal synthesis methods. The crystalline phase resulting from these synthesis routes can be brookite, anatase, or rutile depending on the nature of the precursor solutions, temperature, and pH [13, 38, 48, 49]. Rapid crystallisation generally favours the formation of anatase owing to its aforementioned more flexible construction. The α-PbO_2 structured TiO_2 (II), which is obtained conventionally from high-pressure treatment of rutile, also can be prepared by sulphuric acid leaching of partially reduced ilmenite at atmospheric pressure [50].

5. Stability of TiO_2 Phases

It is accepted widely that rutile is the equilibrium phase of TiO_2 at all temperatures [51, 52]. However, there is some ambiguity regarding the equilibrium phases present at high pressures. The transformations of anatase and brookite to rutile are irreversible as these are transformations from a metastable to a stable phase. In contrast, it has been reported that, at high pressures (2-6 GPa), an equilibrium phase boundary exists between rutile and TiO_2 (II). This implies that rutile may transform to TiO_2 (II) at these reported pressure-temperature conditions [3, 5, 7, 53, 54]. The proposed equilibrium boundary is shown in Figure 2. While this phase transformation is reported to be reversible, other studies suggest that the transformation to TiO_2 (II) is irreversible [54, 55].

Dachille *et al.* [53] reported phase equilibria boundaries between anatase, brookite, rutile, and TiO_2 (II), which were shown later by Jamieson and Olinger [57] to represent irreversible phase transition boundaries. These are shown in the behavioural diagrams of the different titania forms in Figure 3 and Figure 4.

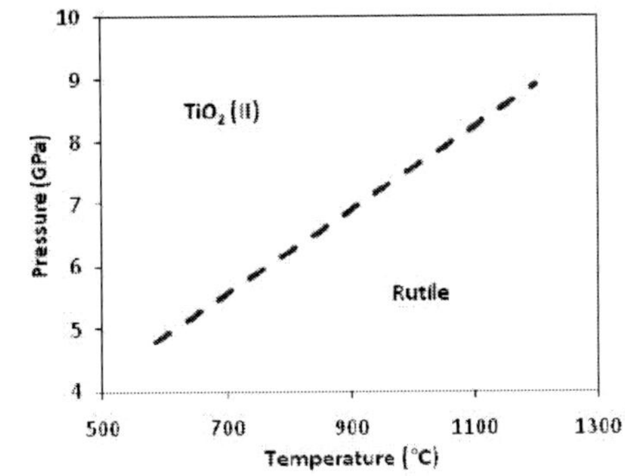

Figure 2. Reported Phase Equilibrium Boundary between Rutile and TiO_2 (II) at High Pressures [3, 56].

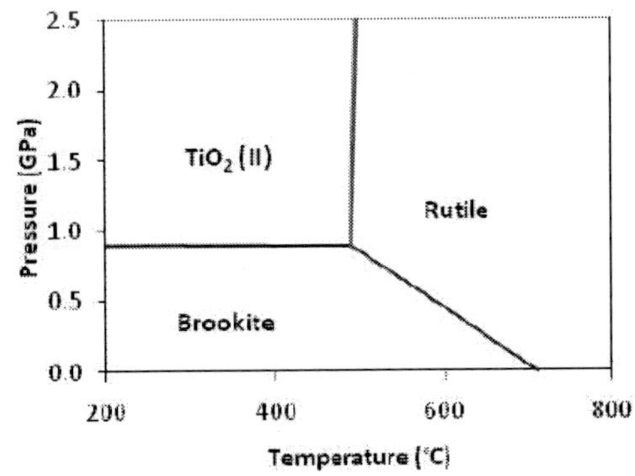

Figure 3. Behavioural Phase Transformation Diagram of Brookite, Rutile, and TiO_2 (II).

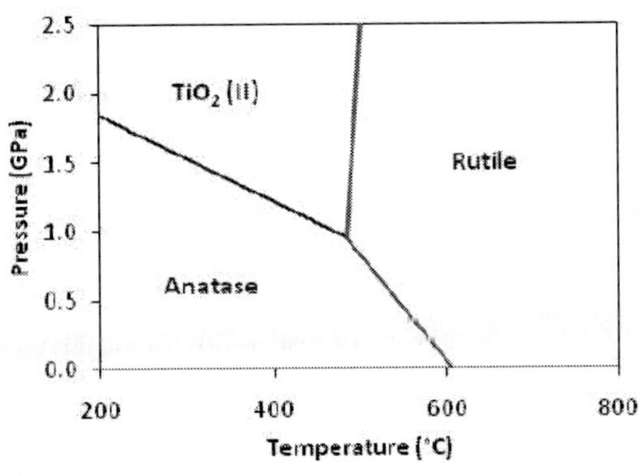

Figure 4. Behavioural Phase Transformation Diagram of Anatase, Rutile, and TiO_2 (II).

A summary of the reported thermodynamic data at standard temperature and pressure (STP) for the anatase to rutile phase transformation is shown in Table 2. The negative free energy values show that the phase transformation is feasible under these conditions leading to the formation of a more stable equilibrium product.

The theoretical lattice free energy variations in the transformations of anatase, rutile, and TiO_2 (II) were reported by Jamieson and Olinger [57] and are shown in Figure 5. Here, it can be seen that anatase shows the highest lattice free energy at all positive pressures (negative pressures are hypothetical). At some undetermined pressure, TiO_2 (II) is considered to become more stable than rutile at room temperature.

Table 2. Thermodynamic Data at STP for Anatase to Rutile Phase Transformation

$\Delta H_{A\rightarrow R}$ (kJ·mol^{-1})	$\Delta S_{A\rightarrow R}$ (J·K^{-1}mol^{-1})	$\Delta G_{A\rightarrow R}$ (kJ·mol^{-1})	Source
-1.70	0.56	-1.87	Smith *et al.*, 2009 [51]
-2.72	0.39	-2.85	Mitsuhashi and Kleppa, 1979 [58]
-12.6	--	--	Winkler, 1977 [9]
-11.72	0.42	-11.84	JANAF tables, 1971 [59]
-5.19	0.42	-5.32	Navrotsky and Kleppa, 1967 [52]

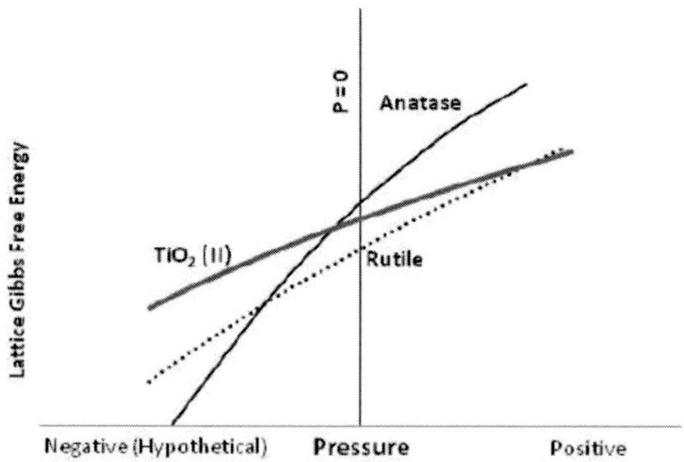

Figure 5. Theoretical Lattice Gibbs Free Energy for Anatase, Rutile and TiO_2 (II) as a Function of Pressure at Room Temperature [57].

5.1. Anatase to Rutile Phase Transformation

The process of the anatase to rutile transformation involves: (1) the nucleation of rutile at point defects, oxygen vacancies, secondary phase inclusions, particle surfaces, and/or at (112) twin interfaces in anatase and (2) the subsequent consumption of the anatase phase by the growing rutile phase. In the absence of point nucleation sites, rutile forms as parallel laths or bands in anatase [45, 60, 61]. The formation of rutile from the metastable phase anatase (or brookite) is reconstructive and so takes place through atomic rearrangement involving the breaking of two of the six Ti-O bonds in the TiO_6 octahedra [62]. As anatase (or brookite) transforms to rutile, significant grain growth takes place, resulting in lower surface area. This is one of the reasons why rutile often exhibits poorer photocatalytic performance in comparison with the metastable anatase and brookite phases.

5.2. Factors Affecting the Stability of Anatase

Although rutile is the equilibrium phase at all temperatures, anatase transforms to rutile only when sufficient thermal energy is present to facilitate the rearrangement of Ti-O bonds to form rutile; thus, anatase is kinetically stabilised. The energy requirement for rutile formation can be reached more readily in the presence of increased levels of oxygen vacancies in the TiO_2 lattice as these ease the rearrangement of ions. This lowering of the energy requirement allows the anatase to rutile transformation to occur more rapidly and at lower temperatures. Conversely, the kinetic stabilisation of anatase may be retained at higher temperatures if the atomic rearrangement is restricted by the lack of oxygen vacancies and/or the presence of constraining lattice defects, such as dopants.

5.2.1. Effect of Temperature on Anatase to Rutile Transformation

Rutile has been observed to form from anatase at various temperatures. The reasons for the multitude of observed transition temperatures can be understood from consideration of the broad range of sample characteristics and analytical methods. All TiO_2 samples exhibit different levels of impurities and morphologies, while analytical methods differ in the heating cycle, heating apparatus, and the methods used for phase analysis. For these reasons, anatase to rutile transformation temperatures have been reported over the broad range of 400º-1200°C.

Studies that utilised differential thermal analysis (DTA) to determine the anatase to rutile transformation temperature generally report an inaccurately high value for this temperature [63-65]. This is because such studies generally report exothermic peak maxima, which correspond to the temperature at which the rate of the anatase to rutile transformation is maximal. This value depends strongly on the heating rate and generally is found at temperatures significantly higher than that of the onset of the phase transformation.

The different phases of TiO_2 exhibit distinct X-ray diffraction (XRD) patterns, as shown in Figure 6. Consequently, XRD is an important tool in the phase analysis of TiO_2 and can be used to assess the anatase to rutile transformation temperature.

XRD can be used to perform quantitative phase analysis of anatase + rutile mixtures using the method of Spurr and Myers [66] according to Equation 1:

$$X_A = \left(1 + 1.26\frac{I_R}{I_A}\right)^{-1} \quad (1)$$

where X_A is the phase fraction of anatase (assuming $X_A = 1 - X_R$ and no amorphous phases or other crystalline phases are present) and I_A and I_R are the intensities of the anatase (101) peak at ~25.2° (2θ) and the rutile (110) peak at ~27.35° (2θ), respectively. The relationship between peak intensities and phase content as described by Equation 1 is shown graphically in Figure 7.

The method of Spurr and Myers [66] may yield inaccurate result if the anatase phase and/or rutile phase are present in preferred orientation. Another drawback of XRD methods is the associated difficulty in determining a distinct temperature for the phase transformation. As anatase is kinetically stabilised, no unique onset temperature can be defined. Rather, the temperature below which the rate of transformation is immeasurably slow depends on the heating rate, heating duration, and sensitivity of the measurement method. Further, unlike DTA, XRD analysis generally is not carried out at high temperatures *in situ* and so samples are analysed subsequently to a certain heat treatment, which allows only the determination of a temperature onset interval in which the transformation begins to take place.

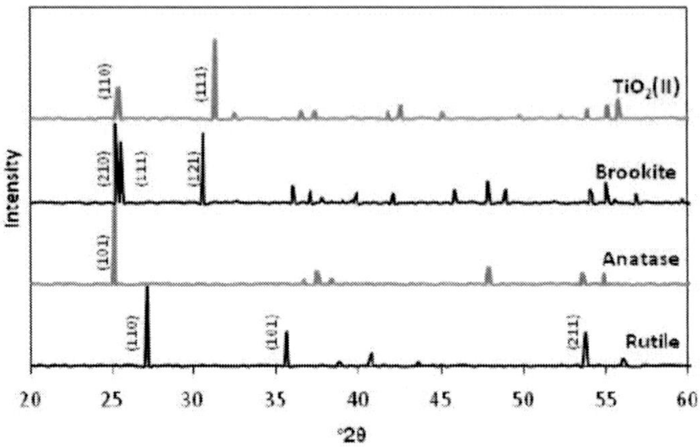

Figure 6. X-Ray Diffraction Patterns of Rutile, Anatase, Brookite, and TiO_2 (II).

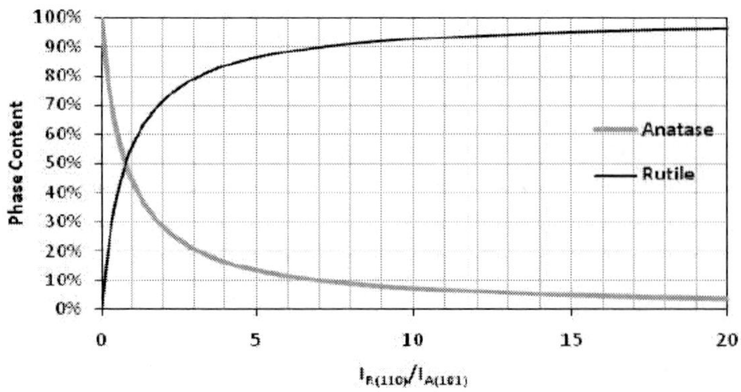

Figure 7. Phase Content in Anatase/Rutile Mixtures as a Function of Peak Intensities.

An examination of more recent studies that have utilised XRD for phase analysis reveals that, when using conventional firing durations (1-24 h), pure anatase begins to transform to rutile in the range 550°-650°C in air at atmospheric pressure. The onset temperature intervals shown in Figure 8 represent those between the highest soak temperature at which rutile was not detected (left end-point of range) and the lowest soak temperature at which rutile formation was detected (right end-point of range) by XRD. These data do not take into account the differing firing durations employed by the various studies.

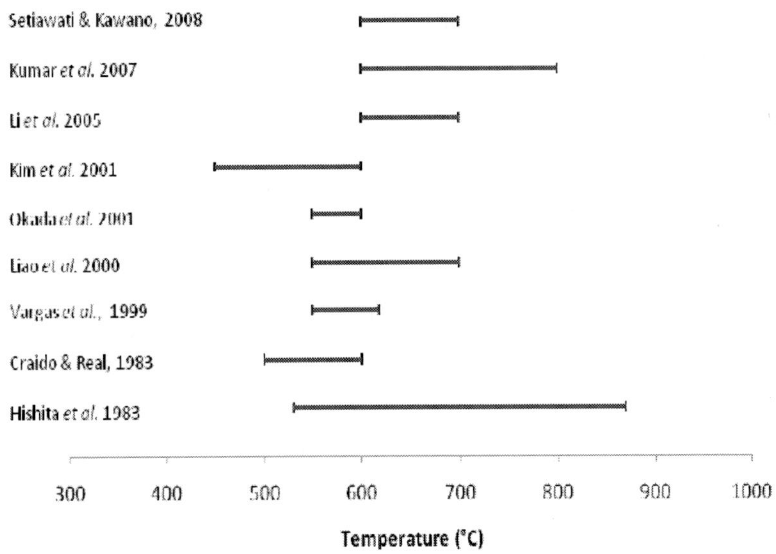

Rutile Formation in Anatase-Onset Temperature Intervals.

Figure 8.Summary of Reported Intervals of Onset Temperatures for the Anatase to Rutile Phase Transition [47, 64, 67-73].

An alternative method of phase analysis for titania is the use of laser Raman microspectroscopy. This method has been used for on-stream analysis in the manufacturing of TiO_2 pigments [74, 75]. The advantages of this method are the rapidity of the analyses, the lack of preferred orientation effects, and the potential for *in situ* phase analysis during heat treatment. The drawbacks of this analytical technique are the difficulty in quantitative interpretation of results and the weak signal from rutile relative to the strong one from anatase.

5.2.2. Effect of Pressure on Anatase to Rutile Transformation

From consideration of studies involving pressure-temperature effects, it is clear that, at higher pressures, the transformation of anatase to rutile can take place at lower temperatures [53, 57, 70, 76]. This is likely a result of the higher density of rutile and the volume contraction that occurs when anatase transforms to rutile, which favours the formation of this phase at high pressures, following Le Chatelier's principle. At ambient temperatures, rather than forming rutile, anatase has been reported to transform directly to TiO_2 (II) phase at elevated pressures [77, 78]. This is consistent with the transformation reaction boundaries reported by Dachille *et al.* [53] and later clarified by Jamieson and Olinger [57] (shown in Figures 3 and 4).

5.3. Rutile/Anatase Formation from Brookite

Studies involving the transformation of brookite to anatase are considerably less common than reports of the anatase to rutile phase transformation. Apparently, the only study of the direct transformation from brookite to anatase reported that this occurred at low temperatures in the range 200°-500°C [37], with anatase transformation to rutile at higher temperatures. In contrast, other studies have reported the formation of rutile directly from brookite at ~700°C [53, 58, 79], as suggested by Figure 3.

The presence of a secondary brookite phase present with anatase enhances the kinetics of rutile formation and lowers the temperature at which the phase transformation occurs in comparison with single-phase anatase. This is reported to occur since brookite/anatase interfaces act as nucleation sites for rutile formation [80].

5.4. Effects of Dopants on Anatase to Rutile Transformation

5.4.1. Cationic Dopants

Lowering the temperature and enhancing the rapidity at which the rutile forms from anatase or brookite can be achieved by imparting higher levels of oxygen vacancies in the precursor material. These can be achieved through doping with small and low-valence cations, which are small enough to diffuse into the TiO_2 lattice and substitute for Ti^{4+} species, thereby forming oxygen vacancies for charge compensation. Further enhancement of rutile formation from anatase can be achieved through the use of a reducing or otherwise oxygen-deficient atmosphere. Again, it is likely that reducing atmospheres accelerate the transformation by enhancing oxygen vacancy formation. Conversely, the substitution of Ti^{4+} by cations of higher valence leads to a reduction in oxygen vacancy levels. Also, cationic dopants of large radii are too large to enter the TiO_2 lattice substitutionally and thus are likely to be relegated to the grain boundaries. Both of these latter effects inhibit rutile formation from the metastable phases.

Inhibition of the formation of rutile from anatase by the presence of intentional dopants or unintentional impurities is a more common observation than the enhancement of the phase transformation by dopants. The presence of silica, zirconia, and/or alumina brings about a strong inhibitory effect on the anatase to rutile transformation [68, 81-83]. This is relevant in explaining the increased thermal stability of commercially produced anatase powders, which often exhibit higher levels of silica impurities. The use of anatase-stabilising oxide dopants has been exploited to fabricate thermally stable anatase. The stabilisation of anatase by dopants generally takes places through a structure-stuffing effect. This occurs when dopant atoms occupy cation lattice positions, interstices, and grain boundaries in anatase and so restrict the rearrangement necessary for rutile formation [73]. The effects of different dopant compounds on the promotion or inhibition of rutile formation have been reviewed in detail [10] and are summarised in Figure 9, which shows the promoting and inhibiting effects of dopants by plotting the Shannon [84] ionic radii against the cation valence.

It can be seen that dopants of low valence and small ionic radii are more likely to enhance the formation of rutile from anatase. Although low radius and valence appear to be necessary for a dopant to have an enhancing effect on the rutilation of anatase, oxidation and reduction effects are also likely to play an important role. The latter explains the strong anatase to rutile transformation-promoting effects of iron Fe^{3+}, copper Cu^{2+}, and manganese Mn^{4+} that have been reported, despite their not lying in the bottom left of Figure 9.

Transition metals of variable valence are reported to enter the titania lattice and create oxygen vacancies through reduction effects [85], as described by Equation 2:

$$2M^{n+} + O^{2-} \rightarrow 2M^{(n-1)+} + \frac{1}{2}O_2 + V_O \quad (2)$$

where M denotes a transition metal atom and V_O denotes an oxygen vacancy. Since transition metal ions of higher valence tend to have smaller ionic radii than those of lower valence, the former's diffusion into the TiO_2 lattice is likely to be enhanced. Once substituted for Ti^{4+}, the reduction of such species gives rise to new oxygen vacancies and so enhances rutile formation from metastable anatase (or brookite).

5.4.2. Anionic Dopants

While the effects of cationic dopants on the thermal stability of the titania phases have been reported widely, the effects of anionic dopants have been reported to a lesser extent. It has been observed that doping by Cl^- and F^- inhibits rutile formation from anatase [86, 87]. It is likely that most anionic dopants have an inhibitory effect on the formation of rutile from anatase (or brookite). Since anions have considerably larger ionic radii than cations, they can be expected to hinder the reconstructive phase transformation. Further, dopant anions may occupy oxygen vacancies and thereby restrict atomic rearrangement. The substitution of oxygen by halogen ions may give rise to the presence of Ti^{3+} species, through charge compensation. These lower valence titanium ions exhibit larger ionic radii than that of Ti^{4+}, further stabilising the anatase phase.

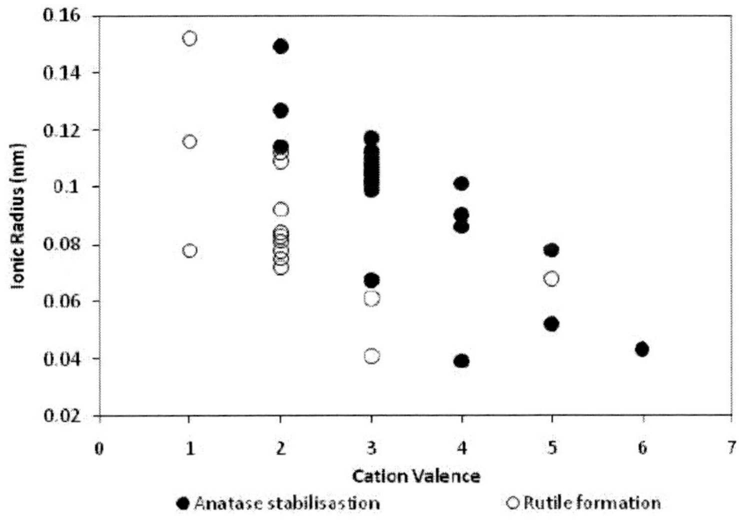

Figure 9. Effect of Cationic Dopants on Anatase to Rutile Transformation.

5.5. Surface Energy Changes during Phase Transformations of Nanoparticles

Despite a lower bulk free energy, rutile crystallographic planes generally exhibit higher surface energies in comparison to anatase and brookite planes. This is likely to be a result of the denser atomic packing on the rutile planes and associated increased density of unsaturated terminating bonds. Consequently, nanosize particles of anatase, and possibly brookite, below a certain size-threshold may exhibit lower total free energy in comparison to rutile particles of the same size [88-92]. This threshold has been reported to be between 11 and 45 nm and is highly dependent on the crystallite shape and surface acidity (surface-terminating groups) of the material.

As significant grain growth is associated with the formation of rutile from anatase at elevated temperatures, the comparison of anatase and rutile crystallites of similar sizes is relevant only for nanocrystalline anatase and rutile, which can be crystallised using hydrothermal or solvo-thermal methods. It also has been reported that rutile may transform to anatase through modification of the surface acidity [93]. However, as this study involved atomic-scale layers of titania, the conventional concept of phase transformation is not applicable. In general, the high surface area of nanocrystalline anatase gives rise to more nucleation sites for rutile formation and thus a faster overall transformation occurs compared to anatase of larger grain size [45].

Summary

While rutile is the only equilibrium phase of TiO_2 at atmospheric pressure, metastable and high-pressure phases also occur, with the anatase phase being particularly common. Since the different phases of titania exhibit different properties, an understanding of the stability of the different phases and the transformations between them is important to TiO_2 in photocatalysis, pigmentary applications, and other uses.

While rutile is preferred for most pigmentary uses, anatase, brookite, and mixtures of these phases are advantageous in photocatalytic applications. The anatase and brookite phases transform to rutile through thermal treatment. The transformation of the metastable phases to the equilibrium rutile phase does not occur at a distinct onset temperature and the kinetics of the phase transformation and the temperature at which it becomes observable are highly dependent on the levels of oxygen vacancies and nucleation sites in the TiO_2 lattice.

Dopants and atmospheric conditions can be controlled to enhance or inhibit rutile formation from the metastable phases. This occurs through the modification of oxygen vacancy levels in the TiO_2 lattice and the restriction/easing of structural rearrangement. Cationic dopants of low valence and small ionic radii have a significant enhancing effect on rutile formation while cations of higher valence and/or larger radius as well as anions tend to extend the thermal stability of the metastable phases.

Despite the lower lattice free energy of rutile, a lower density of surface-terminated bonds in anatase causes nanocrystallites of this phase to have a lower total free energy in comparison with rutile. This effect is dependent on the nature of the surface-terminating groups and the morphology of the nanocrystallites.

References

[1] Pottier, A., Chanéac, C., Tronc, E., Mazerolles, L., Jolivet, J. P.: Synthesis of brookite TiO_2 nanoparticles by thermolysis of $TiCl_4$ in strongly acidic aqueous media. *Journal of Materials Chemistry,* 2001, 11, 1116-1121

[2] Yanqing, Z., Erwei, S., Suxian, C., Wenjun, L., Xingfang, H.: Hydrothermal preparation and characterization of brookite-type TiO_2 nanocrystallites. *Journal of Materials Science Letters,* 2000, 19, 1445-1448.

[3] Dubrovinskaia, N. A., Dubrovinsky, L. S., Ahuja, R., et al.: Experimental and Theoretical Identification of a New High-Pressure TiO_ {2} Polymorph. *Physical Review Letters,* 2001, 87, 275501.

[4] Sato, H., Endo, S., Sugiyama, M., Kikegawa, T., Shimomura, O., Kusaba, K.: Baddeleyite-type high-pressure phase of TiO_2. *Science,* 1991, 251, 786.

[5] Muscat, J., Swamy, V., Harrison, N. M.: First-principles calculations of the phase stability of TiO2. *Physical Review B,* 2002, 65, 224112-224116.

[6] Gerward, L., Olsen, S.: Post-rutile high-pressure phases in TiO_2. *Journal of Applied Crystallography,* 1997, 30, 259-264.

[7] Withers, A. C., Essene, E. J., Zhang, Y.: Rutile/TiO_2 II phase equilibria. *Contributions to Mineralogy and Petrology,* 2003, 145, 199-204.

[8] Mo, S. D., Ching, W.: Electronic and optical properties of three phases of titanium dioxide: Rutile, anatase, and brookite. *Physical Review B,* 1995, 51, 13023-13032.

[9] Winkler, J.: Titanium Dioxide. Vincentz Network GmbH, Hannover, 1977.

[10] Hanaor, D., Sorrell, C.: Review of the anatase to rutile phase transformation. *Journal of Materials Science,* 2011, 46, 1-20.

[11] Peters, G., Vill, V. (1989) Landolt-Börnstein numerical data and functional relationships in science and technologyVerlag, Berlin.

[12] Li, J. G., Ishigaki, T., Sun, X.: Anatase, brookite, and rutile nanocrystals via redox reactions under mild hydrothermal conditions: Phase-selective synthesis and physicochemical properties. *The Journal of Physical Chemistry C,* 2007, 111, 4969-4976.

[13] Bokhimi, X., Pedraza, F.: Characterization of brookite and a new corundum-like titania phase synthesized under hydrothermal conditions. *Journal of Solid State Chemistry,* 2004, 177, 2456-2463.

[14] Kogel, J. E., Trivedi, N. C., Barker, J. M.: Industrial minerals and rocks: commodities, markets, and uses. Society for Mining, Metallurgy, and Exploration, Englewood 2006.

[15] Fisher, J., Egerton, T. A. (2001) Kirk-Othmer Encyclopedia of Chemical TechnologyJohn Wiley and Sons, Inc.

[16] Koleske, J. V.: Paint and coating testing manual: *fourteenth edition of the Gardner-Sward handbook.* ASTM, 1995.

[17] Volz, H. G.: Pigments, Inorganic. Wiley-VCH Verlag GmbH and Co, Weinheim, 2006.

[18] Mackey, T. S.: Upgrading ilmenite into a high-grade synthetic rutile. *JOM,* 1994, 46, 59-64.

[19] Byrne, J. A., Eggins, B. R., Brown, N. M. D., McKinney, B., Rouse, M.: Immobilisation of TiO_2 powder for the treatment of polluted water. *Applied Catalysis B, Environmental,* 1998, 17, 25-36.

[20] Ding, Z., Lu, G. Q., Greenfield, P. F.: Role of the crystallite phase of TiO2 in heterogeneous photocatalysis for phenol oxidation in water. *J. Phys. Chem. B,* 2000, 104, 4815-4820.

[21] Dominguez, C., Garcia, J., Pedraz, M., Torres, A., Galan, M.: Photocatalytic oxidation of organic pollutants in water. *Catalysis today,* 1998, 40, 85-101.

[22] Pozzo, R. L., Baltanas, M. A., Cassano, A. E.: Supported titanium oxide as photocatalyst in water decontamination: state of the art. *Catalysis Today,* 1997, 39, 219-231.

[23] Abe, R., Sayama, K., Domen, K., Arakawa, H.: A new type of water splitting system composed of two different TiO_2 photocatalysts (anatase, rutile) and a IO_3^-/I^-shuttle redox mediator. *Chemical physics letters,* 2001, 344, 339-344.

[24] Fujihara, K., Ohno, T., Matsumura, M.: Splitting of water by electrochemical combination of two photocatalytic reactions on TiO2 particles. *J. Chem. Soc. Faraday Trans.,* 1998, 94, 3705 – 3709.

[25] Ni, M., Leung, M., Leung, D., Sumathy, K.: A review and recent developments in photocatalytic water-splitting using TiO2 for hydrogen production. *Renewable and Sustainable Energy Reviews,* 2007, 11, 401-425.

[26] Fujishima, A., Honda, K.: Electrochemical photolysis of water at a semiconductor electrode. *Nature,* 1972, 238, 37-38.

[27] Mills, A., Wang, J.: Simultaneous monitoring of the destruction of streraic acid and generation of carbon dioxide by self cleaning semiconductor films. *Journal of Photochemistry and Photobiology A,* 2006, 182, 181-186.

[28] Sam, E. D., Urgen, M., Tepehan, F. Z., Gunay, V.: Self cleaning photoactive TiO2 coatings on SLS glasses by sol-gel dip-coating. *Key Engineering Materials,* 2004, 264, 407-410.

[29] Ireland, J. C., Klostermann, P., Rice, E. W., Clark, R. M.: Inactivation of Escherichia coli by titanium dioxide photocatalytic oxidation. *Applied and Environmental Microbiology,* 1993, 59, 1668.

[30] Maness, P. C., Smolinski, S., Blake, D. M., Huang, Z., Wolfrum, E. J., Jacoby, W. A.: Bactericidal activity of photocatalytic TiO_2 reaction: toward an understanding of its killing mechanism. *Applied and Environmental Microbiology,* 1999, 65, 4094.

[31] Sclafani, A., Herrmann, J. M.: Comparison of the photoelectronic and photocatalytic activities of various anatase and rutile forms of titania. *Journal of Physical Chemistry,* 1996, 100, 13655 – 13661.

[32] Hadjiivanov, K., Klissurski, D. G.: Surface Chemistry of Titania. *Chemical Society Reviews,* 1996, 25, 61-69.

[33] Fox, M. A., Dulay, M. T.: Heterogeneous Catalysis. *Chem. Rev.,* 1992, 93, 341-357.

[34] Hurum, D. C., Agrios, A. G., Gray, K. A., Rajh, T., Thurnauer, M. C.: Explaining the enhanced photocatalytic activity of Degussa P25 mixed-phase TiO_2 using EPR. *J. Phys. Chem. B,* 2003, 107, 4545-4549.

[35] Torimoto, T., Nakamura, N., Ikeda, S., Ohtani, B.: Discrimination of the active crystalline phases in anatase–rutile mixed titanium (iv) oxide photocatalysts through action spectrum analyses. *Physical Chemistry Chemical Physics,* 2002, 4, 5910-5914.

[36] Li, G., Chen, L., Graham, M. E., Gray, K. A.: A comparison of mixed phase titania photocatalysts prepared by physical and chemical methods: The importance of the solid–solid interface. *Journal of Molecular Catalysis. A, Chemical,* 2007, 275, 30-35.

[37] Bakardjieva, S., Stengl, V., Szatmary, L., et al.: Transformation of brookite-type TiO_2 nanocrystals to rutile: correlation between microstructure and photoactivity. *Journal of Materials Chemistry,* 2006, 16, 1709-1716.

[38] Kominami, H., Ishii, Y., Kohno, M., Konishi, S., Kera, Y., Ohtani, B.: Nanocrystalline brookite-type titanium (IV) oxide photocatalysts prepared by a solvothermal method:

correlation between their physical properties and photocatalytic activities. *Catalysis Letters,* 2003, 91, 41-47.

[39] Ohtani, B., Handa, J., Nishimoto, S., Kagiya, T.: Highly active semiconductor photocatalyst: Extra-fine crystallite of brookite TiO2 for redox reaction in aqueous propan-2-ol and/or silver sulfate solution. *Chemical physics letters,* 1985, 120, 292-294.

[40] Sharma, R. K., Bhatnagar, M. C., Sharma, G. L.: Effect of Nb metal ion in TiO_2 oxygen gas sensor. *Applied surface science,* 1996, 92, 647-650.

[41] Ashkarrana, A. A., Mohammadizade, M. R.: Superhydrophilicity of TiO_2 thin films using TiCl4 as a precursor. *Materials Research Bulletin* 2007, 43, 522-530.

[42] Bischoff, B. L., Anderson, M. A.: Peptization process in the sol-gel preparation of porous anatase (TiO_2). *Chemistry of materials,* 1995, 7, 1772-1778.

[43] Kim, H. S., Campbell, S., Gilmer, C.: Charge trapping and degradation in high-permittivity TiO_2 dielectric films. *Electron Device Letters, IEEE,* 1997, 18, 465-467.

[44] Andreev, A., Diachenko, A.: Conditions for the production of pigmentary titanium dioxide of rutile and anatase modifications by ilmenite processing with ammonium fluoride. *Theoretical Foundations of Chemical Engineering,* 2009, 43, 707-712.

[45] Hyoung, G. L., Zuo, J. M.: Growth and phase transformation of nanometer sized titanium oxide powders produced by the precipitation method. *Journal of the American Ceramic Society,* 2004, 87, 473-479.

[46] Gopal, M., Moberly Chan, W. J., De Jonghe, L. C.: Room temperature synthesis of crystalline metal oxides. *Journal of Materials Science,* 1997, 32, 6001-6008.

[47] Li, G., Li, L., Boerio-Goates, J., Woodfield, B. F.: High purity anatase TiO_2 nanocrystals: near room-temperature synthesis, grain growth kinetics, and surface hydration chemistry. *J. Am. Chem. Soc,* 2005, 127, 8659-8666.

[48] Aruna, S. T., Tirosh, S., Zaban, A.: Nanosize rutile titania particle synthesis via a hydrothermal method without mineralizers. *Journal of Materials Chemistry,* 2000, 10, 2388-2391.

[49] Cheng, H., Ma, J., Zhao, Z., Qi, L.: Hydrothermal Preparation of Uniform Nanosize Rutile and Anatase Particles. *Chemistry of Materials,* 1995, 7, 663-671.

[50] Grey, I. E., Li, C., Madsen, I. C., Braunshausen, G.: TiO_2-II. Ambient pressure preparation and structure refinement. *Materials Research Bulletin,* 1988, 23, 743-753.

[51] Smith, S. J., Stevens, R., Liu, S., et al.: Heat capacities and thermodynamic functions of TiO_2 anatase and rutile: Analysis of phase stability. *American Mineralogist,* 2009, 94, 236.

[52] Navrotsky, A., Kleppa, O. J.: Enthalpy of the Anatase-Rutile Transformation. *Journal of the American Ceramic Society,* 1967, 50, 626-626.

[53] Dachille, F., Simons, P. Y., Roy, R.: Pressure-Temperature Studies of Anatase, Brookite, Rutile and TiO_2-II. *The American Mineralogist,* 1968, 53, 1929-1939.

[54] Mammone, J., Sharma, M.: Raman study of rutile (TiO_2) at high pressures. *Solid State Communications,* 1980, 34, 799-802.

[55] Murrray, J. L., Wriedy, H. A.: The O-Ti System. *Bulletin of Alloy Phase Diagrams,* 1987, 8, 148-165.

[56] Meng, D., Wu, X., Fan, X., et al.: High Pressure Response of Rutile Polymorphs and Its Significance for Indicating the Subduction Depth of Continental Crust. *Acta Geologica Sinica(English Edition),* 2008, 82, 371-376.

[57] Jamieson, J., Olinger, B.: Pressure Temperature Studies of Anatase, Brookite, Rutile and TiO_2 II: A discussion. *Mineralogical Notes,* 1969, 54, 1477-1482.

[58] Mitsuhashi, T., Kleppa, O. J.: Transformation Enthalpies of the TiO_2 Polymorphs. *Journal of the American Ceramic Society,* 1979, 62, 356-357.

[59] Stull, D. R., Prophet, H.: JANAF thermochemical tables. *NBS STP NO 37, WASHINGTON DC, 1971, 1141 P,* 1971.

[60] Gouma, P. I., Mills, M. J.: Anatase to Rutile Transformation in Titania Powders. *Journal of the American Ceramic Society,* 2001, 84, 619-622.

[61] Hanaor, D., Triani, G., Sorrell, C.: Morphology and photocatalytic activity of highly oriented mixed phase titanium dioxide thin films. *Surface and Coatings Technology,* 2011, 205, 3658-3664.

[62] Riyas, S., Krishnan, G., Mohandas, P. N.: Anatase-Rutile transformation in doped titania under argon and hydrogen atmospheres. *Advances in Applied Ceramics,* 2007, 106, 255 – 264.

[63] Yang, J., Huang, Y. X., Ferreira, J. M. F.: Inhibitory effect of alumina additive on the titania phase transformation of a sol--gel-derived powder. *Journal of Materials Science Letters,* 1997, 16, 1933-1935.

[64] Kumar, K. N. P., Fray, D. J., Nair, J., Mizukami, F., Okubu, T.: Enhanced anatase to rutile phase transformation without exaggerated particle growth in nanostructured titania-tin oxide composites. *Scripta Materialia,* 2007, 57, 771-774.

[65] Shannon, R. D., Pask, J. A.: Kinetics of the anatase-rutile transformation. *Journal of the American Ceramic Society,* 1965, 48, 391-398.

[66] Spurr, R. A., Myers, H.: Quantitative Analysis of Anatase-Rutile Mixtures with an X-Ray diffractometer. *Analytical Chemistry,* 1957, 29, 760 – 762.

[67] Setiawati, E., Kawano, K.: Stabilisation of anatase phase in the rare earth; Eu and Sm doped nanoparticle TiO_2. *Journal of Alloys and Compounds,* 2008, 451, 293-296.

[68] Kim, J., Song, K. C., Foncillas, S., Pratsinis, S.: Dopants for synthesis of stable bimodally porous titanis. *Journal of the European Ceramic Society,* 2001, 21, 2863 – 2872.

[69] Okada, K., Yamamoto, N., Kameshima, Y., Yasumori, A.: Effect of Silica Additive on the Anatse to Rutile Phase Transition. *Journal of the American Ceramic Society,* 2001, 84, 1591-1596.

[70] Liao, S. C., Colaizzi, J., Chen, Y., Kear, B. H., Mayo, W. E.: Refinement of Nanoscale Grain Structure in Bulk Titania. *Journal of the American Ceramic Society,* 2000, 83, 2163-2169

[71] Vargas, S., Arroyo, R., Haro, E., Rodriguez, R.: Effect of cationic dopants on the phase transition temperature of titania prepared by the Sol-gel method. *J. Mater. Res.,* 1999, 14, 3932-3937.

[72] Craido, J., Real, C.: Mechanism of the Inhibiting Effect of Phosphate on the Anatase to Rutile Transformation Induced by Thermal and Mechanical Treatment of TiO_2. *J. Chem. Soc. Faraday Trans.,* 1983, 79, 2765-2771.

[73] Hishita, S., Mutoh, I., Koumoto, K., Yanagida, H.: Inhibition mechanism of the anatase-rutile phase transformation by rare earth oxides. *Ceramics International,* 1983, 9, 61-67.

[74] Clegg, I. M., Everall, N. J., King, B., Melvin, H., Norton, C.: On-line analysis using Raman spectroscopy for process control during the manufacture of titanium dioxide. *Applied Spectroscopy,* 2001, 55, 1138-1150.

[75] Capwell, R., Spagnolo, F., DeSesa, M.: A Rapid Determination of Low Concentrations of Anatase in Rutile TiO_2 Pigments by Raman Spectroscopy. *Applied Spectroscopy,* 1972, 26, 537-539.

[76] Liao, S., Mayo, W., Pae, K.: Theory of high pressure/low temperature sintering of bulk nanocrystalline TiO_2. *Acta materialia,* 1997, 45, 4027-4040.

[77] Sekiya, T., Ohta, S., Kamei, S., Hanakawa, M., Kurita, S.: Raman spectroscopy and phase transition of anatase TiO_2 under high pressure. *Journal of Physics and Chemistry of Solids,* 2001, 62, 717-721.

[78] Arlt, T., Bermejo, M., Blanco, M., Gerward, L.: High pressure polymorphs of anatase TiO2. *Physical Review B,* 2000, 61, 14414-14416.

[79] Rao, C., Yoganarasimhan, S., Faeth, P.: Studies on the brookite-rutile transformation. *Transactions of the Faraday Society,* 1961, 57, 504-510.

[80] Ovenstone, J., Yanagisawa, K.: Effect of hydrothermal treatment of amorphous titania on the phase change from anatase to rutile during calcination. *Chem. Mater,* 1999, 11, 2770-2774.

[81] Li,: Inhibiting effect of $AlPO_4$ and SiO_2 on the anatase to rutile transformation reaction: An x-ray and laser Raman study. *J. Mater. Res,* 1992, 7, 3349.

[82] Reidy, D. J., Holmes, J. D., Morris, M. A.: Preparation of a highly thermally stable titania anatase phase by addition of mixed zirconia and silica dopants. *Ceramics International,* 2006, 32, 235-239.

[83] Akhtar, M. K., Pratsinis, S. E., Mastrangelo, S. V. R.: Dopants in Vapor-Phase Synthesis of Titania Powders. *Journal of the American Ceramic Society,* 1992, 75, 3408-3416.

[84] Shannon, R. D.: Revised effective ionic radii and systematic studies of interatomic distances in halides and chalcogenides. *Acta Crystallographica Section A: Crystal Physics, Diffraction, Theoretical and General Crystallography,* 1976, 32, 751-767.

[85] Mackenzie, K. J. D.: Calcination of titania V. Kinetics and mechanism of the anatase-rutile transformation in the presence of additives. *Transactions and Journal of the British Ceramic Society,* 1975, 74, 77-84.

[86] Yu, J. C., Yu, J., Ho, W., Jiang, Z., Zhang, L.: Effects of F-doping on the photocatalytic activity and microstructures of nanocrystalline TiO2 powders. *Chem. Mater.,* 2002, 14, 3808-3816.

[87] Takahashi, Y., Matsuoka, Y.: Dip-coating of TiO_2 films using a sol derived from Ti (O-i-Pr) 4-diethanolamine-H $_2$ O-i-PrOH system. *Journal of Materials Science,* 1988, 23, 2259-2266.

[88] Reidy, D. J., Holmes, J. D., Morris, M. A.: The critical size mechanism for the anatase to rutile transformation in TiO_2 and doped TiO_2. *Journal of the European Ceramic Society,* 2006, 26, 1527 – 1534.

[89] Zhang, H., Banfield, J. F.: Understanding polymorphic phase transformation behavior during growth of nanocrystalline aggregates: Insights from TiO_2. *J. Phys. Chem. B,* 2000, 104, 3481-3487.

[90] Zhang, Z., Wang, C., Zakaria, R., Ying, J.: Role of Particle Size in Nanocrystalline TiO_2 Based Photocatalysts. *Journal of Physical Chemistry B,* 1998, 102, 10871-10878.

[91] Barnard, A., Curtiss, L.: Prediction of TiO_2 nanoparticle phase and shape transitions controlled by surface chemistry. *Nano letters,* 2005, 5, 1261-1266.

[92] Barnard, A. S., Zapol, P., Curtiss, L. A.: Anatase and rutile surfaces with adsorbates representative of acidic and basic conditions. *Surface science,* 2005, 582, 173-188.

[93] Saponjic, Z. V., Dimitrijevic, N. M., Tiede, D. M., et al.: Shaping Nanometer Scale Architecture Through Surface Chemistry. *Advanced Materials,* 2005, 17, 965-971.

In: Rutile: Properties, Synthesis and Applications
Editor: It-Meng (Jim) Low

ISBN 978-1-61942-233-9

Chapter 3

RUTILE NANOSTRUCTURES: SYNTHESIS, CHARACTERIZATION AND POTENTIAL APPLICATION IN PHOTOCATALYTIC PROCESSES

Mirjana I. Čomor[1*], Nadica D. Abazović[1], Ivana A.Janković[1], Dragana J. Jovanović[1], Milovan Stoiljković[1], Daniela V. Šojić[2] and Biljana F. Abramović[2]

[1]Vinča Institute of Nuclear Sciences, University of Belgrade, 11000 Belgrade, Serbia
[2]Department of Chemistry, Faculty of Sciences, University Of Novi Sad, Trg D. Obradovića 3, 21000 Novi Sad, Serbia

ABSTRACT

Titanium dioxide is one of the most studied semiconductors because of its wide range of potential technological applications including pigments, cosmetics, photovoltaic cells and catalysis. TiO_2 has three most commonly encountered crystalline polymorphs: anatase, brookite and rutile. The possible application of TiO_2 critically depends on crystalline structure, size and shape of the particles. Rutile has the highest density and refractive index and is widely employed in pigments and cosmetic industry. TiO_2 is transparent in the visible light region; its band gap is 3.0 eV for rutile and 3.2 eV for anatase crystalline phase. By doping or sensitization, it is possible to improve the optical activity of TiO_2 and to move its absorption threshold into the visible light region.

The subject of this chapter is different rutile nanostructures of pure and Fe^{3+}- doped titania which were synthesized by co-precipitation in acidic aqueous solution. Obtained materials were characterized by structural (XRD), morphological (TEM and HRTEM), optical (UV/vis absorption, reflection, and photoluminescence), and analytical (XPS and ICP-OES) techniques. Also, photo-catalytic activity measurements were performed, using photo-degradation of specific herbicides as model systems. Applied synthetic procedure resulted in formation of flower like agglomerates of 200-400 nm in diameter. Agglomerates represent assembly of nano-rods consisted of "chains" of spherical particles, 5-7 nm in diameter, most likely interconnected through the so-called oriented attachment or grain-rotation induced-grain coalescence (GRIGC) process. The presence of dopant iron ions didn't distract morphology of rutile nano-particles, nor agglomerates, but significantly changed optical properties of titania compared to pure material. The photoluminescence of pure and Fe-doped rutile nano-particles is characterized with several well-resolved peaks extending in the visible spectral region. Photon energy up-conversion from rutile TiO_2 nano-particles was also observed at low excitation intensities. The energy of up-converted photoluminescence spans the range of emission of normal photoluminescence. The explanation of photon energy up-conversion involves mid-gap energy levels originating from oxygen vacancies.

INTRODUCTION

Since its commercial production in the early twentieth century, titanium dioxide (TiO_2) has been widely used as a pigment [1] and in sunscreens [2, 3], paints [4], ointments, toothpaste [5] etc. In 1972, Fujishima and Honda [6-8] discovered the phenomenon of photo-catalytic splitting of water on a TiO_2 electrode under ultraviolet (UV) light. Since then, enormous efforts have been devoted to the research of TiO_2 material, which has led to many promising applications in areas ranging from photo-voltaics and photo-catalysis to photo-electrochromics and sensors [9-12]. These applications can be roughly divided into "energy" and "environmental" categories, many of which depend not only on the properties of the TiO_2 material itself but also on the modifications of the TiO_2 material host (e.g., with inorganic and organic dyes) and on the interactions of TiO_2 materials with the environment [13].

An exponential growth of research activities has been seen in nano-science and nanotechnology in the past decades [14-18]. Our research group contributed also in this dynamic field with a significant number of publications dealing with various nano-crystalline semiconductor systems such as metal halides [19-25], metal chalcogenides [20, 23, 25-39] and oxides [23, 25, 28, 40-50]. New physical and chemical properties emerge when the size of the material becomes smaller and smaller, and down to the nanometer scale. The movement of electrons and holes in semiconductor nano-particles is primarily governed by the well-known quantum confinement, and the transport properties related to phonons and photons are largely affected by the size and geometry of the materials [14-17]. The specific surface area and surface-to-volume ratio increase dramatically as the size of a material decreases [14, 51]. The high surface area brought about by small particle size is beneficial to many TiO_2-based devices, as it facilitates reaction/interaction between material and interacting media, which mainly occurs on the surface or at the interface and strongly depends on the surface area of the material. Thus, the performance of TiO_2-based devices is largely influenced by the sizes of the TiO_2 building units, apparently at the nanometer scale.

Literature data show, that optical properties of TiO_2 significantly depend on the method of its synthesis and purification, the presence of dopants and defects in it, its surface states and the way of its post synthetic treatment. Although methods such as flame synthesis [52, 53], ultrasonic irradiation [54, 55], chemical vapor synthesis [56] and hydrothermal methods [57-59] have been widely used, sol-gel processes have been shown to be especially versatile synthesis procedures [60-66]. Sol-gel reactions allow good control from the molecular precursors to the final product, which makes possible the low-temperature tailoring of materials to have high purity, high homogeneity, and desired size and morphology. Most of these sol–gel methods are based on the hydrolysis of an alkoxide or halide precursors and subsequent condensation to the inorganic framework. However, in most cases, sol-gel derived precipitates at low temperatures are amorphous, and ultrasonic irradiation, subsequent hydrothermal processing or calcination is necessary to induce crystallization.

From crystallographic point of view, TiO_2 crystallize in three main forms: anatase (space group, $P4_2/mnm$), rutile (space group, $I4_1/amd$) and brookite (space group, *Pbca*). Two other existing structural forms, known as columbite and baddeleyite, are stabilized only under high pressure [67]. In general, structure and microstructure (shape, size, surface morphology, and anisotropy) determines physical and chemical properties of material. The most stable phase is rutile and it is usually obtained after annealing at temperature above 500 °C [68]. TiO_2 is transparent normally in the visible light region; its band gap is 3.0 eV for rutile and 3.2 eV for anatase crystalline phase. By doping or sensitization, it is possible to improve the optical activity of TiO_2 and to move its absorption threshold into the visible light region.

While microstructure parameters strongly affect properties of nano-sized materials, non-stoichiometry can also have a profound influence on some physical properties, such as electronic transport properties [69, 70]. In TiO_2, non-stoichiometry may be due to oxygen vacancies and excess titanium on interstitial sites or Ti vacancies. Ion doping (or partial ion substitution) of TiO_2 can influence a deviation from stoichiometry by forming oxygen vacancies and appearance of cation deficit with a change in cation valence. For example, in TiO_2 doped by *3d* ions an increase in the photo-catalytic activity has been observed [71]. By incorporation of Fe^{3+} in host TiO_2 the light absorption of pure TiO_2 is significantly shifted from UV towards visible spectral region [46, 47]. Crystalline structure, size and shape of the particles strongly influence the possible application of TiO_2. Although for photo-catalytic applications anatase displays higher efficiency than rutile, a combination of anatase and rutile seems to be the best [72]. However, due to its wide band gap (3.0 – 3.2 eV) which requires UV radiation for excitation, commercial photo-catalytical application of TiO_2 is still moderate. Further, fast recombination of photo-generated e^--h^+ pairs lowers the quantum yield below 10%. The photo-catalytic efficiency of titanium dioxide could be enhanced by doping with various elements [13, 44, 46, 71, 73-75], e.g. Cr^{3+}, Ni^{2+} or Fe^{3+} ions. Incorporating these ions in titania crystal structure could shift the absorption threshold above 450 nm, resulting in absorption in the visible part of the spectrum [71]. Moreover, the shape of titanium dioxide nano-crystals greatly influences e^--h^+ recombination rate. For example, in nano-rods charge carriers are free to move along the length of the elongated nano-crystal, which is not the case for nano-spheres. Therefore, charge carriers delocalization is higher and, consequently, probability of their recombination is lower [76].

This chapter is focused on synthesis and characterization of nano-particles and elongated TiO_2 nano-crystals, nano-rods with rutile crystalline structure, pure and doped with Fe^{3+} ions. First of all, a detailed study of materials composition, structure and microstructure (particle size, morphology, micro-strain, and defects concentration), as they provide important insights for better understanding of their properties, essential for applications. To achieve this, we carried out HRTEM, Raman spectroscopy, x-ray diffraction, as well as SQUID magnetic measurements. Optical properties (absorption, reflection and luminescence) of pure and Fe-doped TiO_2 nano-particles were obtained and correlated with materials composition and microstructure. Photo-degradation of selected herbicides in the presence of synthesized catalysts was chosen as a probe reaction for evaluation of photo-catalytic activity of prepared samples. The relationship between optical properties (PL) and photo-activity of samples is discussed.

PREPARATION AND STOICHIOMETRY OF PURE AND FE-DOPED RUTILE TiO_2 SAMPLES

All chemicals used for synthesis were of p.a. purity and were used without further purification. Triply distilled water was used for aqueous solutions. TiO_2 nano-particles were synthesized by procedure developed by Abazović et al. [42, 46, 47], modifying a "low temperature dissolution-reprecipitation process". Briefly, 5 ml of $TiCl_4$ (Fluka) pre-chilled to -20 °C was slowly added to 200 ml of water, under vigorous stirring at 5 °C. The reaction vessel was kept at low temperature for 2 hours, then it was sealed and under continuous stirring the sol was kept at 45-50 °C [42, 46,47,49] or 60- 70 °C [50] for 16 hours. When samples doped with iron ions were synthesized, an appropriate amount of $FeCl_3$ (Aldrich) was dissolved in pre-chilled water before addition of $TiCl_4$. The rest of the procedure was the same as described for pure rutile samples. The obtained dispersions were dialyzed against water to remove chloride ions and the resulting precipitates were separated by centrifugation and dried in vacuum at room temperature.

The concentration of iron ions in dried samples was determined by different techniques: by inductively coupled plasma optical emission spectroscopy (ICP-OES Spectroflame ICP, 2.5 kW, 27 MHz) [46, 50], by scanning electron microscopy using energy dispersive spectrometer as detector (SEM/EDS, JEOL JSM-35) [49] and by X-ray photoelectron spectroscopy (XPS VG Scientific ESCALAB 210 spectrophotometer) [46]. Results and sample numeration which will be used throughout this chapter are presented in Table 1.

The Fe concentrations obtained from ICP-OES and SEM/EDS measurements are comparable, show expected trend and are much lower compared to stoichiometric (initial) values. Obviously the major part of the initial Fe ions were omitted from the precipitated TiO_2 nanoparticles, or just adsorbed on the surface of the particles. Those ions were excluded from the dispersions by dialysis process, prior to centrifugation and drying. XPS is analytical technique which is mainly useful for analyzing surfaces.

Table 1. Sample numeration and real concentration of Fe-dopant in synthesized rutile TiO_2

S. No	Initial Fe at.%	Fe at % in TiO_2 matrix		
		ICP-OES	SEM/EDS	XPS
1.	0.0	0.03±0.03	0.00	0
2.	2.5	0.20-0.22	0.13	0
3.	5.0	0.43-0.47	0.19	0
4.	10.0	1.00-1.05	1.27	0
5.	20.0	1.70-1.75	1.48	0.2

This technique could reveal the direct presence of iron ions only in sample 5 (initial 20% of Fe ions), only sample with the highest doping concentration of Fe ions had dopants observable at the surface, the other samples had dopant ions in the interior of the nano-particles. We have examined three areas of the XPS spectrum of our samples: the Ti 2p region between 455 and 470 eV (Figure 1a), the O 1s region between 525 and 535 eV (Figure 1b) and the Fe $2p_{3/2}$ region near 710 eV (Figure 1c).

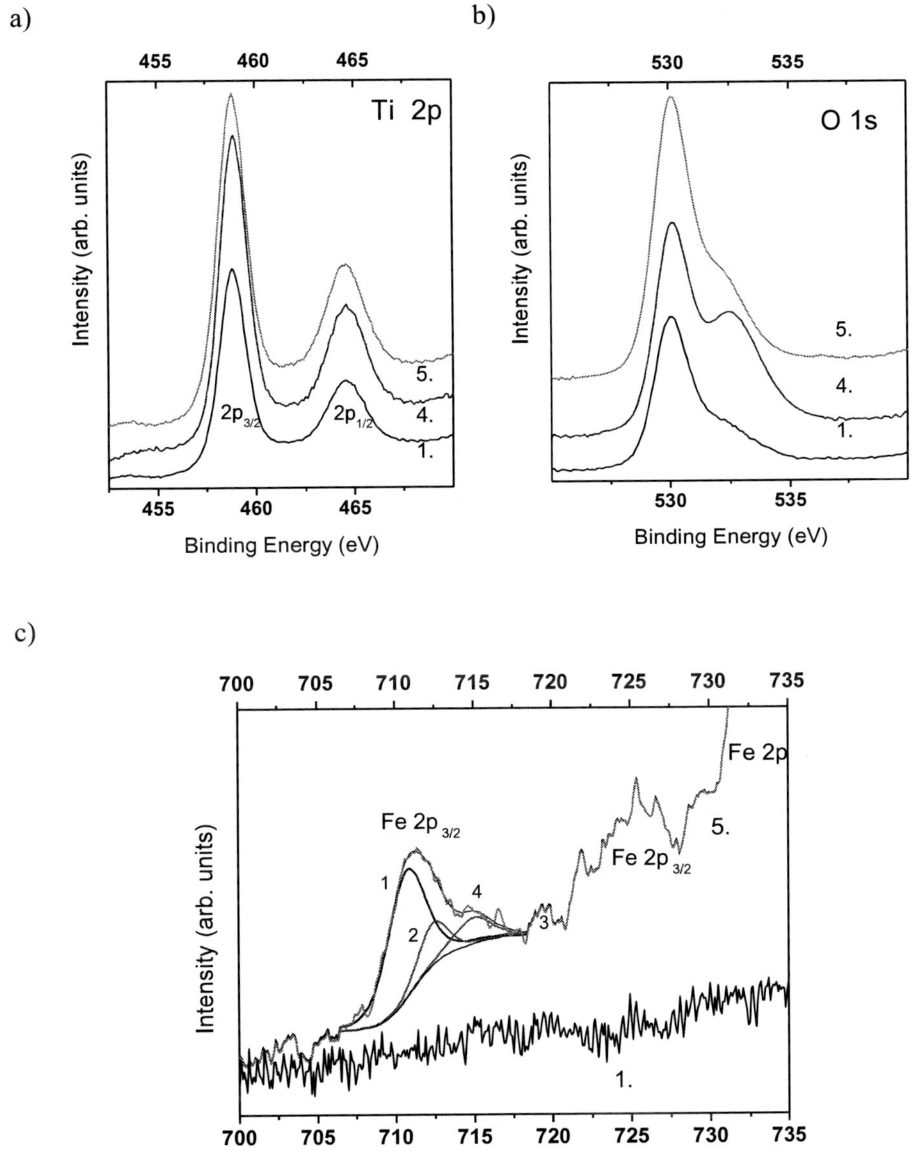

Figure 1. XPS spectra of pure (1.) and Fe doped TiO_2 (4. and 5.) samples: Ti 2p energy region (a), O 1s energy region (b), and Fe 2p energy region together with deconvolution of the Fe $2p_{3/2}$ (c) [46].

It must be noted that in spectra of all samples peak at around 284.6 eV was found (not presented in Figure 1), corresponding to carbon impurities, arising probably from the background of XPS test or the residual precursors. The Ti $2p_{1/2}$ and Ti $2p_{3/2}$ spin-orbital splitting photoelectrons for all samples are located at binding energies of 464.6 and 458.9 eV, respectively, as can be seen in Figure 1a.

These data are in excellent agreement with data obtained by Zhu et al. [77]. It seems that these energies are not affected by doping with iron ions. It is possible that due to the low concentration of iron, the shift of Ti 2p peaks could have been below the detection limit. No

Ti^{3+} species were observed in XPS. The absence of peak broadening of Ti $2p_{3/2}$ signals (FWHM was about 1.4 eV for all samples) may also indicate the presence of Ti^{4+} species only [78], and good crystallization for all samples [79]. Figure 1b shows the O 1s core level spectra of pure and doped TiO_2. The peak at 530.1 eV for all samples is due to O^{2-} ion in the TiO_2 lattice. In all spectra at higher binding energy a shoulder is observed (the spectrum of sample 4 shows a second peak located at a binding energy of 532.6 eV), which can be attributed to the surface hydroxyl groups of chemisorbed water molecules on the titania [79, 80] or this is the way in which presence of Fe ions influences the TiO_2 matrix. The atomic weight ratio of O to Ti in samples increases with doping: from 1. pure TiO_2: 1.955; 4. Fe-TiO_2: 2.3; to 5. Fe-TiO_2: 3. The XPS spectrum of Fe is very weak (Figure 1c) due to the low doping level. The presence of Fe ions was detected only in sample 5. (20 at.% initial) Fe-TiO_2 sample, estimated concentration being about 0.2 %. The Fe ions could not be detected in sample 4. Fe-TiO_2 due to detection limit of instrument, but their presence can be indirectly assumed from O/Ti ratio (2.3), as mentioned before. The binding energies from 711.0 to 711.8 eV and from 725.4 to 726.0 eV should be assigned to $2p_{3/2}$ and $2p_{1/2}$ of Fe^{3+}, respectively. These data exhibit a small shift compared to those in Fe_2O_3 (711.3 eV for $2p_{3/2}$) probably indicative of more positively charged surface due to diffusion of the Fe^{3+} in TiO_2 lattice and possible formation of the Fe-O-Ti bond in the sample [81]. Also it can be seen in Figure 1c that Fe $2p_{3/2}$ peak can be decomposed in several contributions corresponding to different oxidation states of iron. The main contribution can be attributed to Fe^{3+} ions (peak 1, binding energy at 710.9 eV) and the other to Fe^{2+} ions (peak 2, binding energy at 712.8 eV). These statements can be supported with the presence of characteristic satellite peaks at ~720 eV for Fe^{3+} (peak 3) and at ~ 716 eV for for Fe^{+2} (peak 4), well established for Fe 2p spectra for Fe_2O_3 and FeO, respectively [82].

Morphology and Microstructure by TEM and HRTEM

Transmission electron microscopy was used in order to characterize size and morphology of the nano-crystallites. Figures 2-4 show typical TEM and HRTEM images of pure (sample 1.) and Fe doped TiO_2 (samples 2. and 5.), respectively.

In all TEM images with a lower magnification (Figure 2a, Figure 3a and Figure 4a), the flower-like agglomerates with sizes of 200-400 nm can be observed, for pure and Fe-doped TiO_2 samples. Agglomerates are bundles of TiO_2 nano-rods with length of about 80 ± 20 nm and diameters of 5-10 nm (aspect ratio of about 10). Rod-like shapes of nano-crystallites were observed, typical for rutile phase of TiO_2.

Also, Fourier transform (similar to a local micro-difraction) in Figures 2b, 3b, and 4b indicated that nano-particles are crystalline. HRTEM shows the presence of slightly curved surface of the nano-rods (for example, arrows above nano-rod, Figures 2 b and 4b) and the presence of substructure inside the nano-rods.

Images revealed that nano-rods consist of spherical crystalline nano-particles with a diameter of 5-10 nm, one of them being attached to the lower part of the nano-rod (Figure 4b, pointed with arrow).

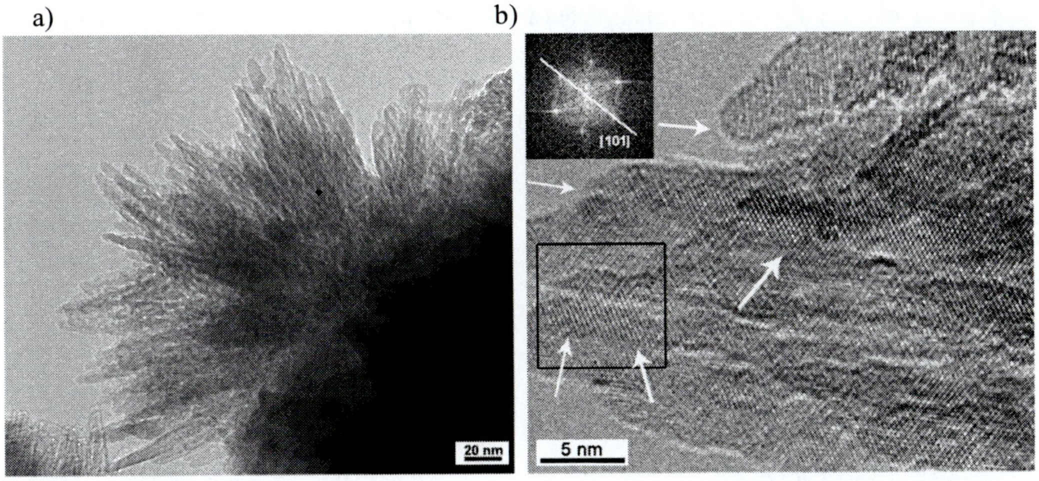

Figure 2. Typical TEM (a) and HRTEM (b) images of pure TiO_2 (sample 1), Fourier transform in the inset of part (b). Measurements were carried out on a transmission electron microscope TECNAI G2 300F, Eindhoven, the Netherlands, operated at 300kV [47].

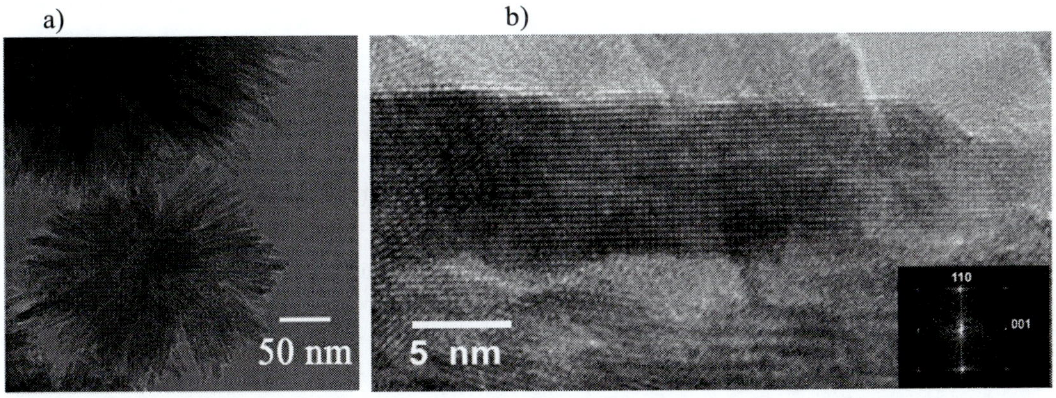

Figure 3. Typical TEM image of Fe-TiO_2 (sample 2) (a), and HRTEM image of longitudinal section of a nano-rod, with Fourier transform in the inset. Measurements were carried out using TOPCON 002B, operating at 200 kV [50].

It has been reported lately that oxides with rutile-like structure form nano-rods which consist of spherical crystalline particles aggregated along a specific crystallographic direction. [84]. Oriented attachment, involves spontaneous self-organization of adjacent particles so that they share a common crystallographic orientation, followed by joining of these particles at a planar interface. Bonding between the particles reduces the overall energy by removing the surface energy associated with dangling bonds, resulting in coherent grain-grain boundary. This mechanism is relevant when particles are free to move in solutions or when particles have abundant surface-bound water, which was the characteristic of our synthetic procedure and subsequent treatment (prolonged aging at elevated temperature). This process can also explain the differences between the directions of elongation of undoped and doped TiO_2 nano-rods synthesized in this work and literature data. Formation of nano-rods took place on the expense of initially formed nano-spheres whose surface energy is highly sensitive to synthetic procedure and defines further organization in nano-rods.

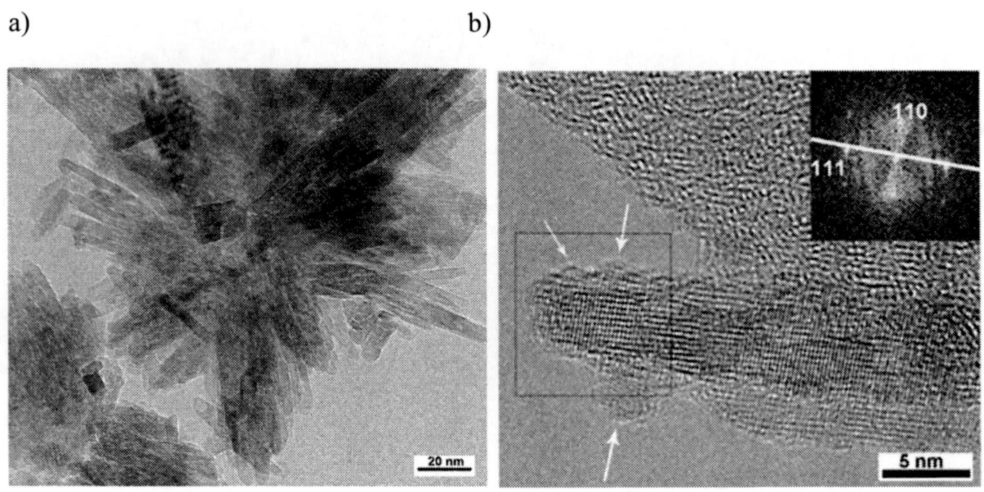

Figure 4. TEM image (a) and HRTEM image (b) of Fe-TiO_2 (sample 5), Fourier transform in the inset of part (b). Measurements were carried out on a transmission electron microscope TECNAI G2 300F, Eindhoven, the Netherlands, operated at 300kV [47].

The interplanar distance of about 3 Å that can be measured in Figure 3b, corresponds to (001) planes of rutile structure, which indicates that preferred particle growth direction is [001]. There are a lot of literature data devoted to grain growth process of TiO_2 [83-88]. Their findings are consistent that rutile's inherent growth is along [001] because that surface has the highest energy among surfaces considered. No crystalline or amorphous phase containing Fe (pure Fe or oxide) was detected from the HRTEM examinations.

Structure Analysis

The XRD measurements of TiO_2 powders pure and doped with of Fe ions revealed once again that all samples have rutile crystal structure, without any detectable dopant related phase (Figure 5). This result is in agreement with literature data by Yin et al. [89], and can be explained by the fact that the radius of Ti^{4+} (0.68 Å) and Fe^{3+} (0.64 Å) are almost the same. Consequently, Ti^{4+} ions can be replaced with Fe^{3+} causing no distortion of rutile crystal lattice [13]. The diameter of the particles was calculated using the Scherrer diffraction formula, equation (1), relating the diffraction angular width (β) at the half height of the (101) diffraction peak, to the domain size:

$$D = \frac{k\lambda}{\beta \cos\theta} \tag{1}$$

where *k* is a geometrical factor taken to be 0.89, $\lambda = 1.541$ Å is the X-ray wavelength, and θ is the diffraction angle of the most prominent peak for rutile structure ($2\theta \sim 27.4°$, (101) peak).

We found that Fe-doped TiO_2 crystal domains are in the range from 5 to 7 nm in diameter. The sizes of the crystalline domain are in a good agreement with dimension of the isolated particle in the Figure 4b.

Microstructure analysis of 1-4 samples (assignment given in Table 1) was performed using Synhrotron X-ray technique, at room temperature at the 6-ID-D beam line at the Advanced Photon Source at Argonne National Laboratory, Argonne, IL. Diffraction data were collected using the rapid acquisition pair distribution function (RAPDF) technique that benefits from two-dimensional (2D) data collection, details as given by Kremenović et al. [50]. The analysis of obtained XRPD patterns, Figure 6, shows that all samples predominantly crystallized in rutile form with small amount of anatase phase that could be noticed only in pure TiO2 (sample 1) and Fe-doped TiO2 –sample 4. The patterns were analyzed by FullProf program [90] started initially with standard Rietveld refinement procedure.

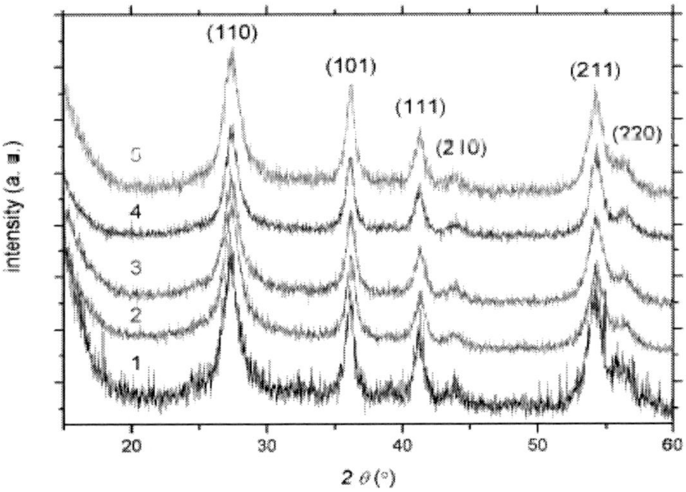

Figure 5. X-ray diffraction patterns of pure (1) and Fe-doped TiO_2 nano-particles (2-5). stoichiometry of the samples is given in Table 1. XRD characterization of the samples was performed using Philips PW1710 (Almelo, the Netherlands) diffractometer.

However, an attempt to fully implement Rietveld refinement in order to refine all crystal structure parameters was unsuccessful in this case; this was attributed to the difficulties in integrated intensity calculations for highly anisotropic particles. To avoid these problems, further analysis was done by using FullProf in profile matching mode with the chosen "needle like" crystallite size line broadening model [90]. The refinement based on this model has shown very good agreement with experiment (Figure 6).

The profile of diffraction peaks was described by Thompson-Cox-Hastings approximation of Voigt function comprising the size and strain calculations based on integral breadths [91, 92]. The instrumental Full Width at Half Maximum (FWHM) was derived from XRPD pattern of standard CeO_2 powder sample for microstructure analysis [93].

The results obtained for cell parameters and microstructure (crystallite size and average microstrain) are summarized in Table 2. It was not possible to calculate average particle size due to the fact that FWHMs of certain reflections were similar to the instrumental FWHM, so in these directions (close to [001]) it was not possible to extract the actual particle size (these values are indicated in Table 2. as "resolution limited").

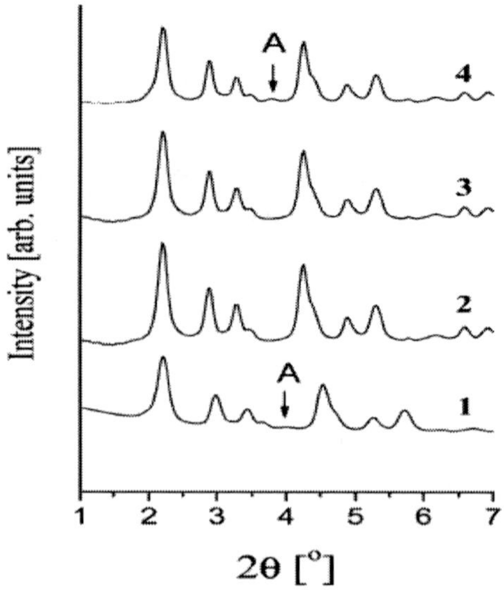

Figure 6. Part of synchrotron X-ray diffraction patterns of pure (1) and Fe-doped TiO_2 (2-4) samples. A denotes anatase contribution [50].

Thus, instead of the average size values, dimensions along three directions are stated in Table 2: [110] direction, for which the smallest size value was found in all samples, indicating the direction corresponding to the cylinder diameter (assuming cylindrical particle shape as observed by TEM); the second one is size along direction [001], the value of which was too large to be correctly evaluated by XRD, indicating the preferential particle growth along [001] direction; the third direction listed in Table 2 is [103], for which the size value was found to be the largest among all non-orthogonal directions to [110]. This value can be used as an indicator of crystallite length along [001] since the real value along [001] cannot be precisely determined.

Table 2 shows that the isotropic micro-strain increases uniformly with the iron content increase. However, crystallite size changes non-uniformly with the Fe content increase. It was shown that smallest crystallites were formed in samples 2 and 3 (doped medium concentrations of iron, Table 1).

Table 2. Unit Cell parameters, Average Strain and Crystallite Size for Pure and Fe-doped TiO_2 Nano-crystals

Unit cell/strain/size	Sample No. 1	2	3	4
a [Å]	4.6055 (3)	4.6149 (3)	4.6139 (3)	4.6078 (3)
c [Å]	2.9543 (3)	2.9536 (3)	2.9545 (2)	2.9569 (2)
Average strain 10-4	20.26 (2)	25.36 (3)	28.1 (2)	30.842 (2)
Size [Å]				
Along [110]	102	51	53	68
Along [001]	23000a	16000a	22000a	12000a
Along [103]	488	245	246	326

[a]Resolution limited. These micrometer- scale values are out of XRD measurement scope and should not be regarded as reliable [50].

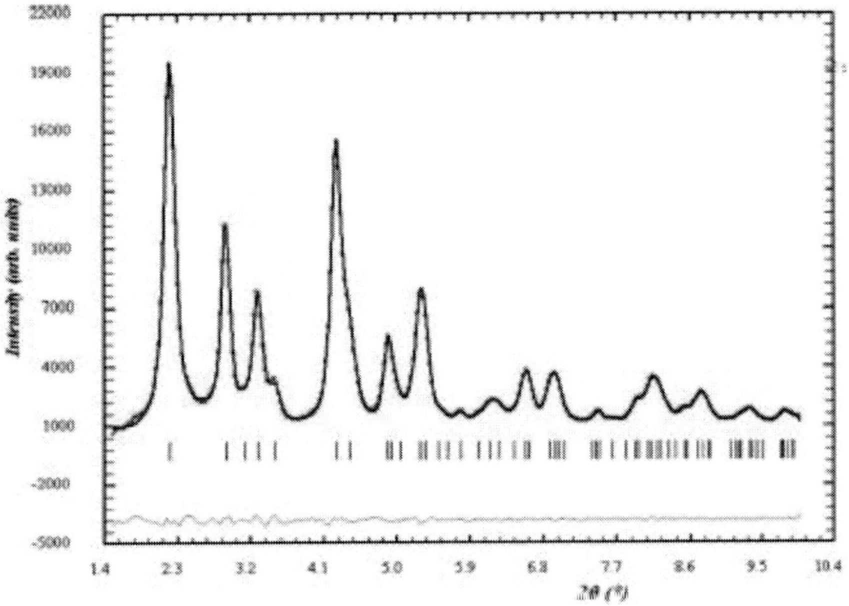

Figure 7. Rietveld fit for Fe-doped TiO_2, sample 2 (R_{wp} ~ 3%).The difference (bottom) curves between experimental and calculated values is offset for clarity. Reflection positions for rutile are represented by vertical ticks [50].

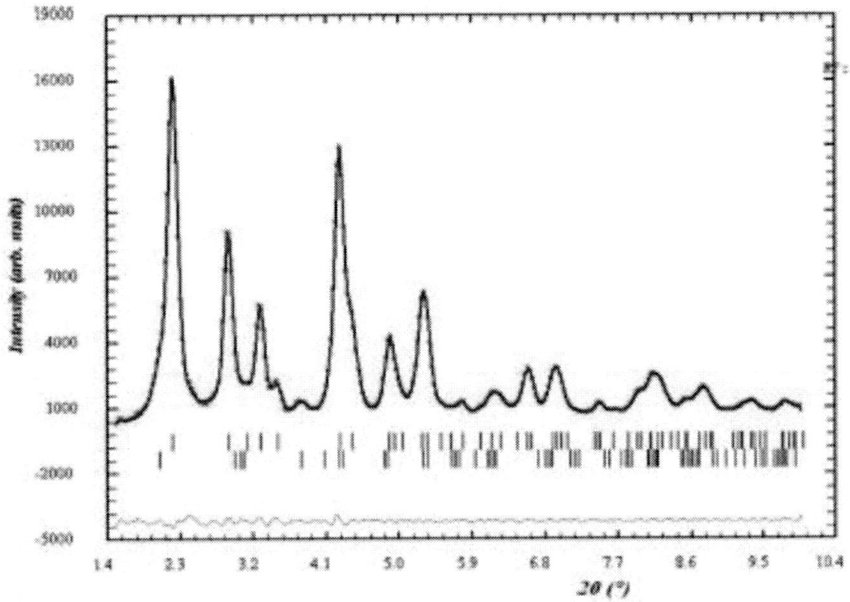

Figure 8. Rietveld fit for Fe-doped TiO_2 , sample 4 (R_{wp}~ 3%).The difference (bottom) curve between experimental and calculated values is offset for clarity. Reflection positions are represented by vertical ticks (first row denotes rutile phase and second row anatase phase) [50].

As expected, the largest crystallite size was found in pure TiO_2. However, further increase of iron content above (sample 4) did not induce further decrease of size, on the contrary, sample 5 showed the largest particle size among all studied doped samples.

The amount of anatase phase was also found not to be in regular correspondence with the iron content. The diffraction patterns of samples doped with medium concentration of Fe-dopant samples 2 and 3 did not show the presence of anatase detectable by x-ray diffraction, and these were refined as single phase patterns, Figure 7. On the other hand, pure TiO_2 and highest Fe-dopant concentration, sample 4 (1.05 at. % Fe doped) gave diffraction patterns with notable presence of anatase phase. They were refined as two-phase patterns (for illustration, refined pattern of sample 4 is shown in Figure 8) and the amount of anatase ratio is estimated to 4.6% and 6% for pure and Fe-doped sample 4 respectively.

The crystal structures of rutile and anatase are in agreement with literature [67]. Due to very small amount of Fe it was not possible to refine reliably Fe content and its distribution in doped samples. Irregular unit cell parameter change with Fe concentration increase could indicate large crystallite macro-strain or random substitution of Ti by Fe. Large crystallite size anisotropy is in accord with HRTEM results. Rather small strain anisotropy indicates absence of dislocations and stacking faults, again in accord with HRTEM results. However, presence of point defects in Fe-doped TiO_2 could be expected.

Optical Properties of Rutile Nanostructures

Influence of doping on UV/Vis spectral properties of TiO_2 is clearly evident in Figure 9, and it is manifested by change of the colour of doped powders from light to dark yellow. DRS of all doped samples contain shoulder at 500 nm, with intensity increasing with Fe^{3+} concentration [77]. Reflectance data presented in Fig. 9 were converted to F(R) values using the Kubelka-Munk equation. The optical band gap energy of doped TiO_2 samples (calculated using Tauc's plot approximation) decrease with increasing Fe concentration - from 3 eV for pure TiO_2 to ~ 2.3 eV for highest dopant concentration (sample 5). This indicates that photo-catalytic activity of TiO_2 could be also influenced by Fe-doping. The tail in the visible spectral region can be observed for all samples, as a consequence of light scattering or presence of oxygen vacancy states between the valence and conduction bands in TiO_2 band structure. The presence of oxygen vacancies can be further confirmed by the substructure of PL spectra of samples, Figure 10.

Considering photoluminescence (PL) spectroscopy as highly sensitive tool to study the photo-physics of the photo-generated species we have investigated electronic properties of TiO_2 rutile nano-particles. We found out that synthesized rutile TiO_2 particles are emissive with well resolved PL spectra extending in visible spectral region due to the presence of oxygen vacancies.

Photon energy up-conversion (UC) in semiconductors (the observation of an emission at energies higher than the excitation energy) has attracted much attention recently. Such UC effects have been reported in bulk semiconductors [94, 95], heterostructures [96- 102], and self-assembled quantum dots [103, 104]. For UC in bulk semiconductors and heterostructures, the proposed mechanism involves either two-step photo-exitation process [98- 100, 102] or Auger recombination that ionizes carriers to the higher band gap region where they undergo radiative recombination [96, 97, 101]. Recently, photon energy UC has been observed in colloidal InP [105], CdSe [105], CdTe [106], and core-shell CdSe/ZnS quantum dots [106, 107], as well as Si nano-crystals [108]. In the case of colloidal quantum dots, the proposed mechanism is based on surface states and phonon processes [105- 107]. PL and UCPL spectra of pure (a) and doped (b) rutile TiO_2 nano-particles as well as their deconvolutions, are presented in Figure 10. PL spectra (Figure 10a λ_{exc}= 250 nm and Figure10b λ_{exc}= 300 nm) presents broad emission in the spectral range from 350-410 nm (> 3eV) as well as the presence of well-resolved peaks /shoulders at 2.91, 2.80, 2.70, 2.56 and 2.34 eV. We assigned high-energy peaks (> 2.9 eV) to band edge luminescence of TiO_2 particles, while lower-energy peaks/shoulders (at 2.8, 2.7, 2.56, and 2.34 eV) are induced by the presence of the oxygen vacancies [42, 46]. The PL spectra of pure (a) and doped (b) rutile nano-particles with excitation energies smaller than band gap energy of TiO_2 are also presented in Figure 10, (λ_{exc}= 600 nm/ E_{exc}= 2.07 eV). UC-PL spectra have the same positions of peaks/shoulders and similar intensities compared to normal PL spectrum. It was found that the lowest excitation energy for which the UC-PL still can be detected is 1.89 eV.

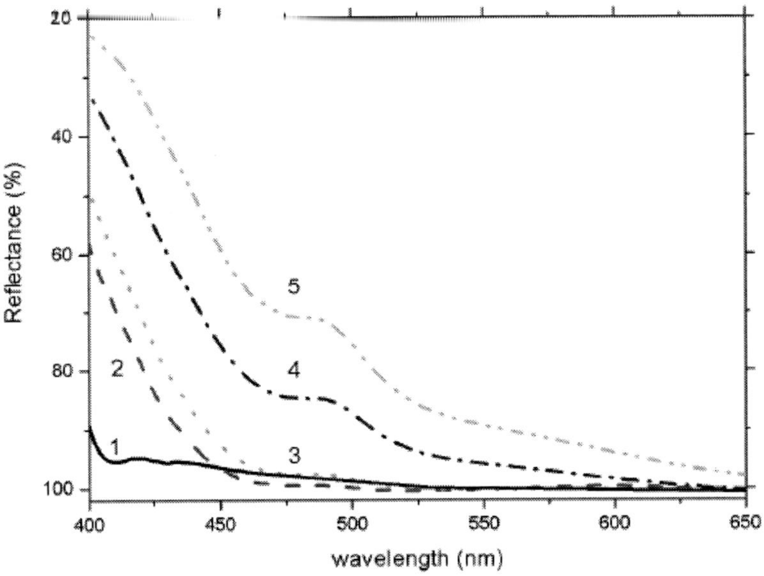

Figure 9. Diffuse reflectance spectra of pure (1) and Fe doped TiO_2 nanoparticles (2-5, Table 1.), Evolution600 UV/Vis spectrophotometer equipped with DRA-EV-600 Diffuse Reflectance Accessory [49].

Doping of TiO_2 with iron ions is also beneficial for introducing oxygen vacancies in/on the crystal lattice or surface of TiO_2 [109]. The observed UC-PL most likely originates from cascade excitation involving mid-gap trap states, i.e. from the valence band to intra-band trap state located near conduction band and subsequent thermally activated de-trapping to the conduction band, followed by recombination from the conduction band or mid-gap energy levels to the valence band. When oxygen vacancies are on or near the particle surface, they can favor adsorption of water and the formation of surface hydroxyl groups (we have many hydroxyl groups on the surface of samples observed by XPS, Figure 1b), which can promote photocatalytic activity, that is the subject of the next paragraph.

Photocatalytic Activity of Pure and Fe-Doped Rutile Nanoparticles

In a view of the former paragraph, we studied efficiency of pure and Fe-doped rutile TiO_2 and Degussa P25 TiO_2 for comparison, as a photocatalyst in degradation of the herbicides RS-2-(4-chloro-o-tolyloxy) propionic acid (mecoprop, MCPP, $C_{10}H_{11}ClO_3$) and (4-chloro-2-methylphenoxy) acetic acid (MCPA, $C_9H_9ClO_3$) (Figure 11) using UV or visible light for irradiation. These herbicides were chosen because of their wide use for selective control of many annual and some perennial weeds and because of their frequent presence in drinking water [110, 111].

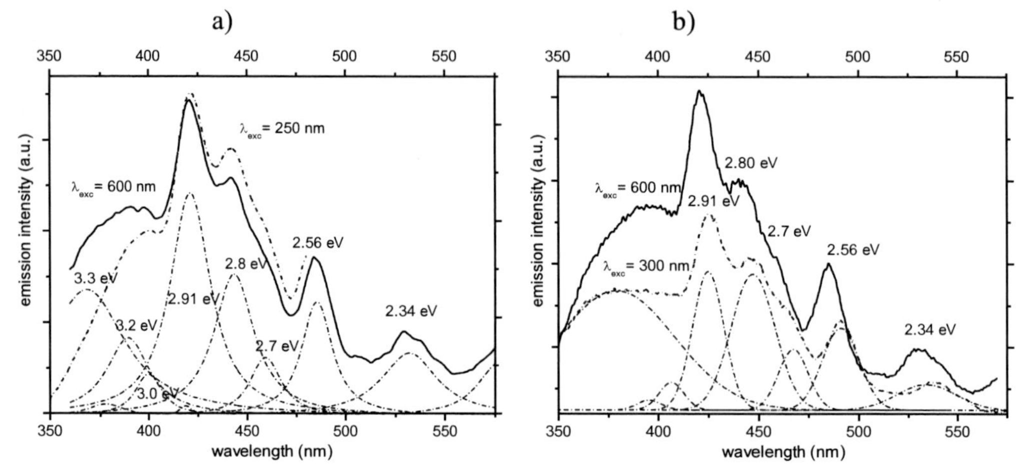

Figure 10. Typical PL and UC-PL spectra and their deconvolutions of pure rutile (a) and (b) Fe doped rutile, sample 1 nano-particles [42, 46]. (Perkin Elmer LS-3b).

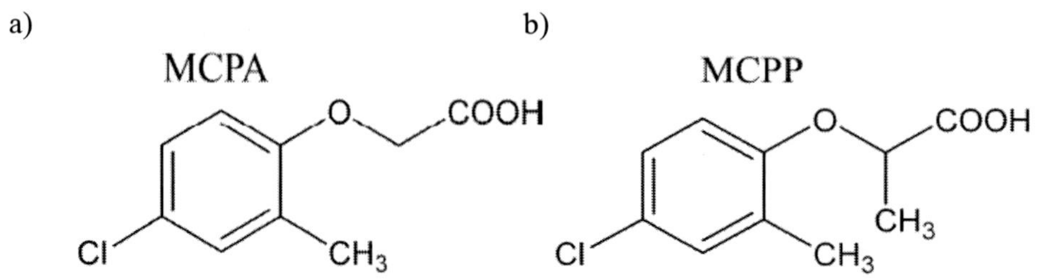

Figure 11. Structures of the herbicides.

Photocatalytic degradation was carried out in a cell described previously [46]. Aqueous suspensions of catalyst containing herbicide were sonicated for 15 min before irradiation, to make the catalyst particles uniform. The suspension thus obtained was thermostated at 40 ± 0.5 °C in a stream of O_2 and then irradiated with UV (125 W medium-pressure mercury lamp- Philips, HPL-N, emission bands at 304, 314, 335 and 366 nm, with a maximum at the 366 nm) or visible light (50 W halogen lamp Philips and a 400 nm cut-off filter). During irradiation, the mixture was stirred at a constant speed. All experiments were carried out at a natural pH ~2.8. For the kinetic studies of herbicides photodegradation aliquots (0.25 cm^3 to keep total volume change below 10%) of the reaction mixture were taken at the beginning of the experiment and at regular time intervals and diluted to 10.00 cm^3 with doubly water.

The suspensions containing photocatalyst were filtered through Millipore (Millex-GV, 0.22 μm) membrane filter. The spectra of obtained solutions were recorded in the wavelength range from 200 to 400 nm on a Secomam Anthelie Advanced 2 spectrophotometer in 1 cm quartz cells. Kinetics of the herbicides degradation was monitored at 228 nm.

The obtained kinetic curves are presented in Figures 12 and 13. As can be seen in Figures 12a and 13a, the best photocatalyst was Degussa P25 for degradation for both herbicides, MCPA and MCPP, when UV light was used. We expected that result because of special properties of Degussa P25 due to its mixed anatase/rutile phase composition which significantly reduce recombination probability [72]. A small amount of anatase phase, that was discovered by synchrotron X-ray diffraction (4.6 % in pure rutile and 6 % in sample 4), was not enough for significant increase of photocatalytic efficacy. Among the photo-catalysts that are the subject of this chapter, the highest efficacy using UV light, showed pure rutile TiO_2 in degradation of both herbicides. Fe-doping of rutile TiO_2 reduces its photo-catalytic efficacy of degradation of both herbicides; the lowest photocatalytic efficacy was obtained for both herbicides using the Fe-doped samples with the highest dopant concentration (sample 5). When visible light was used, Figures 12b and 13b, the best efficacy showed pure rutile TiO_2, and the worst was obtained using sample with highest Fe-dopant concentration (sample 5) for both herbicides similar to results obtained using UV light irradiation. Literature data concerning influence of Fe doping on TiO_2 photocatalytic activity, showed critical dependence on the method of preparation and choice of compound whose degradation was followed [75, 112,113]. Doping of TiO_2 with iron ions is also beneficial for introducing oxygen vacancies in/on the crystal lattice or surface of TiO_2 [109]. When oxygen vacancies are on or near the particle surface, they can favor adsorption of water and the formation of surface hydroxyl groups (we have many hydroxyl groups on the surface of samples observed by XPS), which can promote photocatalytic activity. Since it didn't happen in our experiments we assumed that oxygen vacancies are mainly inside the nano-particles, their existence was proved by photoluminescence measurements (Figure 10).

To interpret the influence of the presence and concentration of Fe^{3+} ions on the photocatalytic activity of the Fe-doped rutile TiO_2 it is necessary to take into account a number of factors [114]. Firstly, Fe^{3+} ions can enhance the intensity of absorption in the UV/vis light region and make a red shift in the band gap transition of the Fe-doped rutile TiO_2 samples (Figure 9). This can induce more photogenerated electrons and holes to participate in the photocatalytic reactions. Apart from this effect, Fe^{3+} ions can also serve as a mediator of the transfer of interfacial charges at an appropriate doping concentration.

Besides, the presence of a small amount of Fe^{3+} ions can enhance the activity, but excessive Fe^{3+} ions are detrimental [114]. A most probable explanation of the decrease of photocatalytic efficiency with increase in the Fe^{3+} content might be the trapping of holes on the catalyst surface, which lowers their apparent mobility and results in an increased probability of their recombination with electrons [60]. Also, the increase in the Fe^{3+} concentration may lead to the formation of the aggregates of Fe-oxides (Fe_2O_3 or FeO), causing a further decrease of the photodegradation rate, which is a consequence of lower mobility of the charges and reduced adsorptive properties of Fe oxide compared to TiO_2 [115].

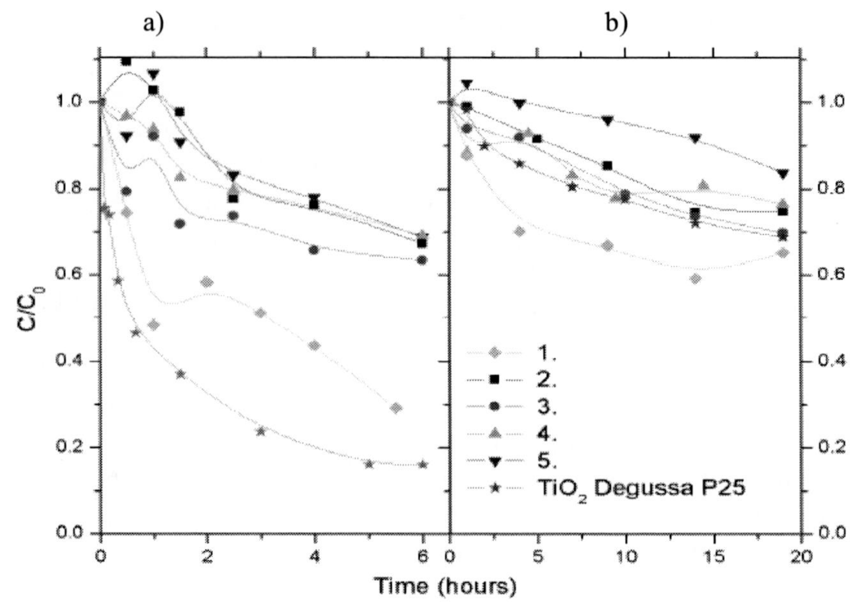

Figure 12. Spectrophotometric monitoring of herbicide MCPA photo-catalytic degradation in the presence of synthesized pure and Fe-doped rutile TiO_2 and Degussa P25: (a) UV and (b) Visible light.

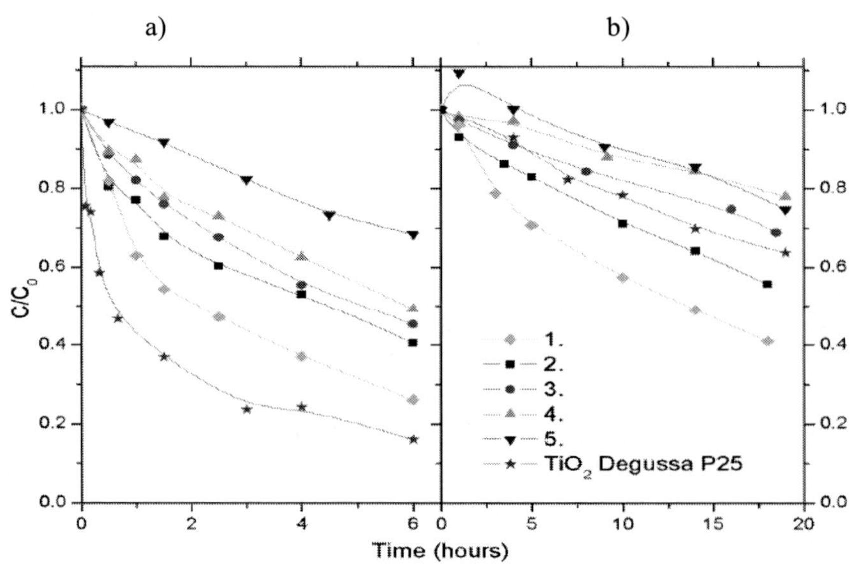

Figure 13. Spectrophotometric monitoring of herbicide MCPP photo-catalytic degradation in the presence of synthesized pure and Fe-doped rutile TiO_2 and Degussa P25: (a) UV and (b) Visible light.

Secondly, the large surface area may be an important factor in certain photocatalytic degradation reactions, since the large amount of adsorbed organic molecules promotes the reaction rate. However, powders with a large surface area are usually associated with large amounts of crystalline defects which favor the recombination of photogenerated electrons and holes, leading to a poor photoactivity [114]. Thirdly, particle size is another important parameter influencing the photo-catalytic efficiency, since the electron–hole recombination rate may depend on the particle size.

Finally, but not the least important, come the effects of the substrate structure and the pH at which the photodegradation process is taking place. The observation that there does not exist an Fe^{3+} concentration at which the photodegradation rate of all tested herbicides would be the highest suggests that the structure of the substrate plays a decisive role in the photocatalytic process. This conclusion is also supported by the finding that the concentrations of MCPA and MCPP on the catalyst with highest Fe^{3+} content (sample 5, Figure 12b and Figure 13b) using visible light, in the first period of photocatalytic degradation and concentration of the MCPA on the catalyst with the smallest and highest Fe^{3+} content (sample 2, Figure 12a), are higher than their initial equilibrium concentrations, which may be ascribed to the desorption of the herbicides, induced by irradiation [77]. It is well known that the surface of TiO_2 is amphotheric, so that at a low pH a significant number of surface sites are protonated, yielding a positively charged surface.

Hence, the maximum adsorption of carboxylic acids will occur at the pH corresponding to the positively charged TiO_2 surface, as the de-protonated acid molecules carry negative charges. Therefore, one can expect efficient adsorption of acids whose pK_a values (MCPP, $pK_a = 3.11 - 3.78$ [116]; MCPA, $pK_a = 3.1$ [117]) are lower than the pH_{ZPC} ($\sim$5.1) of rutile TiO_2 [118]. Also, FTIR measurements [45, 119] confirmed the formation of RS-2-(4-chloro-o-tolyloxy)propionate species after the adsorption of MCPP on the TiO_2 surface. It is obvious that at such high dopant concentration the population of Fe^{3+} ions on the very surface of TiO_2 particles or adjacent to it is relatively high. When the suspension is illuminated, the electrons generated are being rapidly trapped ($Fe^{3+} + e^{-} \rightarrow Fe^{2+}$), which affects the surface properties and yields desorption of MCPA and MCPP. As a consequence, an increase in the concentration of the free herbicide in the solution is observed at the beginning of the photocatalytic experiment. Later on, the degradation becomes more intensive than desorption.

In summary, the results indicated that the rate of MCPA and MCPP degradation is higher on undoped rutile TiO_2 photocatalyst compared to its Fe-doped form, no matter which kind of light was used for irradiation (UV or Vis). A comparison of photocatalytic efficiency of pure and Fe-doped rutile TiO_2 with that of Degussa P25 when UV light was used, shows that Degussa P25 is the most efficient in degradation of applied herbicides. A comparison of the performance of pure rutile TiO_2 and Fe-doped rutile TiO_2 with that of TiO_2 Degussa P25 shows that pure rutile TiO_2 is more efficient in the degradation of MCPA and MCPP when visible light was used, the efficiency of the Fe-doped rutile TiO_2 being similar to or lower than that of TiO_2 Degussa P25.

CONCLUSIONS

Different rutile nanostructures of pure and Fe^{3+}- doped titania were synthesized by co-precipitation in acidic aqueous solution and characterized by various techniques. Analytical techniques (ICP-OES, XPS and SEM/SED) revealed that only small fraction of initial concentration of iron ions are incorporated in TiO_2 nano-particles. Concentrations are about ten times lower that initial concentrations: sample 2 had 0.13-0.22 at. %, initial concentration was 2.5 at.%; similar results are obtained for all samples. XPS technique revealed that at the surface of Fe-doped rutile nano-particles Fe ions can be found only in sample with highest concentration of Fe-ions. In all other samples iron ions are inside of the particles. Pure and Fe-doped samples are characterized by the high concentration of hydroxyl groups of chemisorbed water.

TEM and HRTEM images showed that rutile nano-crystallites were elongated structures with length of 80 ±20 nm and diameters of 5-10 nm, organized in flower-like agglomerates with dimensions of 200-400 nm. Elongated nanostructures are consisted of spherical crystallites with d ~ 5 nm. Nano-rods, nano-cylinders, grow through the mechanism of aggregation called "oriented attachment". Preferential particle grow direction is most probably along [001], but that can be different for different samples (pure or Fe-doped). XRD measurements proved that our samples had pure rutile crystal phase, without any detectable dopant related phases. Also Scherrer's formula gave crystalline domains of about 5-7 nm. Microstructure analysis indicated existence of traces of anatase phase in pure TiO_2 and sample 4, Fe-doped TiO_2. Rietveld and atomic PDF methods were applied on X-ray powder diffraction data. Optical characterization showed that pure rutile TiO_2 nano-particles start to absorb light at ~ 420 nm which corresponds to rutile band gap of 3.0 eV. Fe-doped samples had significantly improved absorption (decreased reflection) in the visible spectral range. The sample with highest Fe-concentration has band gap of ~2.3 eV.

The presence of oxygen vacancies in pure and Fe-doped sample was approved by normal and up-converted photoluminescence measurements. In both cases a series of peaks/shoulders in emission spectra of all samples were observed, in the range from 350-600 nm. Low energy peaks/shoulders (from 2.8-2.34 eV) were assigned to oxygen vacancies. Pure and Fe-doped samples were employed for photocatalytic degradation of two herbicides: MCPA and MCPP using UV or Visible light for irradiation. Obtained results are compared with those obtained using Degussa P25 TiO_2 as commercial and most efficient photocatalyst.

The results indicated that the rate of MCPA and MCPP degradation is higher on pure rutile TiO_2 photocatalyst compared to its Fe-doped form, no matter which kind of light was used for irradiation (UV or Vis). A comparison of photo-catalytic efficiency of pure and Fe-doped rutile TiO_2 with that of Degussa P25 when UV light was used, shows that Degussa P25 is the most efficient in degradation of applied herbicides. A comparison of the performance of pure rutile TiO_2 and Fe-doped rutile TiO_2 with that of TiO_2 Degussa P25 shows that pure rutile TiO_2 is more efficient in the degradation of MCPA and MCPP when visible light was used, the efficiency of the Fe-doped rutile TiO_2 being similar to or lower than that of TiO_2 Degussa P25.

ACKNOWLEDGMENTS

Financial support for this study was granted by Ministry of Education and Science of the Republic of Serbia, Project Numbers: 172056 and 45020.

REFERENCES

[1] Pfaff, G., Reynders, P. *Chem. Rev.* 1999, 99, 1963-1982.

[2] Salvador, A., Pascual-Marti, M. C., Adell, J. R., Requeni, A., March, J. G. *J. Pharm. Biomed. Anal.* 2000, 22, 301-306.

[3] Zallen, R., Moret, M. P. *Solid State Commun.* 2006, 137, 154-157.

[4] Braun, J. H., Baidins, A., Marganski, R. E. *Prog. Org. Coat.* 1992, 20, 105-138.

[5] Yuan, S. A., Chen, W. H., Hu, S. S. *Mater. Sci. Eng. C.* 2005, 25, 479-485.

[6] Fujishima, A., Honda, K. *Nature* 1972, 238, 37-38.

[7] Fujishima, A., Rao, T. N., Tryk, D. A. *J. Photochem. Photobiol. C* 2000, 1, 1-21.

[8] Tryk, D. A., Fujishima, A., Honda, K. *Electrochim. Acta.* 2000, 45, 2363-2376.

[9] Grätzel, M. *Nature* 2001, 414, 338-344.

[10] Hagfeldt, A., Grätzel, M. *Chem. Rev.* 1995, 95, 49-68.

[11] Linsebigler, A. L., Lu, G., Yates, J. T. Jr. *Chem. Rev.* 1995, 95, 735-758.

[12] Millis, A., Le Hunte, S. *J. Photochem. Photobiol. A* 1997, 108, 1-35.

[13] Chen, X., Mao, S. S. *Chem. Rev.* 2007, 107, 2891-2959.

[14] Alivisatos, A. P. *J. Phys. Chem.* 1996, 100, 13226-13239.

[15] Alivisatos, A. P. *Science* 1996, 271, 933-937.

[16] Burda, C., Chen, X., Narayanan, R., El-Sayed, M. A. *Chem. Rev.* 2005, 105, 1025-1102.

[17] Murray, C. B., Kagan, C. R., Bawendi, M. G. *Annu. Rev. Mater. Sci.* 2000, 30, 545-610.

[18] Yin, Y., Alivisatos, A. P. *Nature* 2005, 437, 664-670.

[19] Mićić, O. I., Nenadović, M. T., Rajh, T., Nedeljković, J. M., Vučemilović , M. I. Kinetic and Spectroscopic Properties of Small Colloids of Metal Iodide, u "Trends in Colloid and Interface Science II", V. Degiorgio (Ed.), *Springer-Verlag*, Darmstadt, 1988, p. 20-23.

[20] Rajh, T., Vučemilović, M. I., Dimitrijević, N. M., Mićić, O. I., Nozik, A. J. *Chem.Phys.Lett.*, 1988, 143, 305-308.

[21] Vučemilović , M. I., Mićić, O. I. *Radiation Phys. Chem.* 1988, 32, 79-83.

[22] Nedeljković, J. M., Čomor, M. I., Šaponjić, Z. V., Rajh, T., Mićić, O. I. *Mater. Sci. Forum*, 1996, 214, 41-48.

[23] Dimitrijević, N. M., Nedeljković, J. M., Rajh, T., Čomor, M. I., Šaponjić, Z. V., Mićić, O. I. Charge-transfer Reactions on the Surface of Nanometer-sized Semiconductor Particles, in "Trends in Heat, Mass and Momentum Transfer", Vol. 3, E. Hahne, P. J. Heggs, M. Lebouche, K. Suzuki, I. Tanasawa, B X. Wang, eds., Research Trends, Poojapura Trivandrum, (1997) 1-18.

[24] Čomor, M. I., Nedeljković, J. M. *Chem.Phys.Lett.*1999, *299,* 233-236.

[25] Mićić, O. I., Rajh, T., Čomor, M. I. Influence of Surface Properties of Charge Carrier Dynamics of Quantized Semiconductor Colloids, "Electrochemistry in Colloids and Dispersions", Mackey, R. A.; Texter, J., eds., ACS Series, VCH Publ.Inc., New York, (1992) 457-465.

[26] Vučemilović, M. I., Vukelić, N., Rajh, T. *J.Photochem.Photobiol. A. Chemistry* 1988, 42, 157-167.

[27] Nenadović, M. T., Čomor, M. I., Vasić, V., Mićić, O. I. *J.Phys.Chem.* 1990, 94, 6390-6396.

[28] Mićić, O. I., Rajh, T., Nedeljković, J. M., Čomor, M. I. *Israel J. Chem.*1993, 33, 59-65.

[29] Nedeljković, J. M., Dimitrijević, N. M., Čomor, M. I.; Šaponjić, Z. V., Uskoković, D. P. *Mater. Sci. Forum* 1998, 282-283, 19-26.

[30] Čomor, M. I., Dramićanin, M. D., Rakočević, Z., Zec, S., Nedeljković, J. M. *J. Mat. Sci. Lett.* 1998, 17, 1401-1402.

[31] Čomor, M. I., Nedeljković, J. M. *J. Mat.Sci. Lett.* 1999, 18, 1583-1585.

[32] Čomor, M. I., Babić-Stojić, B., Nedeljković, J. M. *Mat. Sci. Forum.* 2003, 413, 17-21.

[33] Babić-Stojić, B., Milivojević, D., Čomor, M. I., Vodnik, V. *J. Phys.: Condens. Matter.* 2004, 16, 4625-4633.

[34] Kostić, R., Romčević, N., Čomor, M. I., Romčević, M., Grujić-Brojčin, M., Vodnik, V., Nedeljković, J. M. *Mat. Sci. Forum.* 2004, 453-454, 293-298.

[35] Romčević, N., Kostić, R., Romčević, M., Čomor, M. I., Nedeljković, J. M. *J. Phys. D: Appl. Phys.* 2005, 38, 4321-4324.

[36] Romčević, N., Kostić, R., Romčević, M., Grujić-Brojčin, M., Čomor, M. I., Vodnik, V. V., Nedeljković, J. M. *Mat. Sci. Forum.* 2005, 480-481, 237-242.

[37] Kuljanin, J., Čomor, M. I., Djoković, V., Nedeljković, J. M. *Mat. Chem. Phys.* 2006, 96, 67-71.

[38] Kostić, R., Romčević, M., Romčević, N., Klopotowski, L.; Kossut, J., Kuljanin-Jakovljević, J., Čomor, M. I., Nedeljković, J. M. *Opt. Mater.* 2008, 30, 1177-1182.

[39] Abazović, N. D., Kuljanin-Jakovljević, J. Ž., Čomor, M. I. *Russ. J. Phys. Chem.* 2009, 83, 1687-1690.

[40] Romčević, M., Romčević, N., Kostić, R., Klopotowski, L., Dobrowolski, W. D., Kossut, J., Čomor, M. I. *J. Allo. Comp.* 2010, 497, 46-51.

[41] Onjia, A. E., Milonjić, S. K.,Čokeša, D J., Čomor, M., Miljević, N. *Mat. Res. Bul.* 2003, 38, 1329-1339.

[42] Topalov, A. S., Šojić, D. V., Molnar-Gabor, D. A., Abramović, B. F., Čomor, M. I. *Appl. Cat. B: Environmental* 2004, 54, 125-133.

[43] Abazović, N. D., Čomor, M. I., Dramićanin, M. D., Jovanović, D. J., Ahrenkiel, S. P., Nedeljković, J. M. *J. Phys. Chem. B* 2006, 110, 25366-25370.

[44] Abazović, N. D., Ruvarac-Bugarčić, I. A., Čomor, M. I., Bibić, N., Ahrenkiel, S. P., Nedeljković, J. M. *Opt. Mater.* 2008, 30, 1139-1144.

[45] Abazović, N. D., Montone, A., Mirenghi, L., Janković, I. A., Čomor, M. I. *J. Nanosci. Nanotechn.* 2008, 8, 613-618.

[46] Abramović, B. F., Šojić, D. V., Anderluh, V. B., Abazović, N. D.,Čomor, M. I. *Desalination.* 2009, 244, 293-302.

[47] Abazović, N. D., Mirenghi, L., Janković, I. A., Bibić, N., Šojić, D. V., Abramović, B. F., Čomor, M. I. *Nanoscale Res. Lett.* 2009, 4, 518-525.

[48] Abazović, N. D., Čomor, M. I., Zec, S., Nedeljković, J. M., Piscopiello, E., Montone, A., Vittori Antisari, M. *J. Am. Ceram. Soc.* 2009, 92, 894-896.

[49] Janković, I. A., Šaponjić, Z. V., Čomor, M. I., Nedeljković, J. M. *J. Phys. Chem. C,* 2009, 113, 12645-12652.

[50] Šojić, D. V., Despotović, V. N., Abazović, N. D., Čomor, M. I., Abramović, B. F. *J. Hazard. Mater.* 2010, 179, 49-56.

[51] Kremenović, A., Antić, B., Blanuša, J., Čomor, M., Colomban, P., Mazerolles, L., Božin, E. S. *J. Phys. Chem. C* 2011, 115, 4395-4403.

[52] Chen, X., Lou, Y., Dayal, S., Qiu, X., Krolicki, R., Burda, C., Zhao, C., Becker, J. *J. Nanosci. Nanotechn.* 2005, 5, 1408-1420.

[53] Morrison, P. W., Raghavan, R., Timpone, A. J., Artelt, C. P., Pratsinis, S. E. *Chem. Mater.*1997, 9, 2702-2708.

[54] Yang, G., Zhuang, H., Biswas, P. *Nanostruct. Mater.* 1996, 7, 675-689.

[55] Yu, J. C., Yu, J., Ho, W., Zhang, L. *Chem. Commun.* 2001, 1942-1943.

[56] Huang, W. P., Tang, X. H., Wang, Y. Q., Koltypin, Y., Gedanken, A. *Chem. Commun.* 2000, 1415-1416.
[57] Seifried, S., Winterer, M., H. Hahn, M. *Chem. Vap. Deposition.* 2000, 6, 239-244.
[58] Cheng, H., Ma, J., Zhao, Z., Qi, L. *Chem. Mater.* 1995, 7, 663-671.
[59] Wang, C. C., Ying, J. Y. *Chem. Mater.* 1999, 11, 3113-3120.
[60] Aruna, S. T., Tirosh, S., Zaban, A. *J. Mater. Chem.* 2000, 10, 2388-2391.
[61] Zhang, Z. B., Wang, C. C., Zakaria, R., Ying, J. Y. *J. Phys. Chem. B* 1998, 102, 10871-10878.
[62] Scolan, E., Sanchez, C. *Chem. Mater.* 1998, 10, 3217-3223.
[63] Schneider, M., Baiker, A. *J. Mater. Chem.* 1992, 2, 587-591.
[64] Chen, J. Y., Gao, L., Huang, J. H., Yan, D. S. *J. Mater. Sci.* 1996, 31, 3497-3500.
[65] Ivanda, M., Music, S., Popovic, S., Gotic, M. *J. Mol. Struct.* 1999, 481, 645-649.
[66] Zhang, H., Finnegan, M., Banfield, J. F. *Nano Lett.* 2001, 1, 81-85.
[67] Burnside, S. D., Shklover, V., Barbe, C., Comte, P., Arendse, F., Brooks, K., Gratzel, M. *Chem. Mater.* 1998, 10, 2419-2425.
[68] ICSD data base, Version 2009-2, NIST and FIZ (136 different entries for TiO_2)
[69] Diebold, U. *Surf. Sci. Rep.* 2003, 48, 53-229
[70] Bak, T., Nowotny, J., Nowotny, M. K. *J. Phys. Chem. B* 2006, 110, 21560-24567.
[71] Nowotny, M. K., Bak, T., Nowotny, J., Sorrell, C. *Phys. Status Solidi* 2005, 242, R88-R90.
[72] Lee, K., Lee, N. H., Shin, S. H., Lee, H. G., Kim, S. *J. Mat. Sci. Eng. B* 2006, 129, 109-115.
[73] Hurum, D. C., Gray, K. A., Rajh, T., Thurnauer, M. C. *J. Phys. Chem. B* 2005, 109, 977-980.
[74] Litter, M. I., Navio, J. A. *J. Photochem. Photobio. A* 1996, 98, 171-181.
[75] Ding, Y., Han, W. Q., Lewis, L. H. *J. App. Phys.* 2007, 102, 123902.
[76] Wang, C., Bottcher, C., Bahnemann, D. W., Dorhmann, J. K. *J. Mater. Chem.* 2003, 13, 2322-2329.
[77] Cozzoli, P. D., Kornowski, A., Weller, H. J. *J. Am. Chem. Soc.* 2003, 125, 14539-14548.
[78] Zhu, J., Chen, F., Zhang, J., Chen, H., Anpo, M. *J. Photochem. Photobiol. A Chem.* 2006, 180, 196-204.
[79] 78. Zhang, W.,, Li, Y., Zhu, S., Wang, F. *Chem. Phys. Lett.* 2003, 373, 333-337.
[80] Nagaveni, K., Hegde, M. S., Ravishankar, N., Subbanna, G. N., Madras, G. *Langmuir* 2004, 20, 2900-2907.
[81] Erdem, B., Hunsicker, R. A., Simmons, G. W., Sudol, E. D.; Dimonie, V. L., El-Aasser, M. S. *Langmuir* 2001, 17, 2664-2669.
[82] Glisenti, A. *J. Mol. Cat. A: Chem.* 2000, 153, 169-171.
[83] Graat, P., J. Somers, M. A. *Appl. Surf. Sci.* 1996, 100/101, 36-40.
[84] Penn, R. L., Banfield, J. F. *Science.* 1998, 281, 969971.
[85] Leite, E. R., Giraldi, T. R., Pontes, F. M., Longo, E., Beltrán, A., Andrés, J. *Appl. Phys. Lett.* 2003, 83, 1566 1568.
[86] Li, J. G., Ishigaki, T., Sun, X. *J. Phys. Chem. C,* 2007, 111, 4969-4976.
[87] Andersson, M., Kiselev, A., Österlund, L., Palmqvist, A. E. C. *J. Phys. Chem. C* 2007, 111, 6789-6797.
[88] Chu, R., Yan, J., Lian, S., Wang, Y., Yan F., Chen, D. *Solid State Commun.* 2004, 130, 789-792.

[89] Testino, A., Bellobono, I. R., Buscaglia, V., Canevali, C., D'Arienzo, M., Polizzi, S., Scotti, R., Morazzoni, F. *J. Am. Chem. Soc.* 2007, 129, 3564-3575.

[90] Yin, S., Hasegawa, H., Maeda, D., Ishitsuka, M., Sato, T. *J. Photochem. Photobiol. A Chemistry.* 2004, 163, 1-8.

[91] Rodriguez-Carvajal, J. FullProf computer program http://www.ill.eu/sites/fullprof/, 2009.

[92] Rodriguez-Carvajal, J. *Physica B* 1993, 192, 55-69.

[93] Rodríguez-Carvajal, J. Recent Developments of the ProgramFULLPROF. In commission on Powder Diffraction (IUCr) Newsletter 2001, 26, 12-19 (http://journals.iucr.org/iucr-top/comm/cpd/Newsletters/).

[94] Balzar, D., Audebrand, N., Daymond, M. R., Fitch, A., Hewat, A., Langford, J. I., Le Bail, A., Louer, D., Masson, O., McCowan, C. N., Popa, N. C., Stephens, P. W., Toby, B. H. *J. Appl. Crystallogr.* 2004, 37, 911-924.

[95] Johnson, E. J., Kafalas, J., Davies, R. W., Dyes, W. A. *Appl. Phys. Lett.* 1982, 40, 993-995.

[96] Ivanov, V.Yu., Semenov, Yu.G., Surma, M., Godlewski, M. *Phys. Rev. B* 1996, 54, 4696-4701

[97] Seidel, W., Titkov, A., André, J. P., Voisin, P., Voos, M. *Phys. Rev. Lett.* 1994, 73, 2356-2359.

[98] Driessen, F. A. J. M., Cheong, H. M., Mascarenhas, A., Deb, S. K., Hageman, P. R., Bauhuis, G. J., Giling, L. *Phys. Rev. B* 1996, 54, R5263-R5266.

[99] Zeman, J., Martinez, G., Yu, P.Y., Uchida, K. *Phys. Rev. B* 1997, 55, R13428-R13431.

[100] Cho, Y. H., Kim, D. S., Choe, B. D., Lim, H., Lee, J. I., Kim, D. *Phys. Rev. B.* 1997, 56, R4375-R4378.

[101] Schrottke, L., Grahn, H. T., Fujiwara, K. *Phys. Rev. B* 1997, 56, R15553-R15556.

[102] Cheong, H. M., Fluegel, B., Hanna, M. C., Mascarenhas, A. *Phys. Rev. B* 1998, 58, R4254-R4257.

[103] Heimbrodt, W., Happ, M., Henneberger, F. *Phys. Rev. B* 1999, 60, R16326-R16329.

[104] Paskov, P. P., Holtz, P. O., Monemar, B., Garcia, J. M., Schoenfeld, W. V., Petroff, P. M. *Appl. Phys. Lett.* 2000, 77, 812-814.

[105] Kammerer, C., Cassabois, G., Voisin, V, Delalande, C., Roussignol, Ph., Gerrard, J. M. *Phys. Rev. Lett.* 2001, 87, 2074011-2074014.

[106] Poles, E., Selmarten, D. C., Mićić, O. I., Nozik, A. J. *Appl. Phys. Lett.* 1999, 75, 971-973.

[107] Rakovich, Yu.P., Filonovich, S. A., Gomes, M. J. M., Donegan, J. F., Talapin, D. V., Rogach, A. L., Eychmuller, A. *Phys. Status Solidi B* 2002, 229, 449-452.

[108] Rusakov, K. I., Gladyshchuk, A. A., Rakovich, Yu.P., Donegan, J. F., Filonovich, S. A., Gomes, M. J. M., Talapin, D. V., Rogach, A. L., Eychmüller, A. *Opt. Spectrosc.* 2003, 94, 859-863.

[109] Diener, J., Kovalev, D., Heckler, H., Polisski, G., Künzner, N., Koch, F., Efros, Al.L., Rosen, M. *Opt. Mater.* 2001, 17, 135-139.

[110] Serpone, N. *J. Phys. Chem. B.*2006, 110, 24287-24293.

[111] N. Gray (Ed.), Pesticides and Organic Micropollutants, Drinking Water Quality, John Wiley and Sons, Chichester, 1996, pp. 132-148.

[112] Donald, D. B., Cessna, A. J., Sverko, E., Glozier, N. E. *Environ. Health Perspect.* 2007, 115, 1183-1191.

[113] Reddy, E. P., Sun, B., Smirniotis, P. G. *J. Phys. Chem.* 2004, 108, 17198-17205.
[114] Teoh, W. Y., Amal, R., Mädler, L., Pratsinis, S. E. *Catal. Today* 2007, 120, 203-213.
[115] Zhou, M., Yu, J., Cheng, B. *J. Hazard. Mat.* 2006, 137, 1838-1847.
[116] Bahnemann, D. W. *Israel J. Chem.* 1993, 33, 115-130.
[117] Clausen, L., Fabricius, I. *J. Environ. Qual.* 2001, 30, 858-869.
[118] C. D. S. Tomlin, Editor, The Pesticide Manual (11th ed.), The British Crop Protection Council, The Bath Press, Bath, UK (1997)
[119] Machesky, M. L., Wesolowski, D. J., Palmer, D. A., Ichiro-Hayashi, K. *J. Colloid Interface Sci.* 1998, 200, 298-309.

In: Rutile: Properties, Synthesis and Applications
Editor: It-Meng (Jim) Low
ISBN 978-1-61942-233-9

Chapter 4

THE FABRICATION OF RUTILE NANOPARTICLES AND HOLLOW MICROSPHERES WITH MESOPOROSITY

Guohua Li*[1]*, Ying He, Xuanhui Zhang and Shujuan Hu
The School Of Chemical Engineering and Materials Science,
Zhejiang University of Technology,
Hangzhou 310032 P.R.China

ABSTRACT

The use of sol spray hydrolysis to synthesize hollow micro-spherical anatase and rutile particles with mesoporosity using $TiCl_4$ as sole precursor is described. The wall of the hollow micro-sphere is constituted of nanoparticles and mesoporosity. Its formation mechanism can be attributed to the spray dry process and the property of $TiCl_4$ aqueous sol. The optical property of hollow sphere mesoporous titania is better than that of nano-anatase and nano-rutile in aqueous suspensions, and related to its crystal phase, specific surface area, crystal size, porosity size and crystallinity of titania, and these characteristics are changing as the calcined temperature increasing. The merit of this approach is template free and without any additives, organic or inorganic. The sol-spray hydrolysis approach can be used to prepare other materials with the same structure at macroscale.

1. THE SYNTHESIS OF RUTILE NANOPARTICLE IN AQUEOUS SOLUTION

1.1. Introduction

Titania exists in three main crystallographic forms at standard pressure: anatase, rutile and brookite. Each structure exhibits different physical properties, such as refractive index, chemical reactivity and photochemical reactivity. Rutile has been extensively studied for its

[1] Corresponding author: nanozjut@zjut.edu.cn Tel: +086-57188320418; Fax: +086-57188320544.

higher relative permittivity, refractive index, ultraviolet ray absorption rate, stronger resistance against powdering [1] and one of the most appropriate candidates for preparing photonic crystals [2]. In recent years, there has been increasing interest in the preparation of nanocrystalline rutile because of its higher specific surface area, superior physical and chemical properties, such as catalytic activity, photocatalytic property and good stability in adverse environment.

Usually, anatase and rutile polymorphs are produced by hydrolysis synthesis [3-7], but the synthesis of monomineralic, well-shaped rutile nanocrystals *via* hydrolysis is much more difficult than that of anatase nanocrystals in aqueous solution [8]. The hydrolysis reaction of $TiCl_4$ in aqueous solution is [1]:

$$TiCl_4+2H_2O \rightarrow TiO_2(\text{rutile}) + 4HCl \quad (A)$$

or

$$TiCl_4+2H_2O \rightarrow TiO_2(\text{anatase}) + 4HCl \quad (B)$$

The Gibbs free energy of reaction (A) and (B) at 298 K is –59.32 and –54.29 $kJmol^{-1}$, respectively [1], therefore the two reactions can take place spontaneously. The difference of the Gibbs free energy between reaction (A) and (B) is only 4.03 $kJmol^{-1}$, and reaction (A) is more thermodynamically favorable than reaction (B) [1]. Reaction (B) is accelerated more rapidly than reaction (A) when hydrolysis temperature increased [1]. This implies that it is difficult to get phase-pure rutile nanocrystal by hydrolysis of $TiCl_4$ in aqueous solution. In order to get phase-pure nanocrystalline rutile, reaction (A) must be promoted and reaction (B) must be depressed. A lot of efforts has been devoted to solve this problem and get phase-pure nanocrystalline rutile powders in aqueous solution, such as using nano-rutile as crystal seed [1], using mixed organic media instead of aqueous solvent [9], adding additives, such as citric and nitric acids [10-11], preparation of intermediates in microemulsion system [12] with the help of alcohol-water solvent, as well as both cationic and anionic surfactants [8]. Though phase-pure rutile nanocrystal could be prepared by these efforts, but other problems, such as impurities into the nanocrystal [13], lower yield [1] and complication of synthesis process [12] aroused. To the best of our knowledge, there has been no report on fabrication of phase-pure rutile nanocrystal by hydrolysis of $TiCl_4$ in aqueous solution only at the temperature below 373 K. Here, we report our efforts in the development of this method.

1.2. Preparation of Rutile Nanoparticle

The fabrication process could be divided into two steps: the first was the preparation of $TiCl_4$ aqueous sol, and the second was the hydrolysis of $TiCl_4$ aqueous solution at 368 K.

In the first step, 150 mL of titanium tetrachloride liquid was added dropwise into a 300 mL of deionized water, while keeping the mixing course under stirring and the system at the temperature below 313 K in an ice-water bath. After titanium tetrachloride had been added, the mixture liquid was stirred for another 2 hours, and then cooled down to room temperature.

In the second step, 50 mL of the above aqueous sol was added 12 hours latter into a three-necked flask, and 100 mL of deionized water was added dropwise into the flask while stirring.

The flask was fixed in an ice-water bath to keep the temperature of the mixture liquid below 313 K. After deionized water had been added, the system was under stirring for another 30 minutes, the temperature of the mixture was raised to 333 K and stirred for 1 hour, and raised to 368 K and stirred for 2 hours. The resultants were filtered and rinsed with deionized water for several times and dried under a vacuum of 0.01Pa at 353 K over night.

1.3. Characterization of Rutile Nanoparticles

X-ray diffraction was performed with a Thermo ARL SCINTAG X′TRA X-ray at room temperature, using quartz monochromator Cu Kα1 radiation source (λ=0.15406nm) under a voltage of 45 kV and a current of 40 mA. The XRD patterns were recorded with a step size of 0.04° from 20° to 80° at the speed of 2.4°•min^{-1}. The average grain size of rutile nanocrystals was estimated according to rutile 110 plane by the Scherrer Formula.

Figure 1 is the XRD patterns of the sample as prepared by sol-hydrolysis of $TiCl_4$ in aqueous solution at 368 K. There are eight diffraction peaks, five of them appear at 2θ=27.40°, 36.12°, 42.80°, 44.00° and 54.28°, which can be assigned to the diffraction peaks of rutile (110), (101), (111), (210), and (211) planes, respectively, according to JCPDS card No 21-1276. The average grain size estimated by the Scherrer Formula based on rutile (110) plane is 7.5 nm. These results indicate that the sample as prepared are composed of phase-pure nanocrystalline rutile.

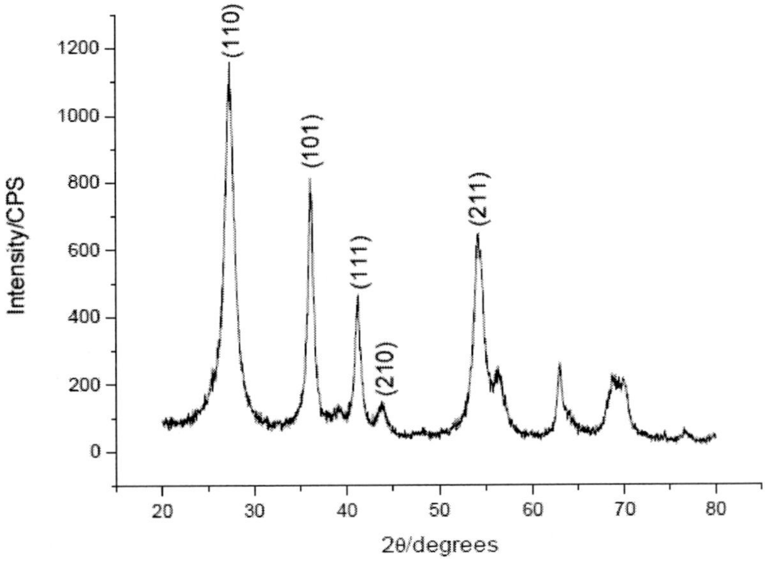

Figure 1. XRD patterns of the sample as prepared.

In order to make it clear, the sample as prepared was characterized by thermogravimetric analysis (TG-DTA). The thermogravimetric analysis (TG-DTA) was performed on RSZ2000 thermogravimetric analyzer at a gas velocity of 100 mL•min^{-1} and at a ramp of 10 K•min^{-1} from room temperature to 1073 K under air atmosphere. TG-DTA curves of the sample as prepared are shown in figure 2.

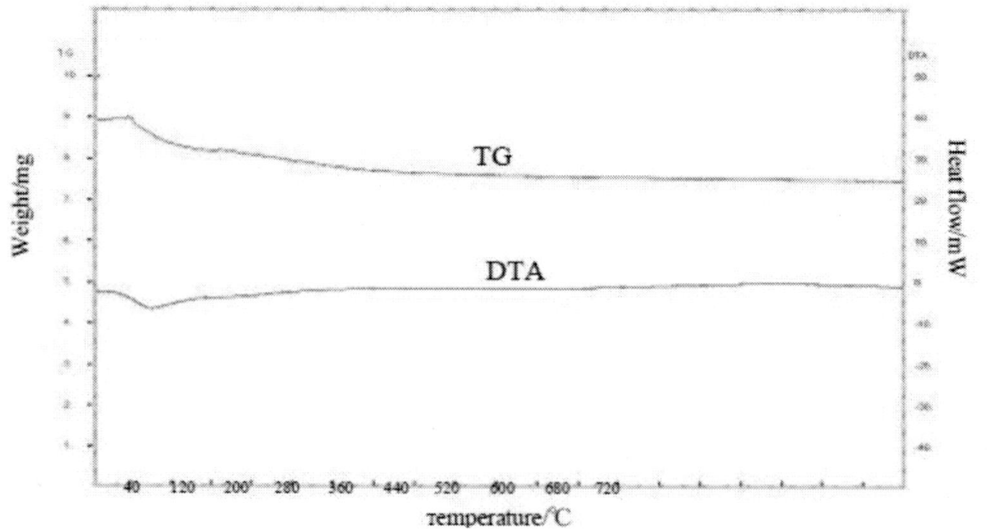

Figure 2. TG-DTA curves of the sample as prepared.

The mass loss of the sample as prepared is 14.95% from room temperature to 1073 K, according to TG curve in figure 2. There is only one endothermic peak at 343 K, with an onset temperature around 323 K, according to DTA curve in figure 2. There are no any other endothermic and exothermic peaks as the temperature considered. The endothermic peak around 343 K is attributed to the evaporation of water adsorbed on the surface of nano-rutile crystals. Based on the above results, one conclusion can be drawn that there is no phase transition from room temperature to 1073 K in the sample as prepared. This indicates that the sample as prepared is phase-pure rutile crystal. This result is in agreement with that of X-ray diffraction analysis.

Figure 3 is the XRD patterns of the sample calcined at 773 K. There are 11 diffraction peaks at 2θ=27.44°, 36.12°, 39.16°, 42.40°, 44.08°, 54.36°, 56.64°, 62.76°, 64.04°, 69.00° and 69.80°, which can be assigned to the diffraction peaks of rutile (110), (101), (111), (210), (211), (220), (002), (310), (301) and (112) planes, respectively, according to JCPDS card No 21-1276. This indicates that the sample calcined at 773 K is of a perfect rutile structure. The average grain size estimated according to the Scherrer Formula based on rutile (110) plane is 28.2 nm.

As there is no phase transition from room temperature to 1073 K, comparison of XRD patterns of the samples as prepared with that of the sample calcined at 773 K is helpful to understand the characteristics of nano-rutile crystal. In figure 1, in addition to the five peaks mentioned above, three other diffraction peaks appear around 2θ=56.40°, 63.00° and 69.36°, respectively.

After calcination at 773 K, as shown in figure 3, the diffraction peak around 2θ=56.40° transforms to 56.64°, which can be assigned to the diffraction peak of rutile (220) plane. The diffraction peak around 2θ=63.00° transforms to 62.76° and 64.04°, which can be assigned to the diffraction peaks of rutile (002) plane and (310) plane, respectively. The diffraction peak around 2θ=69.36° transforms to 69.00° and 69.88°, which can be assigned to the diffraction peaks of rutile (301) plane and (112) plane, respectively.

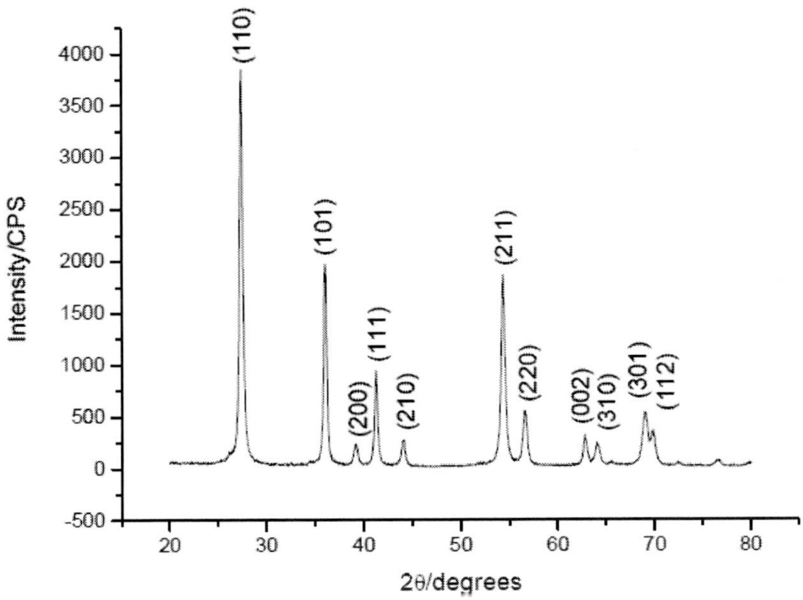

Figure 3. XRD pattern of the sample calcined at 773 K.

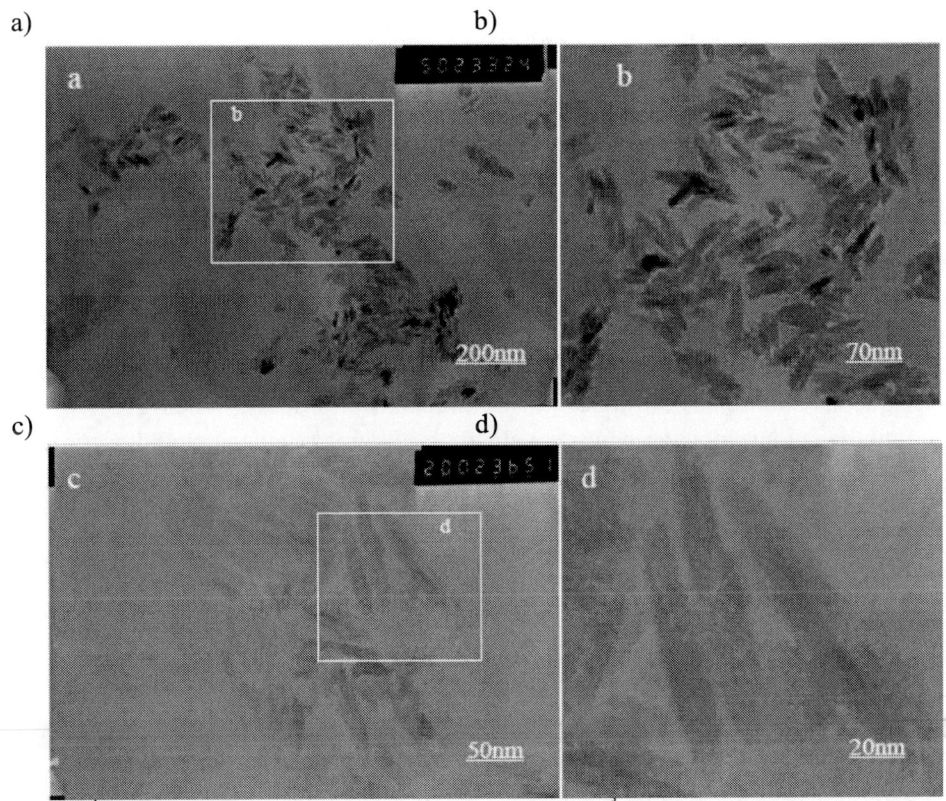

[a] as prepared, [b] the enlargement of local area of a, [c] calcined at 773 K, [d] the enlargement of local area of c.

Figure 4. TEM images of the samples as prepared and calcined at 773 K.

These imply that the diffraction peaks around 2θ=56.40°, 63.00° and 69.36° of the sample as prepared can be attributed to imperfect rutile crystal plane of (220), (002), (310), (301) and (112). Base on these results, one conclusion can be drawn that the sample calcined at 773 K is phase-pure nanocrystalline rutile powders and the sample as prepared is phase-pure nanocrystalline rutile powders too.

TEM images of the samples as prepared and calcined at 773 K are show in figure 4, in which figure 4a and 4b are TEM images of the samples as prepared, and figure 4c and 4d are that of the samples calcined at 773 K. The particles of the sample as prepared are in nanoscale, as shown in figure 4a. In order to see the morphology of the particles more clearly, the local area of figure 4a as shown in figure 4a is enlarged and the result is shown in figure 4b. The morphology of the particles of the sample as prepared is rod-like with a diameter of 10-20 nm, a length of 20-80 nm and an aspect ratio of 2-4, as shown in figure 4b. The morphology of the particles of the sample calcined at 773 K is still rod-like with a diameter of 15-25 nm, a length of 25-105 nm and an aspect ratio of 2-4, as shown in figure 4c and 4d.

The morphologies of rutile nanocrystal of the sample as prepared and calcined at 773 K are rod-like, which are consistent with that of references 2, 15, 16 and 17. From comparison of TEM images of the samples as prepared with that of calcined at 773 K, one can conclude that the morphology and the respect ratio of the samples are almost the same, but the diameter and the length of the particles are increased after calcination at 773 K. This indicates that the calcination temperature has a positive effect on the structure and size of rutile nanocrystals, and has no effect on the aspect ratio of rutile nanocrystals.

1.4. Formation Mechanism of Rutile Nanocrystal

Li *et al* [1] pointed out that when the temperature of hydrolysis of $TiCl_4$ in aqueous solution was above 353 K, no matter whether there was crystal seed or not, anatase was formed as the product because of kinetic effects. In our experiments, phase-pure nanocrystalline rutile was produced by hydrolysis of $TiCl_4$ in aqueous solution at the temperature of 368 K. The result is not in agreement with that of Li *et al* [1] reported.

The fabrication of phase-pure nanocrystalline rutile powders by hydrolysis of $TiCl_4$ at 368 K in aqueous solution is attributed to the preparation process. When $TiCl_4$ was added dropwise into the deionized water continuously, the following reactions would occur [15]:

$$TiCl_4 + H_2O \rightarrow TiOH^{3+} + H^+ + 4Cl \text{ (C)}$$

$$TiOH^{3+} \rightarrow TiO^{2+} + H^+ \text{ (D)}$$

$$TiO^{2+} + H_2O \rightarrow TiO_2 + 2H^+ \text{ (E)}$$

And as the addition of $TiCl_4$, $TiCl_4$ aqueous sol would form automatically. During the aging of $TiCl_4$ aqueous sol, the complex of $[Ti(OH)_nCl_m]^{2-}$(n+m=6) would form [14]. Cheng [14] pointed out that in the complex of $[Ti(OH)_nCl_m]^{2-}$, the value of n and m were concerned with the acidity and the concentration of Cl^- ion in the feedstock, *i.e.*, the higher the acidity and $[Cl^-]$, the larger m was. In our experiment, as the addition of $TiCl_4$, reactions (C), (D) and (E) would advance from left to right, the acidity and the concentration of Cl^- ion in the sol would become higher and higher, thus, in $[Ti(OH)_nCl_m]^{2-}$ complex, the value of m could be larger than n. According to the ligand field theory, $Ti(IV)(3d^0)$ complex ions are all octahedrally coordinated both in solution and crystal. So, it is rational to assume that the structure of $[Ti(OH)_nCl_m]^{2-}$ complex in $TiCl_4$ aqueous sol octahedrally coordinates with four Cl^- ligands on the same quadrangle plane and one OH^- anion at each corner, $[Ti(OH)_2Cl_4]^{2-}$, as shown in figure 5B1. Yin and his colleagues' work showed that room temperature peptizaton of a highly concentrated $TiCl_4$ aqueous solution would give rutile nanoparticles as embryos [11]. In our experiment, the concentration of $TiCl_4$ was 6.69 $molL^{-1}$, it is reasonable to believe that there are some rutile embryos in $TiCl_4$ aqueous sol. And Bahnemann *et al* [16] described the surface of acidic and basic properties of TiO_2 as follows:

for pH<3.5, $TiO_2 + nH^+ \rightarrow TiO_2H_n^{n+}$ (F)

for pH>3.5, $TiO_2 + nOH^- = TiO_2(OH)_n^{n-}$ (G)

As the $TiCl_4$ concentration was 6.69 $molL^{-1}$, the pH value of $TiCl_4$ aqueous sol was smaller than 3.5. It is rational to consider that the surface of rutile embryos is covered by H^+ cations, as shown in figure 5B2.

After $TiCl_4$ aqueous sol was diluted in the second step, the acidity of the diluted aqueous solution became weaker, and its concentration of $TiCl_4$ became smaller too. Cheng *et al* [14] demonstrated that in the complex of $[Ti(OH)_nCl_m]^{2-}$, the value of n and m were related to the acidity and concentration of Cl^- ion in the feedstock, when the acidity in the feedstock was lower, the number of OH ligand in $[Ti(OH)_nCl_m]^{2-}$ was more, and the coordination ability of Cl^- anion was weaker than that of OH^- anion [17]. In this case, some of Cl^- ligands would be replaced by OH^- ligands. Though the structure of $[Ti(OH)_nCl_m]^{2-}$ complex in the diluted $TiCl_4$ aqueous solution was still octahedron, but the detail structure was changed, with four OH^- ligands on the same quadrangle plane and one Cl^- anion at each corner, $[Ti(OH)_4Cl_2]^{2-}$, as shown in figure 5C1. The structure of rutile embryo did not change because of the pH value was still smaller than 3.5, as shown in figure 5C2.

During the hydrolysis in aqueous solution at the temperature of 368 K, the linkage between two $[Ti(OH)_4Cl_2]^{2-}$ complexes and between two rutile embryos were carried out by a dehydration reaction between OH^- ligands through oxolation by means of hydrogen bonds, as shown in figure 5D1 and 5D3, respectively. And the linkage between rutile embryo and $[Ti(OH)_4Cl_2]^{2-}$ complex was carried out by a dehydration reaction between OH ligands through oxolation by means of hydrogen bonds and by escapement of HCl between H ligands of rutile embryos and Cl^- ligands of $[Ti(OH)_4Cl_2]^{2-}$ complex, as shown in figure 5D2. In this way, structure units of $[Ti(OH)_4Cl_2]^{2-}$ complex and rutile embryos were connected with each other, and thus nanocrystals formed, as shown in figure 5E1 and 5E2, in this way nanocrystalline rutile powders formed, as shown in figure 5E3.

As the concentration of $[TiCl_4]$ is high in the aqueous solution, and Cl^- can coordinate to the existing rutile embryos [11], this indicates that some Cl^- ligands coordinate with rutile embryos instead of O^{2-} ligands, and Cl^- cannot form hydrogen bond with H_2O molecule. Thus, the presence of Cl^- ligands in rutile embryo would benefit for the linkage between edges, and obstruct the linkage between planes. In this way, anatase crystallites were suppressed and rutile nanocrystalline could be favored during hydrolyzation, because anatase consists of TiO_6 octahedra with shared faces, while rutile TiO_6 octahedra shares only edges [4]. This is in agreement with the conclusions of Cheng [14].

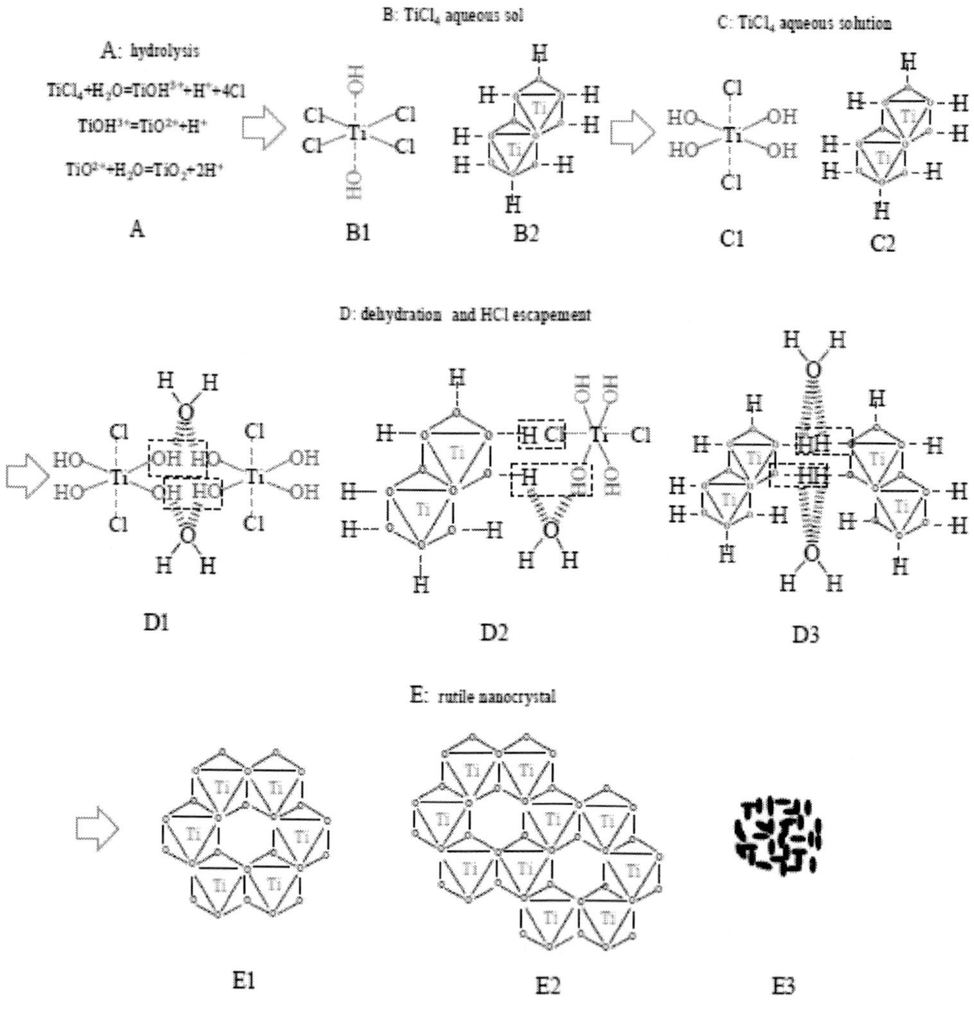

Figure 5. The schematics of rutile formation mechanism of sol-Hydrolysis of $TiCL_4$ in aqueous solution.

Based on the above discussion, a mechanism of rutile formation of hydrolysis of $TiCl_4$ in aqueous solution could be proposed as follows: (1) reactions (C), (D) and (E) advance during hydrolysis, as shown in figure 5A; (2) $[Ti(OH)_2Cl_4]^{2-}$ complex and rutile embryo form in $TiCl_4$ aqueous sol, as shown in figure 5B; (3) $[Ti(OH)_4Cl_2]^{2-}$ complex forms in $TiCl_4$ aqueous solution, and rutile embryo still exists in $TiCl_4$ aqueous solution, as shown in figure 5C; (4)dehydration and HCl escapement during hydrolysis at 368 K, which results in a linkage through edges between two $[Ti(OH)_4Cl_2]^{2-}$ complexes, rutile embryo and $[Ti(OH)_4Cl_2]^{2-}$ complex and two rutile embryos, as shown in figure 5D1, 5D2 and 5D3, respectively; (5)As the linkage advancing, rutile nanocrystals form, as shown in figure 5E.

2. The Transformation Process from Anatase Nanoparticle to Nanorutile during Calcination

2.1. Introduction

Anatase and brookite are stable phase at room-temperature, metastable phase at high temperature and will transform into a thermodynamically more stable rutile phase at high temperature [18]. The transformation from anatase to rutile is metastable-to-stable transformation and there is no an equilibrium temperature of this transformation [19]. The transformation temperature range varies from 673 K to 1473 K, depending on (a) the method of synthesis, (b) the atmosphere, and (c) the presence of foreign ions [20]. The behavior of this phase transformation of anatase as a function of particle size [21-23] and pH of the synthesis gel [24] have been reported. The phase transformation behavior [25] and crystal growth within nanoparticle [26] was investigated with the help of high-temperature X-ray diffraction technique using synchrotron radiation. Several kinetics models have been used to describe the experimental observations [27]. Usually, one model fits one specific study best. Each kinetic model is only related to the specified transformation mechanism [27]. This implies that the models up-to-date do not reveal the essence that determinates the phase transformation. It is interesting and necessary to investigate the course of the phase transformation from anatase to rutile, both from scientific and technological points of view [19].

In rutile and anatase, Ti^{4+} ions occupy the centers of Ti-O octahedrons and are coordinated by six oxygen ions. The numbers of the edges and corners shared by the octahedron vary, and are two and six in rutile, and four and four in anatase, respectively, [28]. As its temperature increasing, the strain energy of a lattice and the octahedrons dislocation of anatase increase, and this type disordered of anatase lattice favors its transformation to rutile structure [21]. But, to the best of our knowledge, there are few investigations that concern the effect of the microstrain energy on the phase transformation from anatase to rutile. Therefore, we prepared the phase pure anatase nanoparticles by a new approach of sol-hydrolysis, and the phase transformation was investigated by in situ XRD. The variation and distribution of the diameter and the microstrain of nano-titania particles were estimated with Win XRD software.

2.2. Preparation of Nanoanatase

The fabrication process can be divided into two steps: the first is the preparation of $TiCl_4$ aqueous sol, and the second is the hydrolysis of $TiCl_4$ aqueous solution at 368K.

In the first step, 100 mL of titanium tetrachloride liquid was added dropwise into a 200 mL of deionized water, while keeping a mixing course under stirring and its temperature below 313 K in an ice-water bath. After titanium tetrachloride had been added, the mixture liquid was stirred for another 2 hours, and then cooled down to room temperature.

In the second step, 50 mL of the above aqueous sol was added 12 hours latter into a three-necked flask, and 100 mL of deionized water was added dropwise into the flask while stirring. The flask was fixed in an ice-water bath to keep the temperature of the mixing liquid below

313 K. After deionized water had been added, the system was under stirring for another 30 minute. At this period, ammonium sulfate was added into the mixture, at a ratio of $(NH_4)_2SO_4$ to $TiCl_4$ is 2:1, the temperature of the mixture was raised to 333 K and stirred for 1 hour, and then to 368 K and stirred for 2 hours. The resultants were filtered and rinsed with deionized water for six times and dried under a vacuum of 0.01Pa at 353 K over night.

2.3. The Transformation Process from Anatase to Rutile

The detail XRD patterns used to estimate crystallite size and microstrain are carried out with a step-scanning mode, a step of 0.02° and counting times of 1second per step. The instrumental broadening was corrected by using an NIST 1976 (α-Al_2O_3) (Standard Reference Material from U.S Department of Commerce National Institute of Standard and Technology (under the same conditions as actual experiments.

All the XRD data obtained in this work were analyzed using a Win XRD software, which is a part of the Thermo X'TRA package supplied by Bruker-AXS Company. The size of crystallite and its distribution, and the lattice microstrains of anatase and rutile were conducted by a line-broadening analysis. An average grain size is obtained by the Scherrer formula without considering the effect of lattice microstrain.

A reflection broadening in the XRD patterns is attributed mainly to three kinds of contributions: the crystallite size and micro-strain of the sample particle, and the instrument itself [29, 30]. In order to separate the effect of the crystallite size from that of the micro-strain on reflection broadening, Fourier transfer was used for the profile analysis of the reflection broadening. In these XRD experiments, the software based on Warren–Averbach Fourier transfer (W–A/FT) method was used to calculate the distribution of crystallite size and micro-strain of anatase.

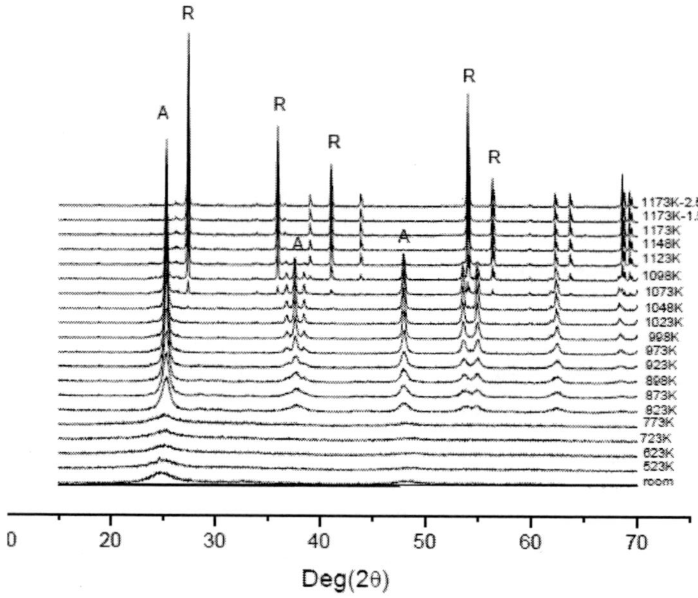

Figure 6. In situ XRD patterns of the sample (A: anatase; R: rutile).

Figure 6 is XRD patterns of the samples treated from room temperature to 1173 K. From room temperature to 773 K, the sample is composed of anatase particles (JCPDS card No 21-1272) and noncrystalline particles; from 823 K to 1023 K, the sample is composed of anatase particles mainly; from 1048 K to 1123 K, the sample is composed of anatase particles and rutile particles (JCPDS card No 21-1276). When the temperature is over 1148 K, the sample is composed of rutile particles.

These results indicate that the transformation process from nano-anatase to nano-rutile can be divided into four steps: from room temperature to 773 K, 823 K to 1023 K, 1048 K to 1123 K and over 1123 K, and the starting point temperature of the phase transformation from anatase to rutile is around 1048 K, and the end point temperature is over 1123 K.

Based on the above results, one conclusion can be drawn that the crystal growth of anatase nanoparticles can be divided into four steps: the stable stage of anatase crystallite is from room temperature to 723 K; the crystal growth stage of nano-anatase is from 773 K to 998 K; the coexistence stage of nanoanatase and nanorutile is from 1023 K to 1073 K; the nano-rutile growth stage is over 1123 K, as the temperature considered in our experiments.

2.4. The Rate Competition Determines the Size Distribution and the Diameter of Anatase Nanoparticles during Calcinations

Figure 7 shows the size distribution of anatase particles. The size of anatase particles is around 10 nm, from 5 nm to 16 nm at 873 K; the size of anatase particles is around 22 nm, from 13 nm to 34 nm at 973 K; the size of anatase particles is around 32 nm, from 17 nm to 51 nm at 1023 K. The size of anatase particles is around 54 nm, from 26 nm to 84 nm at 1048 K; the size of anatase particles is around 42 nm, from 26 nm to 67 nm at 1073 K; the size of anatase particles is around 35 nm, from 17 nm to 51 nm at 1123 K, almost the same with that at 1023.

Based on the above results, one can see clearly that from 873 K to 1048 K, the size of anatase nanoparticles increases as the temperature increases, and its distribution range becomes broader and broader; when the temperature is over 1048 K, the size of anatase nanoparticles decreases as the temperature increasing, and its distribution range becomes narrower and narrower. This phenomenon can be attributed to titania phase transformation. During calcination, the crystal size and the crystallinity of anatase increase along with the calcination temperature.

When the calcination temperature is high enough, rutile phase forms and the rutile phase is transformed from anatase [31], and the transformation from anatase to rutile occurs automatically. When the calcination temperature is over 1048 K, anatase begins to transform into rutile, as shown in figure 6. So, the rate of anatase transforming into rutile is just the rate of rutile growth. Based on the above discussions, the size decrease of nano-anatase implies that the phase transformation rate from anatase to rutile is bigger than the rate of anatase growth. That is the reason why the size of anatase decreases.

Figure 8 shows the size distribution of rutile nanoparticles. The size distribution of rutile nanoparticles is around 25 nm, from 14 nm to 40 nm, at 1048 K; the size distribution of rutile nanoparticles is around 60 nm, from 42 nm to 82 nm, at 1073 K. The average diameter of rutile nanoparticles at 1073 K is over two times bigger than that at 1048 K. These results indicate that when the rutile phase starts to form, the rutile nanoparticles grow rapidly. This conclusion is consistent with that in reference [32].

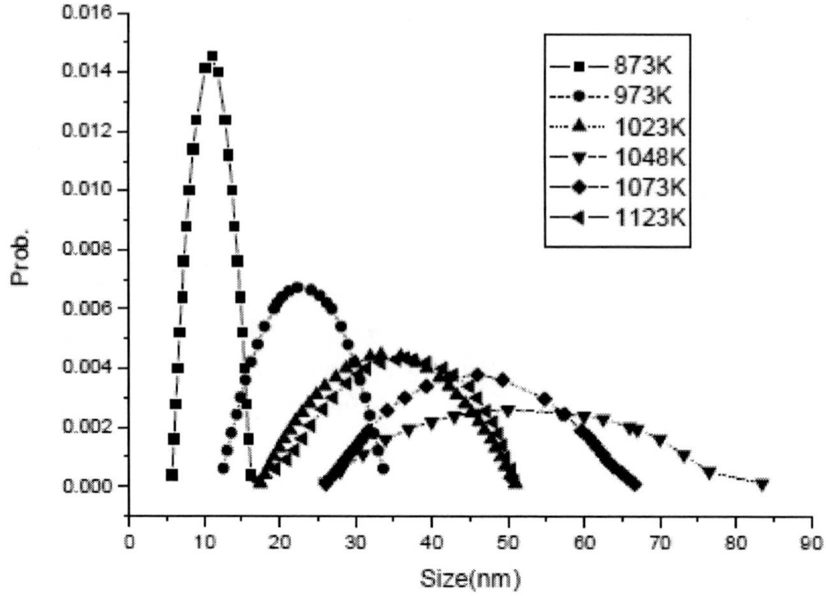

Figure 7. The size distribution of anatase during calcination.

Compared the result of figure 7 with that of figure 8, one can see it clear that the size of rutile nanoparticles is smaller than that of anatase nanoparticles at 1048 K; the size of rutile nanoparticles is bigger than that of anatase nanoparticles at 1073 K.

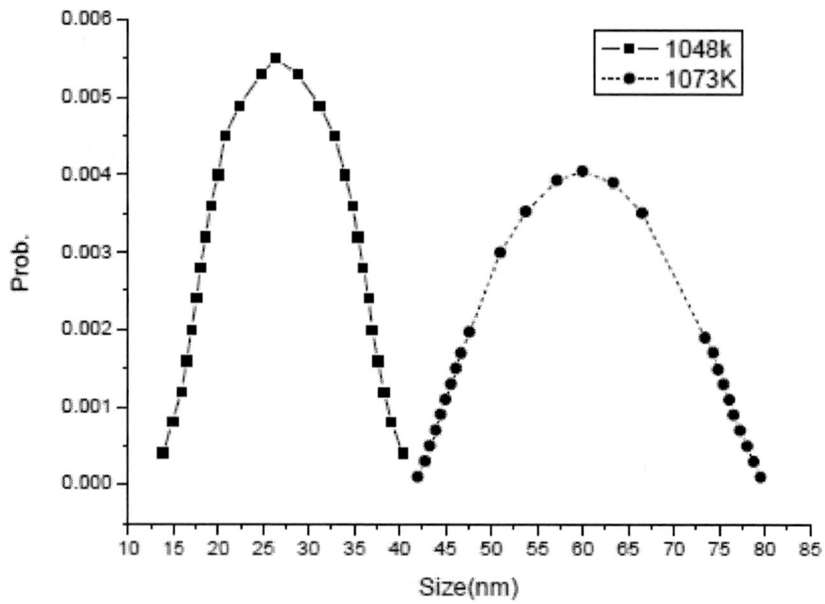

Figure 8. The size distribution of rutile during calcinations.

The rate competition determines the size distribution and the diameter of anatase nanoparticles. During the phase transformation, the size distribution of anatase is the result of the competition between anatase growth rate and the transformation rate from anatase to rutile. When the growth rate is faster than the transformation rate, the size distribution becomes broader, and the average diameter becomes bigger too. When the growth rate is slower than the transformation rate, the size distribution becomes narrower, and the average diameter becomes smaller.

2.5. The Microstrain Determines the Rates of the Anatase Growth and the Transformation

Figure 9 shows the strain of anatase particles during calcination. From figure 9, when the calcination temperature is at 873 K, the strain of anatase particles is weak, it becomes stronger and stronger as the temperature increases, and it is the strongest when the temperature is at 1023 K. It becomes weaker and weaker as the temperature over 1023 K, and becomes the weakest when the temperature is at 1123 K, which is almost the same with that at 873 K.

Based on the above results, one can draw a conclusion that the strain of anatase particles increases along with the temperature increasing from 873K to 1023K, and decreases as the temperature increases over 1023 K. This is consistent with that of the phase transformation and the size distribution of anatase particles. Combined the crystal size with the microstrain of anatase nanoparticles, we can draw a conclusion that when the microstrain of anatase nanoparticles become the strongest, the crystal size of anatase nanoparticles growths rapidly, as shown in figure 7 and 9.

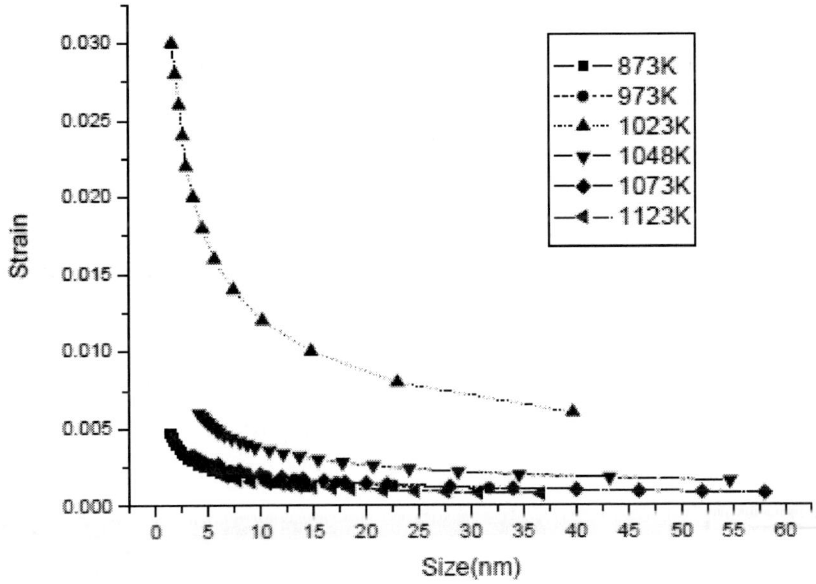

Figure 9. The strain of anatase particles during calcinations.

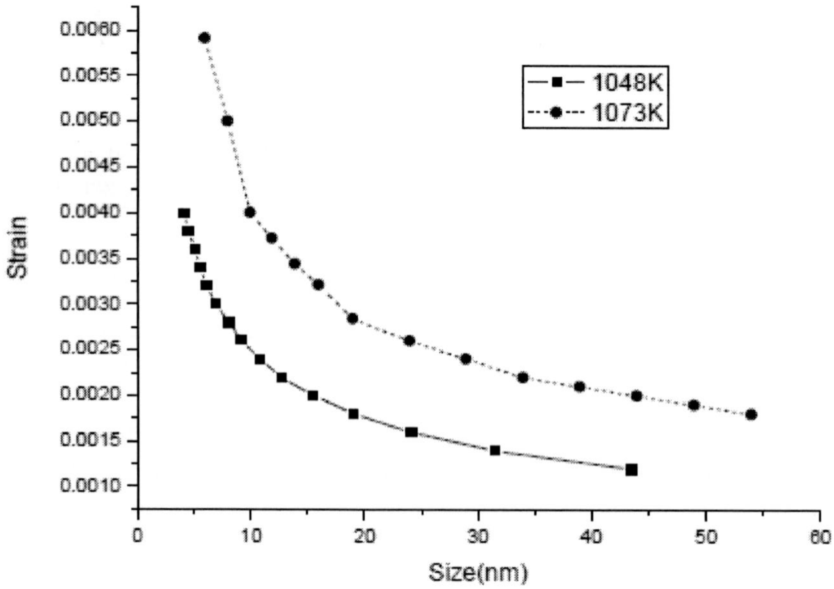

Figure 10. The strain of rutile particles during calcinations.

Figure 10 is the microstrain of rutile nanoparticles at 1048 K and 1073 K. The microstrain of rutile nanoparticles increases from 1048 K to 1073 K.

Compared the result of figure 9 with that of figure 10, one can see clearly that the microstrain of rutile nanoparticles is weaker than that of anatase nanoparticles at 1048 K; the microstrain of rutile nanoparticles is stronger than that of anatase nanoparticles at 1073 K.

When the calcination temperature is over 1048 K, anatase begins to transform into rutile, as shown in figure 6. These results indicate that during the transformation from anatase to rutile, the relative value of the microstrain of nanoparticles changes along with the temperature. Combined the microstrain with the size distribution of nanoparticles, one conclusion can be drawn that when the microstrain of rutile is stronger than that of anatase, the transformation rate of anatase to rutile is faster than the rate of anatase growth, and when the microstrain of rutile is weaker than that of anatase, the transformation rate of anatase to rutile is slower than the rate of anatase growth. This indicates that the microstrain of nanoparticles plays an important role during the phase transformation from anatase to rutile. The microstrain determines the rates of the anatase growth and the transformation.

3. The Fabrication of Rutile Hollow Microspheres with Mesoporosity

3.1. Introduction

Since the photocatalytic splitting of water on TiO_2 was found by Honda and Fujishima in 1972 [33]. The reason why titania can be used in this field has been ascribed to its intrinsic quality. Once titanium dioxide is illuminated by light with energy higher than its band gaps (rutile has a band gap energy of 3.0 eV), some electrons in TiO_2 will jump from the valence band to the conduction band, and the electron (e^-) and electric hole (h^+) pairs will form on the surface of the photocatalyst:

TiO_2+ hv→TiO_2 (e^- + h^+).

The negative electrons and oxygen will combine into O_2^-, the positive electric holes and water will generate hydroxyl radicals [34]. Under an UV-illumination, highly oxidative valence band holes (+2.7 V vs the normal hydrogen electrode at pH7) and various oxygen-containing radical species (e.g., ·OH, O^-_2, HO_2·) generates on the surface of titania [35]. Although titania has wide range of applications, its low absorption in visible region is a drawback from a view point of effective usage of solar energy.

In order to fulfill industrial application of sunlight photocatalysis through titania, a lot of efforts had been done to improve its efficacious absorption of UV light, and enlarge its absorption region of sunlight from UV band to visible band. Doping was one of the efficacious technologies that used widely to improve the photocatalytic activity of titania. Doping transition metal cation, such as Cu, Fe, Ni, Cr, V and Cd into TiO_2 lattice to replace the position of Ti, and doping anion such as S, N, C and P to into TiO_2 lattice to occupy the position of O were two efficacious ways that has been investigated intensively, except for these two kinds of ions, noble metal nanoparticles, such as Pt, Au, Ag and Ru nanoparticles were also used as active content of visible absorption to fabricate noble-titania composite to enlarge absorption region of doped titania systems to visible band of sunlight [36]. These investigations had shown that one of the key factors for improving the photocatalytic efficiency of doped titania system is optimum dopant concentration, because doping concentrations exceeding an optimum value may even reduce photocatalytic activity, and certain dopants (i.e. defects) may weaken the mesostructure of TiO_2 framework, leading to structure collapse at elevated temperatures and reduce photocatalytic activity. But it is not easy to get an optimum value of a dopant and make the dopant to distribute in doped TiO_2 homogeneously.

Hollow sphere of titania is one of promising candidates of solar energy-responsive photocatalysts, because hollow titania particles has a smaller band gap resulting in the absorption in longer wavelength region, and are able to use sunlight more efficiently to degrade organic pollutants in environments [37,40]. And higher energy conversion efficiency and photocatalytic activity are expected using TiO_2 hollow structure as photocatalytic [41-45]. Ren et al showed that a large number of surface hydroxyl groups and a reduced band gap were found in hollow microsphere of mesoporous titania, which should be significant for multi-functionalisation of the surface and photocatalysis applications [38]. Zhang et al [46] showed that a structure with meso- and macro-porosity could enhance the photocatalytic activity of mesoporous titania. These merits are ascribed to phase structure, crystallite size, low density, high surface area, good surface permeability, pore structure and greater light-harvesting capacity of hollow structure of titania. [47-48]. That's the reason why fabrication and application of hollow structure titania is one of hot topics, especially in sunlight photocatalytic area in recent years.

The chemical and physical properties of titania photocatalysts depend on its preparation procedures and conditions. A lot of papers concerning the fabrication of hollow structure of titania had been published in these years, in which the following approaches were used: polystyrene latex particles as template [37, 49-52], surfactant assisted [36, 47, 53,54], LBL

[41], Sol-gel [55], core-shell gel particle as templates [56-57], hydrothermal [58-59], exfoliated layer as precursor [60], sonochemical [48], nanocast [61]. However, almost all the above methods generally require the usage of surfactants, polymers, or inorganic materials as templates or assisted reagent, which have to be removed to create porosity. These materials will cause impurity, such as carbon impurity, or cation impurity on the surface of the product [62], and this impurity will alter the property of titania products [63], for the surface properties of the titania powders plays a vital role for catalysis and photocatalysis [64]. A surface consisting of $_6Ti^{3+}$-OH for coordinative saturation (octahedral site), along with $_4Ti^{4+}$-O (tetrahedral) is essential in photocatalysis [65]. The catalytic property of TiO_2 is also influenced by synthetic methods. TiO_2 synthesized by $TiCl_4$ is found to be three times more active than that synthesized by titanium tetra-isopropoxide, whereas TiO_2 synthesized by titanium sulphate is comparatively much less active [66–68].

Based on the above results, to investigate synthetic strategy for mesoporous TiO_2 with hollow structure that crystallized well and without template residual is valuable and necessary, using $TiCl_4$ precursors. In this section, we report a novel approach to prepare phase pure rutile with nanoparticle frameworks and mesoporosity, using $TiCl_4$ precursors only. The optical property of the samples was investigated.

3.2. Sample Preparation

100 ml of $TiCl_4$ was added dropwise into 200 ml of deionized water, while keeping the mixing course under stirring and the temperature of mixing liquid below 313 K in an ice-water bath. After titanium tetrachloride had added, the mixture solution was stirred for another 2 hours and $TiCl_4$ aqueous sol formed, then cooled down to room temperature to age and peptize. 50 ml of the sol was added into a three-necked flask 12 hours later, and 100 ml of deionized water was added dropwise into the flask while stirring. The temperature of the mixing liquid was kept below 313 K in an ice-water bath. After deionized water had added, the mixing liquid was under stirring for another 30 minutes. The sol was passed into a spray desiccator while stirring with a magnetic force stirrer. During pyrogenation process in the spray desiccator, a velocity of hot air flow was $35m^3•h^{-1}$, a velocity of aqueous solution was 15 $ml•min^{-1}$, the temperature of the hot air at the threshold of the spray desiccator was 473 K, and the temperature of a mixture of gas and powder at the exit of the spray desiccator was 373 K. In this way, a powder precursor was prepared. The powder precursor was then calcined at 553K, 773 K and 1023K for two hours.

3.3. XRD

Figure 11 is the XRD patterns of the samples calcined at 573 K, 773 K and 1073 K for two hours. The crystal phase of the samples calcined at 573 K and 773 K is anatase (JCPDS: 21-1272) only and almost all the diffraction peaks of anatase appear. Diffraction peaks intensity of the sample calcined at 773 K is higher than that calcined at 573 K. These results indicate that the crystallinity of the sample calcined at 773 K is higher than that calcined at 573 K. The sample calcined at 1073 K is phase pure rutile (JCPDS: 21-1276), as figure 11 shown.

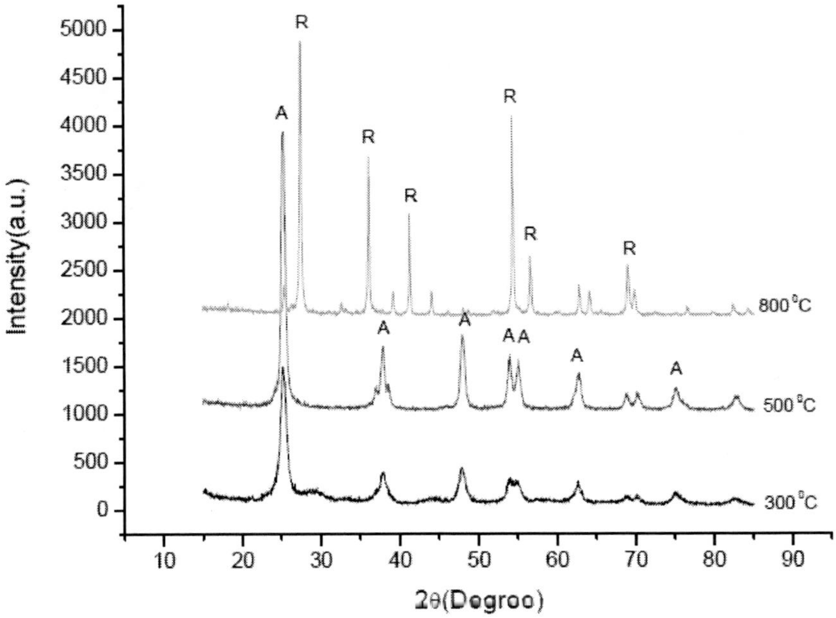

Figure 11. The XRD patterns of the sample calcined at different temperatures fot two hours. A: anatase; B: rutile.

3.4. SEM

Figure 12 is the SEM images of the sample calcined at 573 K, 773 K and 1073 K for two hours. Figure 12a is one of the particles selected from the sample calcined at 573 K. The morphology of this particle is hollow sphere with a diameter around 15 μm, and the wall thickness of the hollow sphere is around 5μm, as showing in figure 12a. Figure 12b shows the surface structure of the broken hollow sphere that showed in figure 12a. The wall of the hollow sphere is constituted of nanoparticles, with a diameter of 10~30 nm, and porosity with a size around 4.5 nm, as showing in figure 12b. Figure 12c shows one of the particles selected from the sample calcined at 773 K. The morphology of this particle is hollow sphere with a diameter around 6 μm, and the wall thickness of the hollow sphere is around 0.4 μm, as showing in figure 12c. Figure 12d is the surface structure of the broken hollow sphere that showed in figure 12c. The wall of the hollow sphere is constituted of nanoparticles, with a diameter of 20~50 nm, and porosity with a size around 8 nm, as showing in figure 12d. Figure 12e is one of the particles selected from the sample calcined at 1073 K. The morphology of this particle is hollow sphere with a diameter around 15 μm, as showing in figure 12e. Figure 12f shows the surface structure of the broken hollow sphere that showed in figure 12e.

The wall of the hollow sphere is constituted of particles, with a diameter bigger than 100 nm, and mesoporosity, as showing in figure 12f. These results indicate that the wall of the hollow microsphere of the sample calcined below 773 K is of nanostructure with mesoporosity. Comparing the images of figure 12b, 12d and 12f, it is clear that the boundary of titania particle becomes more evident as the calcined temperature increasing, and the size of particles becomes bigger and bigger. These phenomena can be ascribed to the crystal growth and the crystallinity increasing as the calcined temperature increased.

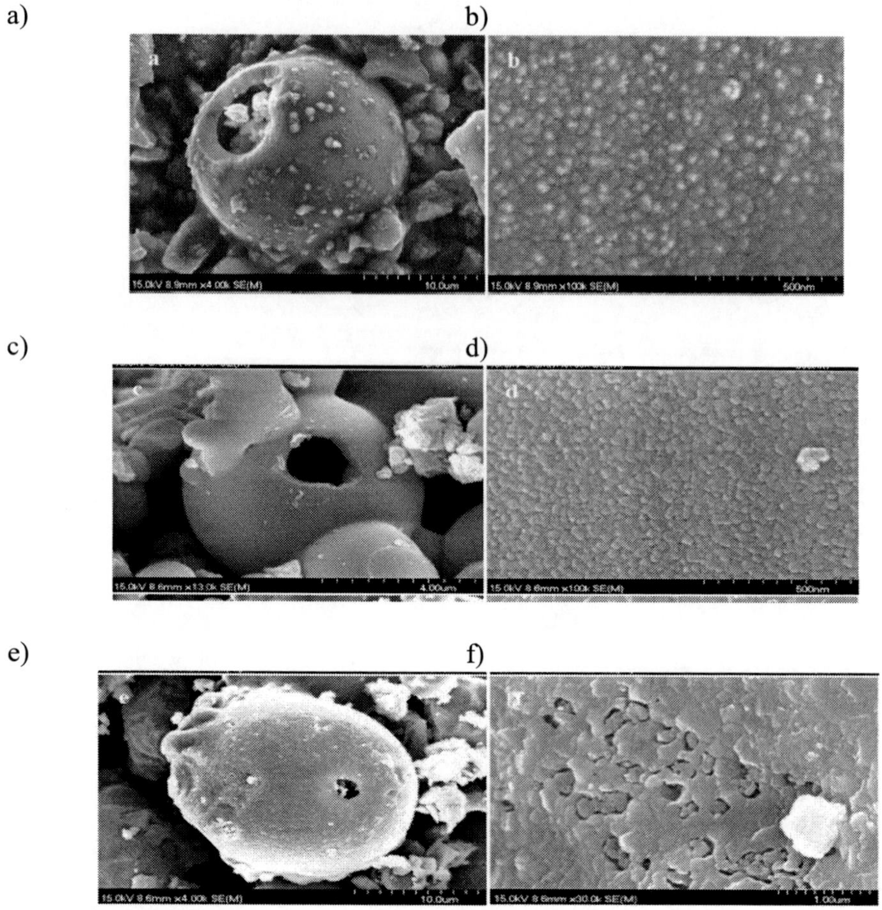

Figure 12. The morphology of the samples calcined ar different temperatures. a and b: 373 K; c and d: 773 K; e and f: 1073 K. a, c and e: morphology; b, d and f: surface structure.

3.5. Specific Surface Area

The specific surface area and pore width of the samples measured by N_2 adsorption-desorption with an ASAP2020 Micromeritics are displayed in table 1.

From table 1, one can see it clearly that the specific surface area of the samples decreasing as the calcined temperature increasing, and the pore average width by BJH desorption increasing as the calcined temperature increasing. These conclusions are consistent with that of SEM results.

3.6. UV Absorption

UV-vis absorption spectra of the suspension of the samples were recorded to investigate their optical properties. The results are shown in figure 13. Nanoanatase and nanorutile were used as comparative samples for UV adsorption, the particle size of nanoanatase is smaller than 10 nm, which prepared by our group and reported in reference 38, and its adsorption for UV is well, with an evident adsorption region of 297~345 nm, as curve A showing in figure

13. The particle of nanorutile is rod-like with a diameter of 10~20 nm and a length of 20~80 nm, which prepared by our group and reported in section 1.2, and there is no evident adsorption for UV, as curve R showing in figure 13. The UV-vis absorption of the sample calcined at 573 K is the best, with an intensive adsorption region from 271 nm to 387 nm, as curve 573 K showing in figure 13. The UV-vis absorption of the sample calcined at 773 K is better, with an evident adsorption region from 200 nm to 387 nm, as curve 773 K showing in figure 13. The UV-vis absorption of the sample calcined at 1073 K is the worst, with a weak adsorption region from 200 nm to 300 nm, as curve 1073 K showing in figure 13. The UV adsorption of the samples calcined at 573 K and 773 K is better than that of nano-anatase and nano-rutile, but that of the sample calcined at 1073 K is much worse than that of nano-anatase and nano-rutile, as figure 13 shown. The results indicate that the optical property of the samples calcined at 573 K and 773 K is better, and that of the sample calcined at 1073 K is worse than that of nano-anatase and nano-rutile. Based on these results, one conclusion can be drawn that the optical property of the samples is connected with their crystal phase.

Table 1. the results of the samples measured by N_2 absorption

Contents	Temperature calcined (K)		
Conditions	573	773	10.73
BET Surface area (m^2/g)	60.35	38.38	19
BJH Desorption average pore width (4v/nm)	4.54	7.98	

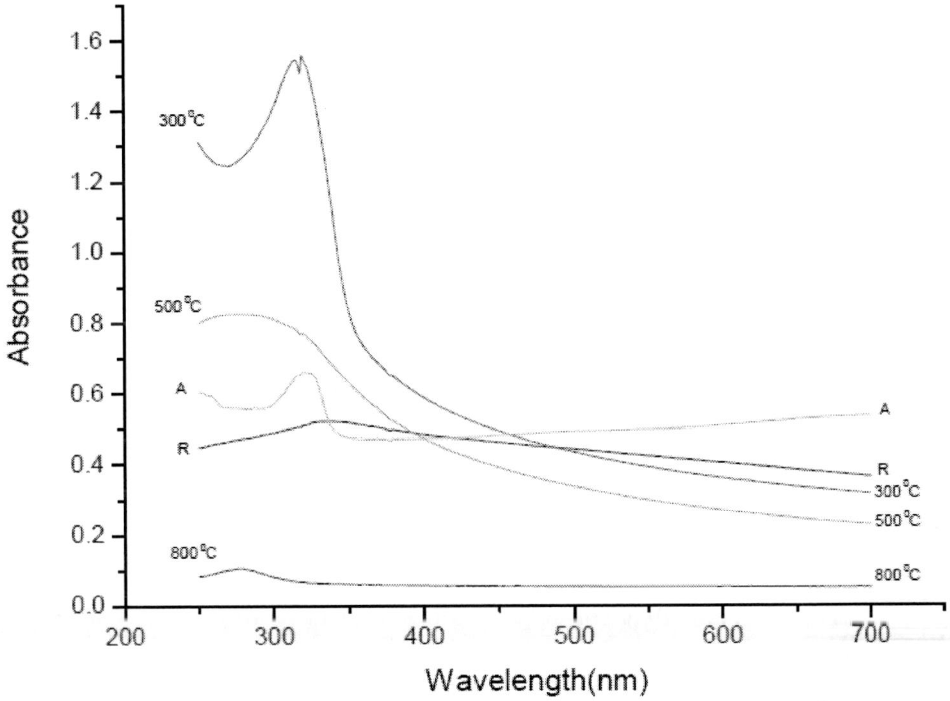

Figure 13. UV-vis spectra of the samples. A: nano-anatase; R: nano-rutile; 300^0C: the temperature that the samle calcined.

3.7. The Formation of the Hollow Structure

The formation of the hollow structure of the sample can be attributed to the process of spray pyrogenation. In our experiments, the spray dryer was an air flow dryer, a BÜCHI B-290, and its working manner was that a flow of hot pressurized air drove a pulverized liquid into a drying spray cylinder to hydrolysis. As $TiCl_4$ aqueous sol with a lot of titania embryos was passed into the spray dryer, a hydrolysis reaction would take place quickly and intensively, as mentioned above, and a conglomerate that composed of titania embryos and liquid, such as water, would form. When this resultant was driven into the spray cylinder to be dried, as the temperature in the spray cylinder was about 473 K, the liquid in the conglomerate evaporated quickly, and titania embryos leave and grew up to lager titania nanoparticles with a volume expansion. Thus, a hollow structure formed.

Conclusions

In summary, hollow micro-spherical anatase and rutile particles with mesoporosity were fabricated using $TiCl_4$ as sole precursor by a sol spray hydrolysis approach. The wall of the hollow micro-sphere is constituted of nanoparticles and mesoporosity. Its formation mechanism can be attributed to the spray dry process and the property of $TiCl_4$ aqueous sol.

The optical property of hollow sphere mesoporous titania is better than that of nano-anatase and nano-rutile in aqueous suspensions, and related to its crystal phase, specific surface area, crystal size, porosity size and crystallinity of titania, and these characteristics are changing as the calcined temperature increasing.

The merit of this approach is template free and without any additives, organic or inorganic. The sol-spray hydrolysis approach can be used to prepare other materials with the same structure at macroscale.

References

[1] Li, Y. Z., Fan, Y. N., Chen, Y. J. *Mater. Chem.* 12 (2002) 1387.
[2] Kuang, D. B., Xu, A .W., Zhu, J. Y., et al. *New J. Chem.* 26 (2002) 819.
[3] Zhang, H., Banfield, J. F., *J. Phys. Chem.* B. 104 (2000) 348.
[4] Yanagisawa, K., Ovenstone, J., *J. Phys. Chem.* B. 103 (1999) 778.
[5] Pottier, A., Chanéac, C., Trone, E., et al. *J. Mater. Chem.* 11 (2001) 1116.
[6] Wu, M., Lin, G., Chen, D., et al. *Chem. Mater.* 14 (2002) 1974.
[7] Zhang, Q.H., Gao, L., Guo, J. K., *Nanostruct. Mater.* 11 (1999) 1293.
[8] Wang, W., Gu, B. H., Liang, L. Y., et al. *J.Phys.Chem.*B. 47 (2004) 952.
[9] Wang, C., Deng, Z .X., Li, Y. D., *Inorg.Chem.* 40 (2001) 5210.
[10] Yin, H. B., Wada, Y. J., Kitamura, T., et al. *J.Mater. Chem.* 11 (2001) 1694.
[11] Yin, H. B., Wada,Y. J., Kitamura, T., et al. *J.Mater. Chem.* 12 (2001) 378.
[12] Andersson, M., Osterlund, L., Ljungström, S., et al. *J.Phys.Chem.*B. 106 (2002) 10674.
[13] Aruna, S. T., Tirosh, S., Zaban, A., *J.Mater.Chem.* 10 (2000) 2388.
[14] Cheng, H., Ma, J., Zhao, Z. G., Qi, L. M., *Chem. Mater.* 7 (1995) 663.

[15] Zhang,Qing-Hong, Gao, Lian, Guo Jing-Kuan, Wuji Cailiao Xuebao *(Chinese Journal of Inorganic Materials*). 15 (2000) 992.

[16] Bahnemann, D., Henglein, A., Spanhel, L., Faraday Discuss. *Chem. Soc.* 78 (1984) 15.

[17] Constable, E. C., Metals and ligand Reactivity: An Introduction to the Organic Chemistry Complexes, New York: VCD Publishers. (1996) 7.

[18] Zhang Heng-Zhong, Jillian, F. Banfield. Understanding polymorphic phase transformation behavior during growth of nanocrystalline aggregates: insights from TiO2 [J]. *J. Phys.Chem.* B 104 (2000) 348.

[19] Krishnankutty-Nair, P. Kumar. Growth of rutile crystallites during the initial stage of anatase-to-rutile transformation in pure titania and in titania-alumina nanocomposites [J]. *Scripta Metallurgica et Materialia*. 32(6) (1995) 873.

[20] Robert, D. Shannon, Joseph, A. Pask. Kinetics of the Anatase-Rutile [J]. *Journal of the American Ceramic Society*. 48(8) (1965) 39.

[21] Laura, E. Depero, Paolo Bonzi, Marcello Zocchi, Cristina Casale, Gennaro De Michele. Study of the anatase-rutile transformation in TiO2 powders obtained by laser-induced synthesis [J]. *J. Mater. Res.* 8 (10) (1993) 2709.

[22] Jong Ho Lee, Wooseok Nam, Misook Kang, Gui Young Han, Ki June Yoon, Moon-Sun Kim, Kenji Ogino, Seizo Miyata and Suk-Jin Choung. Design of two types of fluidized photo reactors and their photo-catalytic performances for degradation of methyl orange [J]. *Appl. Catal. A Gen.* 244 (1) (2003) 49.

[23] Hengzhong Zhang, Jillian, F. Banfield. Thermodynamic analysis of phase stability of nanocrystalline titania [J]. *J. Mater. Chem.* 8(9) (1998) 2073.

[24] Y. Hu, H. L. Tsai, C. L. Huang. Phase transformation of precipitated TiO2 nanoparticles [J]. *Mater. Sci. Eng.* A. 344 (2003) 209.

[25] R. Nicula, M. Stir, C. Schick, E. Burel, High-temperature high-pressure crystallization and sintering behavior of brookite-free nanostructured titanium dioxide: in situ experiments using synchrotron radiation [J]. *Thermochim. Acta*. 403 (2003) 129.

[26] Benjamin Gilbert, Hengzhong Zhang, Feng Huang, Michael, P Finnegan, Glenn, A; Waychunas, Jillian, F. Banfield. Special phase transformation and crystal growth pathways observed in nanoparticles [J]. *Geochem. Trans.* 4 (2004) 20.

[27] Henghong Zhang, Jillian, F.Banfield. New kinetic model for the nanocrystalline anatase-to-rutile transformation revealing rate dependence on number of particles [J]. *American Mineralogist*. 84 (1999) 528.

[28] T. Mitsuhashi, O. J. Kleppa. Transformation enthalpies of the TiO2 ploymorphs [J]. *Journal of the American Ceramic Society*. 62(7-8) (1979) 356.

[29] N. R. Tao, Z. B. Wang, W. P. Tong, M. L. Sui, J. Lu and K. Lu. An investigation of surface nanocrystallization mechanism in Fe induced by surface mechanical attrition treatment [J]. *Acta. Mater.* 50 (2002) 4603.

[30] B. E. Warren. X-ray diffraction [M]. Dover, New York (USA), 1990.

[31] Robert D. Shannon, Joseph A. Pask. Topotaxy in the anatase1-rutile transformation [J]. *The American Mineralogist*. 49 (1964) 1707.

[32] Bin Xia, Huizhong Huang, Youchang Xie, Heat treatment on TiO2 nanoparticles prepared by vapor-phase hydrolysis [J]. *Materials Science and Engineering*. B57 (1999) 150.

[33] K. Honda, A. Fujishima, *Nature* 238 (1972) 37.

[34] C. Srinivasan, N. Somasundaram, *Curr. Sci.* 85 (2003)143.

[35] O. Carp, C. L. Huisman, A. Reller, *Prog. Solid State Chem*. 32 (2004) 33.

[36] Y. S. Shin, B. W. Arey, C. Q. Wang, X. H. S. Li, M. H. Engelhard, G. E. Fryxell, Inorganic *Chemistry Communications* 10 (2007) 642, *and references therein.*

[37] Syoufian, O. H. Satriya, K. Nakashima, *Catalysis Communications* 8 (2007) 755.

[38] T. Z. Ren, Z. Y. Yuan, B. L. Su, *Chem. Phys. Lett*. 374 (2003) 170.

[39] C. Song, D. Wang, G. Gu, Y. Lin, J. Yang, L. Chen, X. Fu, Z. Hu, J. *Colloid Interface Sci*. 272 (2004) 340.

[40] Y. S. Han, G. Hadiko, Y. Tarutani, M. Fuji, M. Takahashi, *Annual Report of the Ceramics Research Laboratory,* Nagoya Institute of Technology 4 (2004) 1–14.

[41] F. Caruso, R. A. Caruso, H. Mohwald, *Science* 282 (1998) 111.

[42] A. Imhof, *Langmuir* 17 (2001) 3579.

[43] T. Nakashima, N. Kimizuka, *J. Am. Chem. Soc*. 125 (2003) 6386.

[44] H. G. Yang, H. C. Zeng, *J. Phys. Chem*. B 108 (2004) 3492.

[45] X. X. Li, Y. J. Xiong, Z. Q. Li, Y. Xie, *Inorg. Chem*. 45 (2006) 3493.

[46] L. Z. Zhang, J. C. Yu, *Chem. Commun*. 20(2003) 2078.

[47] F. B. Li, X. Z. Li, M. F. Hou, K. W. Cheah,W. C. H. Choy, *Appl. Catal.* A 285 (2005) 18.

[48] J. Yu, J. C. Yu, M. K. P. Leung, W. Ho, B. Cheng, X. Zhao, J. Zhao, *J. Catal*. 217 (2003) 69.

[49] I. A. Kartsonakis, P. Liatsi, I. Danilidis, D. Bouzarelou, G. Kordas, *Journal of Physics and Chemistry of Solids,* In Press.

[50] S. J. Ding, C. L. Zhang, M.Yang, X. Z. Qu, Y. F. Lu, Z. Z. Yang, *Polymer* 47 (2006) 8360.

[51] D. Wang, C. Song, Y. Lin, Z. Hu, *Materials Letters* 60 (2006) 77.

[52] X. Cheng, M. Chen, L. Wu, G. Gu, *Langmuir* 22 (2006) 3858.

[53] K. Cassiers, T. Linssen, V. Meynen, P. Van Der Voort, P. Cool, E. F. Vansant, *Chem. Commun.*, 11 (2003) 1178.

[54] M. Collins, C. Spickermann, S. Mann, *J. Mater. Chem*., 13 (2003) 1112.

[55] Y. X. Zhang, G. H. Li, Y. C. Wu, T. Xie, *Materials Research Bulletin* 40 (2005) 1993.

[56] T. H. Kim, K. H. Lee, Y. K. Kwon, *Journal of Colloid and Interface Science* 304 (2006) 370.

[57] Z. Yang, Z. Niu, Y. Lu, Z. Hu, C. C. Han, Angew. *Chem. Int. Ed*. 42 (2003) 1943.

[58] J. Yu, S. Liu, H. Yu, *Journal of Catalysis* 249 (2007) 59.

[59] C. W. Guo, Y. Cao, S. H. Xie, W. L. Dai, K. N. Fan, *Chem. Commun*. 7 (2003) 700.

[60] Masaki Iida, Takayoshi Sasaki, Mamoru Watanabe, *Chem. Mater*. 10 (1998) 3780.

[61] Y. Xia, R. Mokaya, *J. Mater. Chem*. 15 (2005) 3126.

[62] H. Jensena, A. Soloviev, Z. S. Li, E. G. Sogaard, *Applied Surface Science,* 246 (2005) 239.

[63] Sclafani, L. Palmisano and M. Schiavello, *J. Phys. Chem*. 94 (1990) 829.

[64] E. Wahlstrom, E. K. Vestergaard, R. Achaub, A. Ronnau, M. Vestergaard, E. Laegsgaard, I. Stensgaard, F. Besenbacher, *Scince,* 303 (2004) 51.

[65] P. M. Kumar, S. Badrinarayanan, M. Sastry, *Thin Solid Films,* 358 (2000) 122.

[66] C. Wang, J. Y. Ying, *Chem. Mater*. 11 (1999) 3113.

[67] Pottier, C. Chaneac, E. Tronc, L. Mazerroles, J. Jolivet, *J. Mater. Chem.* 11 (2001) 1116.

[68] R. Debnath, J. Chaudhuri, *J. Mater. Sci. Lett*. 10 (1991) 494.

In: Rutile: Properties, Synthesis and Applications
Editor: It-Meng (Jim) Low
ISBN 978-1-61942-233-9

Chapter 5

TITANIUM DIOXIDE UNDER AMBIENT AIR: PREPARATION AND PHOTOCATALYTIC PROPERTIES OF TiO_2 PRECIPITATED AEROSOLS

Valery Zakharenko[1]
Boreskov Institute of Catalysis
Pr. Lavrentieva, 5, 630090, Novosibirsk
Russia

ABSTRACT

The comparative analysis of experimental data over titanium dioxide powders prepared by various ways under ambient air and investigated in conditions close to those of the real troposphere is carried out. The results over TiO_2 prepared by high-temperature heating of anatase, produced by burning of titanium micro particles and grinding of rutile crystal are used for that comparison.

It is shown that Earth's atmosphere ecology can be keep in safe level in the result of photocatalytic and photoadsorption processes proceeding on aerosol particle surface under troposphere solar irradiation. And what is more, it is possibly to clean the atmosphere near a pollution source by the production of titania aerosol in direct place of a contamination.

The conception of oxidative photocatalysis as nature phenomenon are developed. According to that: photocatalytic reaction proceed only under light quanta from the fundamental absorption band of photocatalyst; oxygen photoadsorption is not an limiting step of photocatalytic oxidation reaction; photocatalytic activity is only inherent in metal oxides, the top level of whose valence band is formed by 2p - orbital of oxygen; during of photocatalytic oxidation reaction the oxidized gas-phase molecule interacts with surface lattice oxygen at the moment of its discharge by a free mobile hole. The quantum efficiency of photocatalytic reaction mainly depends upon the bond energy of surface lattice oxygen; the quantum yield is also determined by physicochemical characteristics of photocatalysts. In agreement with the proposed mechanism the photocatalytic oxidation reaction does not require an active surface center.

[1] E-mail address: zakh@catalysis.ru

Keywords: TiO_2, tropospheric conditions, photoadsorption, photodesorption, photocatalytic oxidation, quantum efficiencies.

1. Introduction

To solve the problem of possibility of self-cleaning Earth's atmosphere from harmful chemical compounds under photoinduced heterogeneous processes over aerosol particle surfaces it is necessary to simulate conditions of aerosol particle creation in the Troposphere.

The tropospheric conditions of the gas phase (concentration, composition, temperature, pressure) and solar radiation (intensity, spectral composition) reproduce simply enough into the experimental set-up. Oxygen, water, carbon dioxide, nitrogen oxides are the main components of the gas phase of the atmosphere. The intensity and spectral composition of tropospheric solar radiation are also known [1].

The phase chemical composition of solid aerosol particles is less known. Solid tropospheric aerosol comprises a large number of compounds of certain chemical composition and in crystal form (phase chemical components). The total activity of aerosol particles can be determined by the sum of activities of separate phase components of aerosol particles. The determination of phase compositions of solid aerosol particles is complicated task, which has not yet been achieved, although in some papers the attempts have been described to solve this problem [2, 3].

The elemental compositions of solid aerosol particles of the troposphere qualitatively [4] and for many elements quantitatively [5] correspond to the elemental compositions of the natural rocks of the earth's crust. This makes it possible to assume that the chemical phase compositions of solid aerosol particles mainly correspond to mineralogical compositions of the earth's crust.

The creation of troposphere aerosol particles with fine crystallised structure was the result of a dispersion of various lithosphere minerals over very long time [6]. Thus, aerosol particles are microcrystals with a sufficiently ideal crystalline structure as opposed to particles of the high dispersion adsorbents (photoadsorbents) and catalysts (photocatalysts) produced by the traditional way, i.e. under conditions of preparation TiO_2 with a high specific surface in laboratory and industry. The powder-like samples from minerals can be expected to have properties differing from those of the artificially produced compounds. In accordance with the published data, photocatalytic reactions in semiconductor metal oxides [7], as well as photosorption processes in oxide insulators [8], include a stage of transfer of electrons and holes formed under illumination to the surface. Consequently, the probability that free carriers will reach the surface and change the reactivity of surface atoms of the lattice or adsorbed atoms and molecules depends on the conditions of the carriers transfer in the bands (valence and conductivity bands). At small sizes of particles (high dispersion oxides insulators γ-Al_2O_3 and SiO_2 with a specific surface of 150 and 300 $m^2 \cdot g^{-1}$, respectively), these bands fail to be formed and the photosorption activity relative to halogen-containing organic compounds has not been observed [8].

A significant part of the tropospheric continental aerosol mainly consists of well-crystallised silicate and quartz microparticles. In this case, the photosorption activity relative to halogen-containing organic compounds is observed. As data [9-11] show, efficient photoinduced processes actually proceed on particles of the coastal or desert sand with the participation of organic and halogen organic compounds.

To simulate conditions for producing well crystallised particles in tropospheric aerosol, the powder-like metal dioxides were made through grinding metal-oxide crystals under ambient air. This procedure allows preparation of aerosol particles having the phase chemical composition and the crystal structure close to them for aerosol particles in the Troposphere.

The Earth's atmosphere has the ability to self-clean from dangerous compounds because of the photocatalytic and photosorption processes on the particle aerosol surface in the troposphere. These processes are anticipated to primary affect the intensity of acid rains, concentration of greenhouse and ozone-depleting gases, as well as cleaning up harmful compounds from the atmosphere.

However, in case of a man-caused catastrophe, which accompanies a local emission of pollutants, the decontamination of the atmosphere by natural aerosols may be ineffective. In order to neutralize a drastic pollution of the Troposphere can be used technogenic technologies of the decontamination. One of such technologies is the technology using metal oxide solid aerosols which are active in removal of pollutant compounds and obtainable by combustion under ambient air of appropriate metal particles, for example, aluminum [12], magnesium [13] and etc. It is known that magnesium oxide photosorbs efficiently halogen-containing organic compounds [14] and titanium dioxide photocatalyzes an oxidation breakdown a large number of organic compounds polluting atmosphere [15-16].

The pyrotechnic technology of a metal oxide preparation is mobile, available and inexpensive way of a solid aerosol production near a pollution source. The technical applications may be look like on a firework, lighting flare and the like. It allows locating an environment impact of contaminating compounds. For a development of such technology the study of photochemical properties of the titanium dioxide produced by pyrotechnic mixture combustion under ambient air was carried out.

2. Experimental

2.1. Tropospheric conditions

a). Gas phase composition of troposphere

In spite of significant concentration of molecular oxygen in the gas phase in the troposphere, the tropospheric conditions, in the absence of solar radiation, do not favor oxidation of tropospheric trace gases (CO, SO_2, nitrogen oxides, H_2S, various hydrocarbons, etc.) even in the case that there are aerosol particles in the troposphere, which can serve as catalysts in the oxidation process. In the presence of solar radiation, the situation drastically changes, since radiation quanta initiate many gas-phase and heterogeneous reactions.

The solar radiation is the driving force of chemical reactions in the troposphere. Transmitted through the stratosphere ozone layer is only solar radiation with wavelength $\lambda >$ 300 nm. Gases that are present in the troposphere in the largest amounts (see the Table 1) do not absorb such radiation and, consequently, cannot be activated. The effect of photochemical reactions in the spectral region of solar radiation in the troposphere for gases that are present in very low concentrations (less than 1 ppb) on the chemistry of the troposphere is insignificant, and therefore it is not considered. Among the trace gases (concentration ~ 10 ppb), photochemical active ones are the ozone and nitrogen dioxide, and the concentration of the latter at air pollution can achieve 100 ppb [17]. The concentration profiles for other atmospheric gases can be found in the Internet (http://atmos.iao.ru).

Table 1. Content of gases in unpolluted atmosphere, according to Ref. [1]

Gas	Concentration, $mol\cdot cm^{-3}$, % or fractions		Atmospheric life time
Gases present with the highest concentrations			
N_2	$2\cdot10^{19}$	78%	10^6 years
O_2	$5\cdot10^{18}$	21%	$5\cdot10^3$ years
H_2O	$<6\cdot10^{17}$	<2%	10 days
CO_2	$9\cdot10^{15}$	330 ppm	4 years
Gases absorbing solar radiation in the troposphere			
O_3	10^{11}	10 ppb	
NO_2	$3\cdot10^9$	0,3 ppb	4 days
CH_2O	10^9-10^{10}	0,1-1 ppb	5-10 days
Gases causing acid rains			
NO	10^9	0,1 ppb	4 days
SO_2	10^8-10^9	0,01-0,1 ppb	3-7 days
Gases leading to formation of smog			
Hydrocarbons (without CH_4 and CH_2O)	$<10^{10}$	<1 ppb	
CH_4	$4\cdot10^{13}$	1,5 ppb	3,6 years
Radicals			
OH	$7\cdot10^5$		
HO_2	$2\cdot10^7$		
$O(^1D)$	1 в 100 cm^{-3}		

Anthropogenic pollution of the troposphere with nitrogen dioxide is largely related to oxidation of molecular nitrogen and nitrogen monoxide by air oxygen in high-temperature processes (burning, volcanic activity, and lightning). The background content of ozone in the troposphere, in its turn, is caused by its formation in stratosphere and, probably by the following slow diffusion into the troposphere. In the atmosphere polluted by nitrogen monoxide, the interaction of NO with ozone additionally increases the concentration of nitrogen dioxide. This reaction is the main one in the troposphere in the absence of solar radiation (nighttime). In the presence of solar radiation (daytime) photochemical reaction of ozone and nitrogen dioxide dissociation occur in the troposphere.

Under the effect of radiation having different wavelengths from the solar spectrum in the troposphere the following ozone photodissociation reactions take place [18]:

$$O_3 \rightarrow O(^3P) + O_2(^3\Sigma^-_g), \quad \lambda < 1180 \text{ nm}, \tag{1}$$

$\rightarrow O(^3P) + O_2(^1\Delta_g)$, $\lambda < 611$ nm , (2)
$\rightarrow O(^3P) + O_2(^1\Sigma^+_g)$, $\lambda < 463$ nm , (3)
$\rightarrow O(^1D) + O_2(^3\Sigma^-_g)$, $\lambda < 411$ nm , (4)
$\rightarrow O(^1D) + O_2(^1\Delta_g)$, $\lambda < 310$ nm . (5)

The maximum absorption cross-section in the region near 600 nm is equal to 4.89×10^{-21} cm^2, and the quantum yield for reaction (5) is 0.90.

Photodissociation of nitrogen dioxide in the troposphere occurs under the effect of quanta with the wavelengths $300 < \lambda < 398$ nm by the single reaction [18]:

$NO_2 \rightarrow NO + O(^3P)$, $\lambda < 398$ nm . (6)

The maximum absorption cross section at 400 nm is to 6.76×10^{-19} cm^2 and the quantum yield is about 0.69.

Formation of oxygen atoms at photodissociation of ozone and nitrogen dioxide leads to secondary reactions and production of radicals, whose concentration in an unpolluted atmosphere is given in Table 1. The main secondary reaction is the interaction of atomic oxygen in the singlet D state, which is the product of reaction (4) and (5), with the water molecule yielding formation of the hydroxyls:

$O(^1D) + H_2O \rightarrow 2\,OH$. (7)

The effect of hydroxyls on the chemistry of the troposphere is significant [19-22] and can be illustrated by the following reactions:

(a) formation of the nitric acid:

$OH + NO_2 \rightarrow HNO_3$; (8)

formation of sulfuric acid (oxidation of SO_2):

$2\,OH + SO_2 \rightarrow H_2SO_4$; (9)

(c) interaction with methane:

$OH + CH_4 \rightarrow CH_3 + H_2O$; (10)

(d) formation of hydroperoxide radical:

$OH + O \rightarrow HO_2$; (11)

(e) formation of strong oxidizer – hydrogen peroxide:

$OH + OH \rightarrow H_2O_2$; (12)

(f) formation of hydrogen atoms:

$$OH + CO \rightarrow CO_2 + H. \quad (13)$$

Formation of strong oxidizers (H_2O_2 and HO_2) in reactions (11) and (12) leads to oxidation of hydrocarbons and production of organic peroxide compounds which are the precursor of smog. This process proceeds effectively in the troposphere polluted with car exhausts [22, 23]. Besides, peroxide compounds oxidize nitrogen monoxide, and the ozone concentration increases, since ozone is no longer consumed for oxidation of nitrogen monoxide.

Thus, initiation of the gas-phase chemical reaction in the troposphere is caused by the effect of solar radiation on ozone (unpolluted troposphere) and ozone and nitrogen dioxide (polluted troposphere).

b). Chemical composition of tropospheric solid aerosol

The troposphere is polluted not only with various gaseous compounds, but also with particles of solid-state and liquid-droplet aerosol. Chemical reactions in the volume of liquid-droplet aerosol particles are similar to gas-phase reactions, differing only by the concentration of liquid-phase reagents because of their different solubility in liquids [24, 25]. In the presence of aerosol particles having a well-developed surface [26] and nearby the Earth's surface, chemical reactions by radical mechanism become unlikely [27]. In the presence of solid particles of tropospheric aerosol, photoinduced reactions under solar radiation are possible on the particle surface [28].

The Troposphere contains significant amount of solid particles of various chemical compositions. The entire mass of particles of solid aerosol in the Troposphere exceeds 100 million tons, and their overall common surface attains up 10^{14} m^2 [26]. The determination of phase chemical composition of aerosol particles has not yet realized. However, the elemental composition of tropospheric aerosol is known [4, 5].

c). Solar radiation

Of importance is that the troposphere is fed with the solar radiation of wavelengths $\lambda >$ 300 nm only, since the shorter part of the radiation (the "rigid" UV) is absorbed by atmospheric gases already in their upper layers. However, even the atmosphere penetrating flux of solar photons is enormous.

At sunny noon, the density of the solar tropospheric radiation with wavelengths of the mild or "near" UV region of 300–400 nm corresponds to approximately 10^{13} photon$\cdot$cm$^{-2}\cdot$s^{-1}, while the overall flux of the illumination of the troposphere with the UV radiation of this spectral region attains approximately 10^{31} photon$\cdot$s^{-1}. In case each photon is generating a chemical reaction, the overall value of this enormous flux can be considered as an upper limit of the maximum possible overall rate of photoinduced chemical transformations which is also enormous: approximately 10^7mole$\cdot$s^{-1}. However, the directly photoinduced atmospheric processes at the visible-near UV illumination at such a spectral region should occur with not very high quantum yields, and thus corresponding photoinduced chemical transformations exist only indirectly via photoexcitation of photocatalysts, rather than immediate commonly uncolored reactants present in the atmosphere.

Note that many metal oxides — which are the main components of solid tropospheric aerosols — are known to experience absorption of quanta with wavelengths $\lambda < 400$ nm and even of visible light [29-33].

2.2. Creation of experimental conditions close to tropospheric conditions

To simulate the processes in the Earth's troposphere, the experiments were performed under conditions similar to the natural conditions. This implies (1) the small partial pressure of the gas to be photoadsorbed (10^{-2}–1 Torr) and the tropospheric pressure of air, (2) the small irradiation in the near ultraviolet region, (3) the presence of water vapor in the reaction mixture, and (4) the absence of preliminary hard treatment of the oxide surface, which is known to significantly influence the photoadsorption and photocatalytic properties of samples, for example, at the expense of removal of preadsorbed impurities and a decrease in the number of surface defects.

a). High vacuum installation

A home-built static installation was used for the measurement of the rates of adsorption, desorption, catalytic oxidation in darkness and under irradiation. The experiments were performed in a high vacuum apparatus equipped with facilities of mass-spectrometer, Pirani gage and gas introduction. The base pressure of the apparatus was below about 10^{-7} Torr.

b). Adsorbed water and water vapor in gas phase

The presence of water vapour in the reactor was due to its desorption from the sample surface after connection of the reactor to the installation. The pumping out was performed through an absorption column with a cooling mixture in order for water in the gas phase of the reactor volume and at the oxide surface to be present continuously.

Water vapour did not get into the high-vacuum (measuring) part of the set-up, which allowed us to measure small (in comparison with water vapour) amount of adsorbates and products desorbed from the surface of oxides during the dark and photo processes.

c). Mass-spectrometric checkup of gas phase composition

The chemical composition of the freons during the course of the experiment was monitored with a mass spectrometer via their sampling through a leak valve. The mass spectrometer was made from a monopoly analyzer APDM-1. The kinetics of the observed processes was registered by measuring of the alteration in amplitude of the most intensive peak in the mass spectra of the corresponding substances.

d). Determination of molecule amounts in gas phase

The amounts of O_2, N_2O, NO, CO, ^{13}CO, $^{13}CO_2$, CO_2, halogen-containing organic compounds, and products of their interaction with the titania surface were determined using a

Pirani gauge, which was calibrated with oxygen and carbon monoxide. A necessary correction was introduced for the sensitivity with respect to other gases.

e). Monitoring of adsorbed layer composition

In the tropospheric conditions, the oxides are affected by the gas components of the atmosphere, and primarily by the components having the higher partial pressure. The latter are water, carbon dioxide and nitrogen oxides. So, according to recent studies [34] for magnesia obtained from $Mg(OH)_2$ and exposed by the atmosphere for a long time, the amount of adsorbed compounds attained are: NO and N_2 (from nitrogen oxides) ca.1% of monolayer, CO_2 ca.10% of monolayer, H_2O ca. one monolayer. The total amount of the adsorbed compounds in this situation equals to 5 weight % (according to the data of derivatographic analysis).

Traditionally before performing experiments or using in various chemical processes, the surfaces of adsorbents and catalysts (including photoadsorbents and photocatalysts) are exposed to a rigid pretreatment of various kinds. The main goal of this pretreatment (e.g., in vacuum, in an appropriate gas environment, at a certain pressure and temperature, etc.) is usually to obtain either reproducible results, which are consistent with those of other researchers, or the maximum of adsorption or catalytic (accordingly, photoadsorption or photocatalytic) activity. However, it is essential that the rigid pretreatment of a surface may result in some sufficient changes of chemical composition of the adsorption layer, stoichiometry of the surface and near-surface layers of an adsorbent or a catalyst.

In the natural atmospheric conditions, the aerosol particles are not exposed to any rigid pretreatment of the above mentioned types. So, the state of their surface corresponds approximately to the state of an insulator oxide surface obtained in rather "mild" conditions after long-duration contact with air. Thus, the (photo) catalytic and (photo) adsorption properties of the air-exposed oxide can differ sufficiently from those of pretreated oxides.

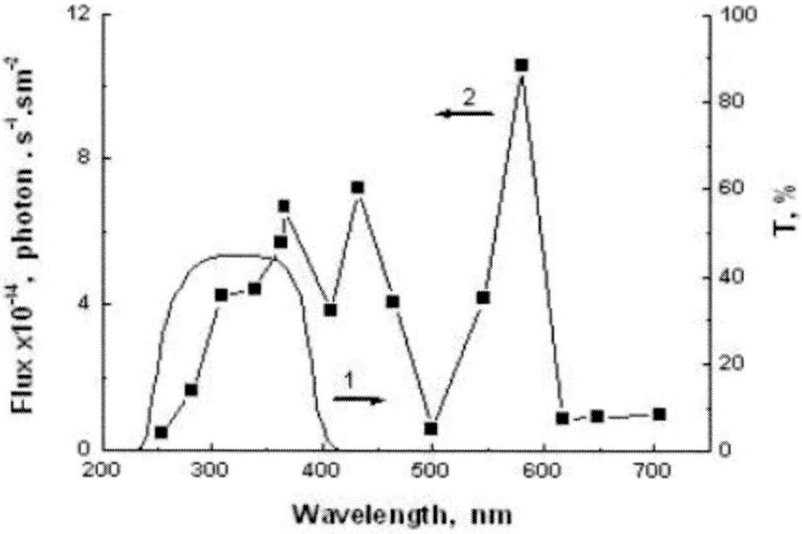

Figure 1. Transmission spectrum (T) of UV filter used (1) and spectrum of the flux of radiation passing through the front wall of the quartz reactor, measured with the interference filter (2).

f). Radiation conditions

To expose the oxides under UV radiation, an OSL-1 illuminator with a high pressure mercury lamp (DRSH-250, power 250 W), including a thermal water filter and UV filter, were used. The transmission spectrum of UV filter is shown in Figure 1, spectrum 1. The total radiation flux density, reaching the reactor's surface and recorded by an RTN-20S thermopile, was equal to ~1 milliwatt $\cdot$ cm^{-2} for this filter.

To separate monochromatic radiation, interference filters were used. The intensity of the flux of quanta having passed through the front wall of the quartz reactor for these filters is shown in Figure 1 (spectrum 2).

The quantum yield of the photoadsorption (photodesorption) was found as the ratio of the quantity of photoadsorbed (photodesorbed) molecules to the number of quanta passed through the reactor's frontal (transparent) wall.

2.3. Preparation of samples under ambient air

a). TiO_2(A) produced by grinding of rutile crystal

The TiO_2 monocrystal of the crystalline structure of rutile produced by the Czochralski method [35] was broken and ground in a corundum mortar. The powder specific surface was 1.6 $m^2 \cdot g^{-1}$. The titanium dioxide, as a water suspension, was applied to the interior wall of the cylindrical quartz reactor and dried in air at room temperature for a week.

b). TiO_2(B) prepared by high temperature treatment of anatase

The titania powder of anatase crystalline modification was heated at 1273 K for two hours in air in order to obtain from the anatase phase (basically) the rutile phase (basically). After that treatment the titanium dioxide was a rutile modification with a specific surface of 8 $m^2 \cdot g^{-1}$. Suspension of titanium dioxide was deposited onto the reactor walls and was dried in air.

c). TiO_2(C) made by combustion of titanium microparticles

Next sample of powdery titanium dioxide was prepared by combustion of pyrotechnic mixture consisting of ammonium perchlorate, hydroxy-terminated polybutadiene (HTPB) and titanium particles (particle size from 60 to 90 μm) under ambient air [36]. The highly dispersed titanium dioxide powder had the specific surface area of 6 $m^2 \cdot g^{-1}$. After a treatment of the oxide surface at temperature 625 K for 30 min in air the specific surface area was decreased to 4 $m^2 \cdot g^{-1}$.

A suspension of TiO_2 in the distilled water was supported on the internal wall of a quartz glass reactor, which was transparent for the light with wavelength longer than 185 nm. The TiO_2 layer was dried at room temperature in air. Then, without any additional treatment, the reactor was sealed to the high vacuum system for the photoadsorption experiments. The reactor was evacuated to 10^{-5} Pa at room temperature for several hours.

2.4. Physical – Chemistry Characteristics of Titania Samples

The X-ray diffraction (XRD) spectrum of the powder-like sample was recorded with an HZG-4C diffractometer using CuK_{α}- radiation. The diffuse reflectance spectra were taken in air on a SPECORD M40 spectrophotometer. In the studies a powder-like magnesium oxide was used as the reference standard. The extended X-ray absorption fine structure (EXAFS) and X-ray absorption near edge structure (XANES) spectra were measured at the EXAFS station of Siberian Center of Synchrotron Radiation. All the spectra were recorded in a transmission mode using two ionization chambers as detectors. The sample was prepared as moist pellet in order to adjust the jump of the K-edge absorption of about 0.5-0.8. The radial distribution of atoms (RDA) function was calculated from the EXAFS spectra in $k^3\chi(k)$ by using Fourier analysis for the intervals of wave numbers k = 2.5-13.8 $Å^{-1}$ [37, 38]. Curve fitting by means of the program EXCURV-92 [39] was realized for $k^3\chi(k)$ in similar wave number intervals and the titanium foil was used as a reference sample. The titania spectra were subjected to transmission electron microscopy (TEM) analysis on a JEOL model JEM-2010 instrument operated at acceleration at 200 V. *TiO_2(A).* Titania microparticles produced by grinding of the rutile crystal have the crystal structure which corresponds to the structure of a crystal obtained under conditions of low growth of big size monocrystal. The experimental results for powder-like samples consisting of such microparticles conform more to fundamental characteristics of titania (band gap, local structure, coordination geometry, lattice constant, and et all.), than to the experimental data for fine powders of titania used as adsorbents (photoadsorbents) and catalysts (photocatalysts). According to the results of the X-ray analysis, the titanium dioxide sample matched the titanium dioxide of the rutile crystalline structure (Figure 2).

The characteristic features of pre-edge and near-edge regions inherent in titanium dioxide of rutile structure are given in Figure 3. Local structure of titanium dioxide: coordination numbers, distances between atoms (Ti-O, Ti-O-Ti) are presented in Figure 4. Obtained data are in a good agreement with X-ray data for TiO_2 (rutile).

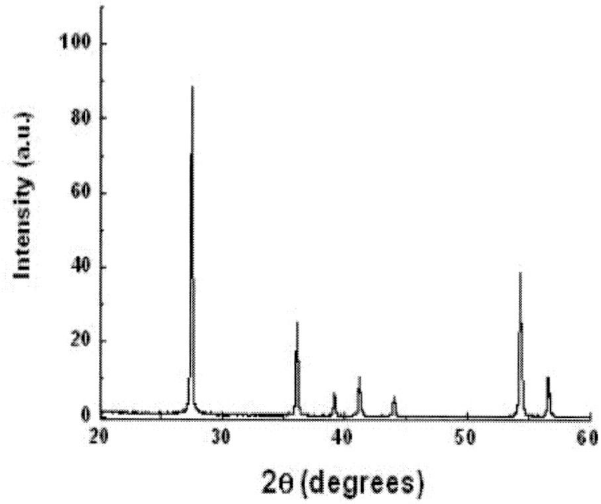

Figure 2. XRD pattern recorded from titania TiO_2(A).

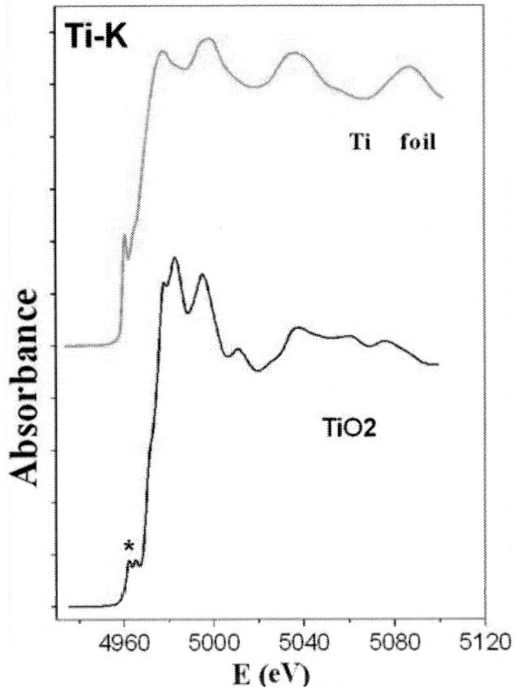

Figure 3. Spectra (Ti-K edge, *-1s-3d pre-edge region) XANES for microcrystal rutile prepared by grinding of monocrystal TiO_2 under ambient air.

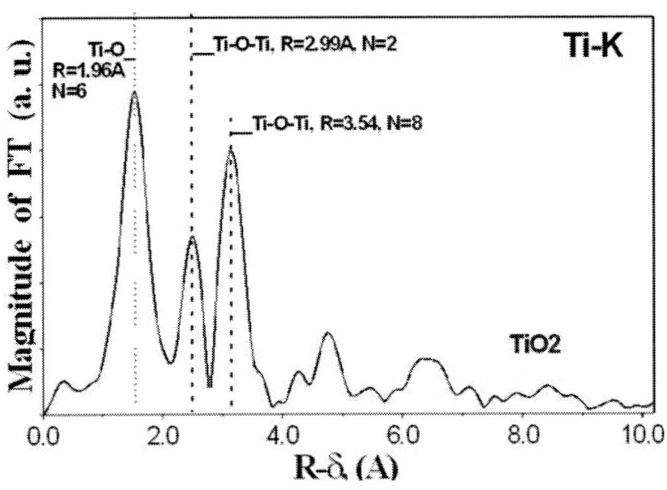

Figure 4. RDA curves for the titanium surrounding in microcrystal rutile prepared by grinding of monocrystal TiO2 under ambient air.

The TEM picture of the titania, produced by grinding of crystal rutile are given in Figure 5. *$TiO_2(B)$*. There are three stable crystalline modifications of titania encountered in the nature in the form of rutile, anatase and brookite crystals. Titania samples of all these crystalline modifications with stoichiometric chemical composition are white powders. The absorption of light with energies below the band gap of titania in rutile modification (Eg = 3.05 eV) is minimal by all such powders (Figure 6, curve 1 and 2).

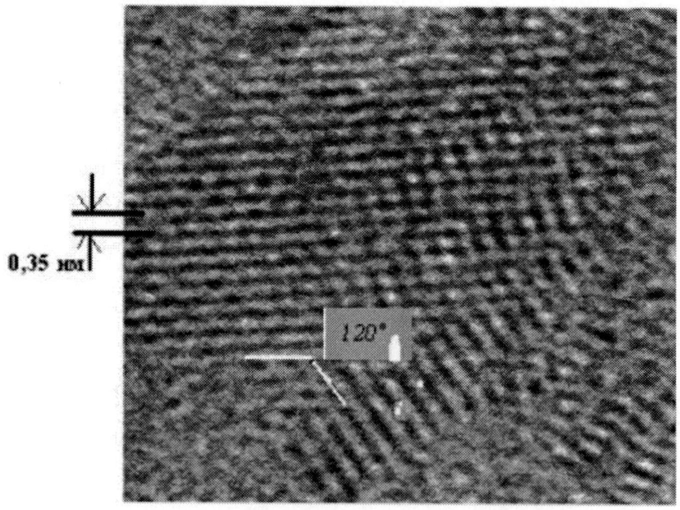

Figure 5. TEM picture of the titania powder produced by grinding of crystal rutile.

This absorption corresponds to the impurity absorption bands for single crystals or surface absorption for finely dispersed powders.

According to these data (Figure 6), there is an inflection of absorption edge curve for powdered titania samples consisting from a mixture of anatase and rutile modifications. Such an inflection of curve is not observed for pure anatase or rutile (Figure 6, curve 1) samples. The absorption band edge of anatase powder is shifted by 0.1-0.2 eV to shorter wavelengths compared to that of rutile powder (3.0 and 2.8-2.9 eV, respectively).

If the titania stoichiometry is changed due to partial reduction of Ti^{4+} to Ti^{3+} and formation of lattice oxygen vacancies during the oxide synthesis, the powders become colored and absorption in the surface band of the oxide appears.

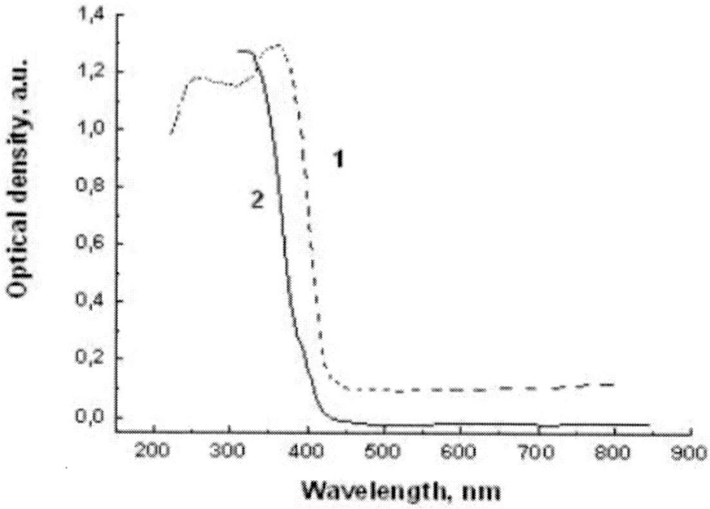

Figure 6. Diffuse reflection spectra of titania samples of different production. 1 – TiO2(A) produced from a monocrystal; 2 – TiO2(B) produced through powder-like titania heating at 1270 K.

$TiO_2(C)$. The crystal structures of rutile and anatase were confirmed by X-ray diffraction of the prepared titanium oxide powders.

The absorption properties of dispersed titania after storage in air for a long period of time and after activation by heating at 625 K in air for 30 min with respect to the light from different parts of the solar spectrum. The experiments with the samples contacted with the air for a long time are especially important for evaluating the prospects of titania prepared by this method for application in processes leading to purification of the environment during various man-caused catastrophes.

The diffuse reflectance optical density spectrum of titania heated in air at 620 K for 30 min is shown in Figure 7 (curve 1). Its spectrum is compared with that of the titania collected immediately after pyrotechnic mixture combustion (curve 2) and sampled from the gas mixture after combustion of the pyrotechnic mixture using a vacuum impactor [36] (curve 3).

Thus, the absorption of titania powders prepared by combustion of pyrotechnic mixture containing titanium microparticles in the region before the TiO_2 intrinsic absorption edge (corresponding to the surface absorption region) depends on the synthesis method and after-treatment of the samples.

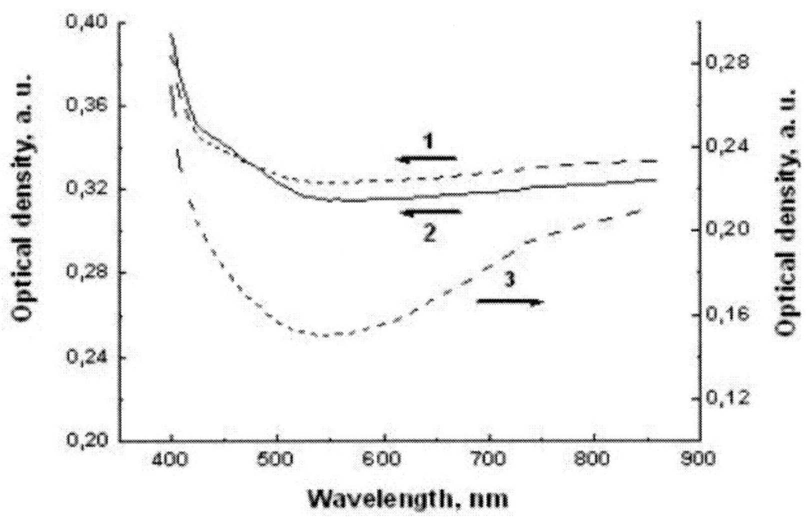

Figure 7. Diffuse reflection spectra for TiO_2 samples. 1 – used in the present work TiO_2 after an air treatment at 620 K during 30 min; 2 – TiO_2 collected immediately after pyrotechnic mixture combustion; 3 – titania sampled using a vacuum impactor.

2.5. Pre-starting procedure in high-vacuum set-up

The samples prepared under ambient air were deposited from water suspension in the form of thin layer on the walls of the reactor made from optical quartz. After soldering to high-vacuum set-up the sample was pumped out at room temperature to evacuate the most part of adsorbed CO_2. For samples TiO_2(A) and TiO_2(C) the pumping out was performed through a water vapour trap with cooling mixture.

The pre-starting procedure for TiO_2 (B) sample was different to the treatment of TiO_2 (A) and TiO_2(C). Titanium dioxide surface was cleaned from the adsorbed organic impurities, carbon dioxide, and water by long-time, high temperature, oxygen-vacuum treatment. The analysis of the products released upon heating from a rutile surface purified by on oxygen-vacuum treatment for 130 hrs shows the absence of CO_2 evolution and occurrence of water and oxygen evolution at about 450 K (H_2O alone) and above 800 K ($H_2O + O_2$). Increasing the time of oxygen-vacuum treatment to 300 hrs leads to the disappearance of the water thermodesorption peak at 450 K and, above 800 K, only small amount of oxygen evolve.

Prior to the oxygen and carbon monoxide photoadsorption measurements, the TiO_2 (B) rutile was heated in oxygen at 700 K and 10^{-1} Torr for 1 hr, cooled to room temperature over 1-2 hrs and then evacuated at 290 K for 10-15 min.

3. Results and Discussion

3.1. Darkness Processes

$TiO_2(A)$. The presence of water vapour in the reactor was due to its desorption from the titania after the connection of the reactor to the set-up. During evacuation, water vapour was frozen in trap cooled down to 173 K. Thus, water vapour was always present in the reactor. After pumping-out of the reactor through this trap, mainly NO evolved from the TiO_2 surface (fourfold greater than CO_2 accumulated inside the reactor).

The dark adsorption was also absent for the TiO_2 (A) sample, because the surface at its formation through grinding the crystal was saturated with oxygen from air. Moreover, the surface reconstruction by means of high temperature treatment in a vacuum or in some reducing atmosphere was not conducted in the experiment.

The dark adsorption of acetone on titania, produced from rutile crystal, is more significant in magnitude than the dark adsorption of freons and reaches 1% of the titania surface monolayer at a pressure of 10^{-2} Torr inside the reactor volume.

The dark adsorption kinetics of acetone is given in Figure 8 (curve 1). For comparison the dark adsorption kinetics of Freon 22 is shown in that figure (curve 2). The mass-spectrometric analysis confirms also that the dark adsorption of acetone proceeds without a displacement (desorption) of different gases from the titania surface.

Our earlier measurements of the amount and composition of gases evolved from the oxide surfaces in the dark (with other metal oxides produced in laboratory or industrial conditions) were characterised mainly by CO_2 desorption. Thus, the amount of carbon dioxide evolved from the MgO surface is equal approximately 10% of the monolayer, and nitrogen oxides (predominantly N_2O) are equal approximately 1% [34].

It can be assumed that nitrogen oxide (II) is produced in a sufficient amount when grinding the TiO_2 crystal in air due to molecular nitrogen oxidation at centres formed at the break of the Ti–O bonds of the titanium dioxide lattice. The NO generation is also possible through the well-known reaction of N_2O decomposition at the electron donor centre of the metal oxide surface [40, 41]. In this case, N_2O is adsorbed from air like the carbon dioxide.

The dark adsorption of Freon 134a (CF3--CH2F) or Freon 22 (CHF_2Cl), insignificant in magnitude (0.03% of the oxide surface monolayer at a pressure of ~10^{-2} Torr inside the reactor volume) is observed for the powder-like titanium dioxide produced from a monocrystal of the rutile crystalline modification.

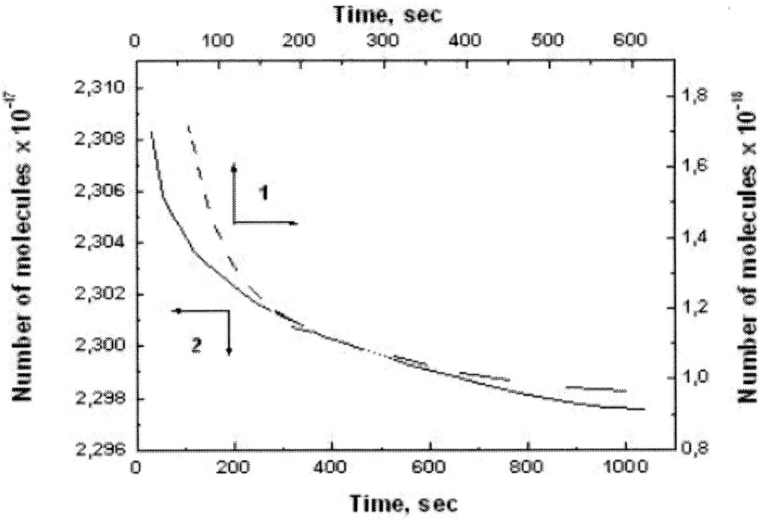

Figure 8. Kinetics of acetone (1) and Freon 22 (2) adsorption on a powder-like sample produced from a TiO_2 monocrystal.

TiO_2(B). Manometric and mass-spectrometric data have shown the absence of the dark adsorption of oxygen on TiO_2(B), which probably is connected with the oxygen preadsorption during the sample cooling in oxygen.

TiO_2(C). When the reactor with deposited TiO_2 was evacuated at room temperature, water and carbon dioxide were the main products released from the surface of the titania powder. During the experiments described below water vapor was always present in the reactor. When the reactor volume was connected to the measurement volume, water vapor was trapped in the reactor volume in a trap with a cooling liquid (ethyl alcohol cooled to 173 K).

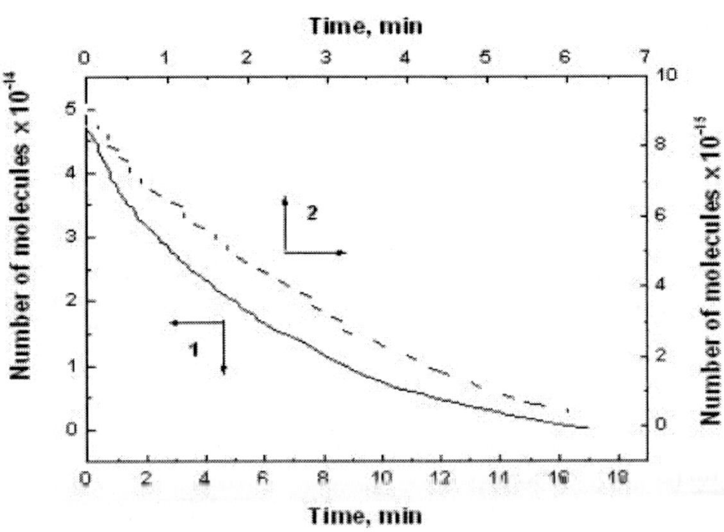

Figure 9. Kinetics of the desorption CO_2 from TiO_2 surface. 1 – decrease of number of CO_2 molecules on TiO_2 surface (dark desorption); 2 – decrease of number of CO_2 (in addition to dark desorption) under illumination of TiO_2 surface through the UV filter (photodesorption).

Prolonged evacuation of the reactor volume for 1 h in high-vacuum installation through a trap with the cooling liquid resulted in partial removal of carbon dioxide adsorbed on the titania surface. As a result, quasi-equilibrium filling of the surface with CO_2 was established. The kinetics of approaching to the equilibrium CO_2 pressure is shown in Figure 9, curve 1. This kinetics is presented as the decrease of the amount of CO_2 molecules adsorbed on the titania surface (desorption). The kinetic curve follows the first-order equation with the time constant $\tau_1 \approx 6$ min. For comparison, the kinetics of CO_2 photodesorption, obtained after the attaining equilibrium of the dark adsorption, is also given in Figure 9, curve 2.

No other gases except for CO_2 and H_2O were observed to be released in measurable amounts in the dark at room temperature from the surface of titania prepared by combustion of the pyrotechnic mixture containing ammonium perchlorate and organic stabilizer under ambient air.

3.2. Heterogeneous Chemical Processes Induced by Radiation

Photoinduced processes can be divided into groups with different radiation absorption processes determining the direction of chemical reactions on the surface of particles of solid-state tropospheric aerosol. This division allows us to separate and characterize experimentally observed photoinduced processes: photoadsorption (photodesorption), photocatalytic reactions, reactions of interaction of gas-phase molecules with the surface of solid-state aerosol under the effect of radiation, and surface photochemical reactions.

Photoadsorption is characterized by the formation of surface electron or hole centers after electron transport in a solid particle under the effect of radiation. The maximum coverage of the surface by photoadsorbed molecules does not exceed 0.1% of the surface monolayer [42].

Photocatalytic reactions are characterized by formation of free mobile electrons in the conduction band of solid particles and, simultaneously, free mobile holes in the valence band after absorption of radiation with the energy exceeding the width of the solid forbidden band. Thus, at photocatalytic reaction of oxidation, the oxidized gas-phase tropospheric molecules (H_2, CO, NO, SO_2, hydrocarbons) interacts with surface lattice oxygen [43].

Particles of troposphere aerosol may include, as a phase, metal oxide [2, 45] that are photocatalysts for reaction of oxidation of various chemical compounds, including organic ones, by oxygen. Such catalysts are ZnO, SnO_2, TiO_2, In_2O_3, V_2O_5, MoO_3, WO_3, ZrO_2 [44, 46, 47]. In these reactions, hydrocarbons are oxidized to water and CO_2, while halogen carbons are oxidized to water, CO_2, and the corresponding acids (reaction of mineralization).

The tropospheric conditions (temperature, pressure, high oxygen concentration, solar radiation) favor photocatalytic reactions. Photocatalytic reaction has low activation energy, they proceed at low partial pressure of the oxidized gas, and quantum yields of the photocatalytic reactions can achieve 30% [44, 46, 47]. As a result of these reactions, organic and halogenorganic substances are removed from the gas-phase troposphere, and CO, SO_2, and nitrogen oxides are oxidized.

Destructive photosorption of halogen containing organic compounds observed in Refs. [43] and [48] is likely an example of a photoinduced reaction from the group reactions, in which the gas-phase molecules interact with the surface of metal oxides. As a result, such an interaction causes replacement of the lattice oxygen of metal oxide by the halogen atom of the gas-phase halogen hydrocarbon molecule.

Insulator oxides (SiO_2, Al_2O_3, MgO, and CaO) have no photocatalytic activity in the spectral region of solar radiation in the troposphere [49, 50, 51]. Their intrinsic absorption band is in the wavelength range shorter than 200 nm [28], while the photocatalytic activity is connected with the formation of mobile electrons and, simultaneously, mobile holes at absorption of radiation by a solid (intrinsic absorption of a solid).

Surface photochemical reactions occur after absorption of radiation by compounds adsorbed on the surface of solid particles as they transform into the excited reactive state [52]. An example of such a reaction is photodecomposition of nitrous oxide adsorbed on the surface magnesia [42, 53]. The surface in this case plays the role of a "carrier" (substrate), changing the energy of excited state of the adsorbed compound, or role of stabilizer of the adsorbed compound in the state different from the state of molecules in the gas phase, for example, for such compounds as CO_3 complex, OH group, etc.

a). Carbon dioxide photodesorption

$TiO_2(A)$. When illuminating the titanium dioxide surface through a UV filter in the presence of Freon 134a or Freon 22 (after saturation of the dark adsorption of these gases), the carbon dioxide photodesorption is observed, while the photoadsorption of the freons is low. The transmission spectrum of the UV filter is given in Figure 1. The CO_2 photodesorption kinetics in the presence of Freon 134a and Freon 22 are shown in Figure 10.

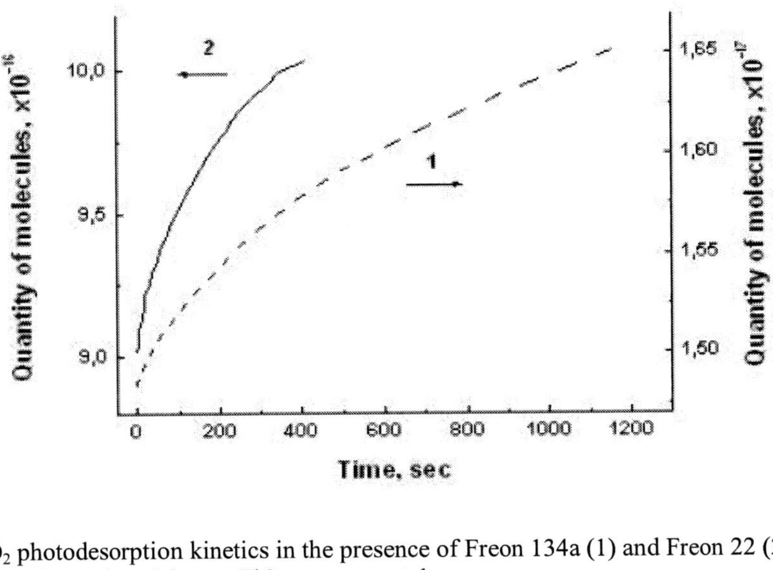

Figure 10. CO_2 photodesorption kinetics in the presence of Freon 134a (1) and Freon 22 (2) on a powder-like sample produced from a TiO_2 monocrystal.

The photodesorption kinetics is characterized by the fast desorption of carbon dioxide in the initial time of illumination (up to 3–5 min), followed by slow desorption with a constant rate. In the first case, probably, desorption of carbon dioxide is connected with classic photodesorption under illumination of the surface under quanta with energy from the fundamental absorption band of TiO_2 [54, 55]. In the second case, desorption is attributed to the photocatalytic reaction of the oxidation oxygen of an oxide of the by adsorbed carbon-containing compounds (for example, carbon oxide from air) followed by reducing the titanium dioxide surface [56]. During this process, as it follows from mass-spectrometric data, the magnitude of mass peak 44 increases by several times (Figure 11, curve 1).

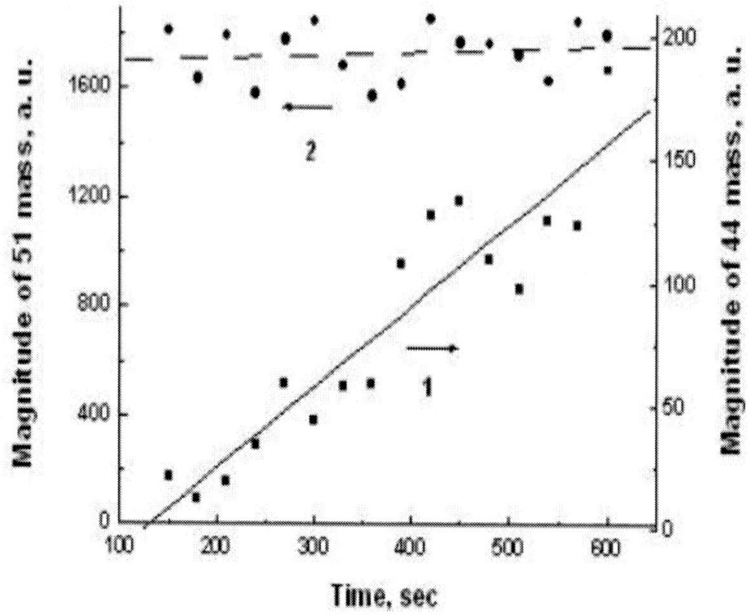

Figure. 11. Kinetics of the variation of peaks 44 (1) and 51 (2) of masses in the mass spectra of CO2 and Freon 22, respectively, under illumination of the TiO2 surface through the UV filter in the presence of Freon.

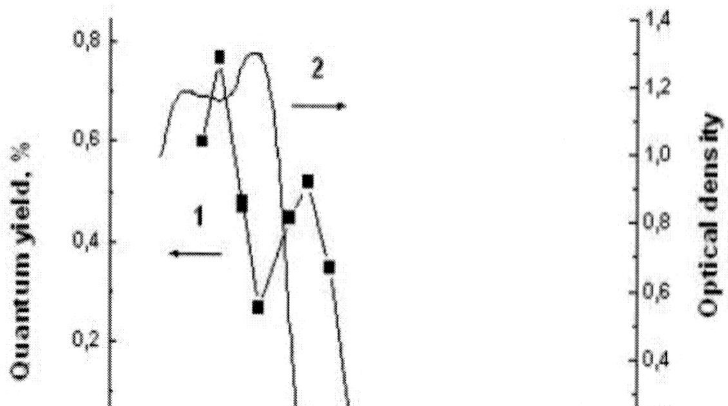

Figure 12. Spectral dependencies for the titanium dioxide sample produced from a monocrystal: (*1*) the quantum yield of CO_2 photodesorption in the presence of Freon 134a and (*2*) the optical density relative to MgO.

Measurements of the CO_2 photodesorption rate under illumination by monochromatic light at different wavelengths (via the use of interference filters) and light flux intensities allowed for the estimation of the quantum yield of the CO_2 photodesorption. The calculation results are given in Figure 12 (curve 1). Curve 2 shows the diffusion reflection spectrum of the powder-like TiO_2 measured relative to magnesium oxide as the reference standard.

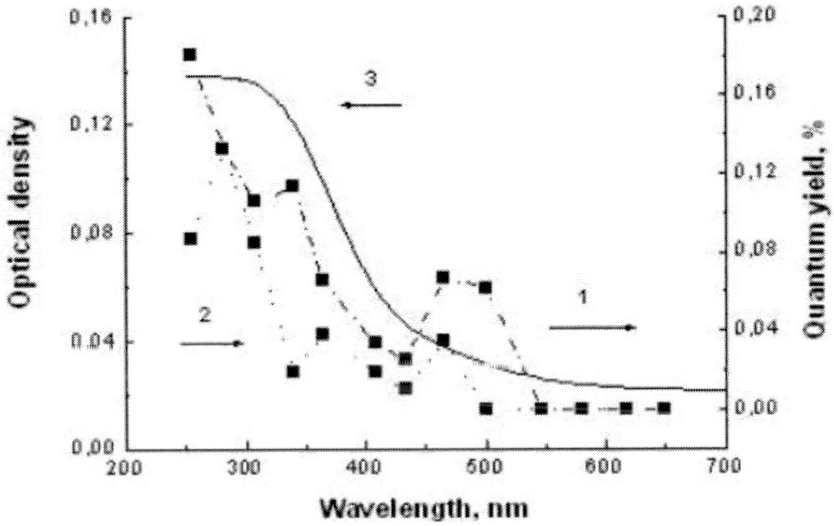

Figure 13. Spectral dependencies for titanium dioxide, prepared by burning of titanium particles under ambient air. 1 – photodesorption CO_2 after first illumination; 2 – photodesorption CO_2 after the long-time illumination through the UV filter; 3 – diffuse reflection spectrum for TiO_2 which is the same as located on the wall of a reactor.

TiO_2(C). Much more processes take place on the titania surface under visible light illumination than in the dark. In eddition to the dark desorption of CO_2, its photodesorption was observed with the first-order kinetics similar to that of the dark desorption ($\tau_2 \approx 5$ min) by 20 times larger amount of desorbed CO_2 (Figure 9, curve 2). The photodesorption kinetics is presented as the decrease in the amount of CO_2 molecules adsorbed on the titania surface. The measurement of the spectral dependencies of the CO_2 photodesorption quantum yield showed that the highest quantum yield was observed in the region of the TiO_2 intrinsic absorption (Figure 13, curves 1 and 2). A diffuse reflectance spectrum of the titania powder used in this study is shown for comparison (Figure 13, curve 3). If the time of the TiO_2 surface illumination was increased to several hours, the CO_2 photodesorption quantum yield decreased.

b). Oxygen photoadsorption

TIO_2(A). The photoadsorption activity of TiO_2(A) produced from a monocrystal of the rutile crystalline modification was compared with that of the TiO_2(B) sample produced from powder-like TiO_2 of the anatase modification. The spectral dependence of the quantum yield of oxygen photoadsorption for the TiO_2 - sample is presented in Figure 14 (curve 1). Of interest are high photoadsorption quantum yields at $\lambda > 400$ nm (corresponding to the TiO_2

surface absorption band), which are close in magnitude to quantum yields in the range of the TiO_2 absorption.

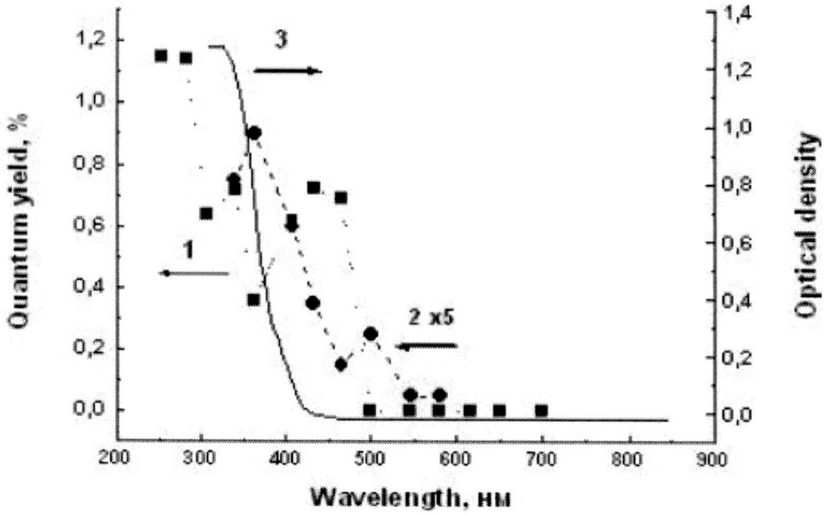

Figure 14. Spectral dependencies for titanium dioxide samples: the quantum yields of O_2 photoadsorption at TiO_2(A) (*1*) and TiO_2(B) (*2*); (*3*) the optical density of TiO_2(B) relative to MgO.

At a wavelength of about 465 nm, the quantum yield is 0.7%; at $\lambda = 340$ nm, it is 0.75% (Figure 14, curve 1). The spectral dependence of the photoadsorption quantum yield for the TiO_2(A) sample is shown by curve 2. For comparison, the diffusion reflection spectrum for TiO_2(B) is also given. The quantum yield for this sample at a wavelength of 465 nm is 0.03% and at 340 nm is 0.15%. Close quantum yields in the surface absorption band were observed also for titanium dioxide produced through the burning in air of a pyrotechnic mixture including titanium microparticles [58].

The absorption in the range of energies of quanta less than the TiO_2 forbidden band can be related to electron transition from deep surface levels of the oxide to the conduction band [59]. At the same time, the oxygen photoadsorption proceeds on the electron localised in the surface trap of the semiconductor oxide [41, 60]. Thus, it can be supposed that the sharp increase of the quantum yield in the range of TiO_2 surface absorption relative to oxygen is connected with the increase of deep surface levels formed at a break of the Ti – O bonds when producing powder-like oxide from a monocrystal.

$TiO_2(B)$. The rate of oxygen photoadsorption was measured at pressure of 10^{-2} Torr. The intensity of monochromatic radiation was $1.9x10^{14}$ photon · s^{-1}. Kinetics curve of oxygen photoadsorption on rutile is given in Figure 15, curve 1. The kinetic curves of oxygen photoadsorption are linear in parabolic coordinates. The bend on the kinetic curve seems to be due to two different mechanisms of oxygen photoadsorption. Mass-spectrometric analysis of oxygen thermodesorption has shown the presence of oxygen only. This is evidence for true oxygen photoadsorption on rutile.

The spectral dependencies of photosorption oxygen for rutile are given in Figure 16. Upon the treatment 800 K the rutile assumes a steady activity to oxygen photoadsorption. The adsorption of water vapours at 370 K for 15 hrs followed by evacuation for 1hr with heating to 770 K restored the photoadsorption activity in sub-band gap absorption region and does not change it in the fundamental absorption band.

The steady-state coverage by photoadsorption oxygen should be small and governed by the depth of the electron traps, while relatively easy recombination should lead to reverse photoadsorption even at low temperatures. Indeed, the storage of rutile in darkness at 20 °C for 10-15 hrs restores its photoadsorption activity. The electron traps are probably anion vacancies, whose concentration decreases with increasing time of storage.

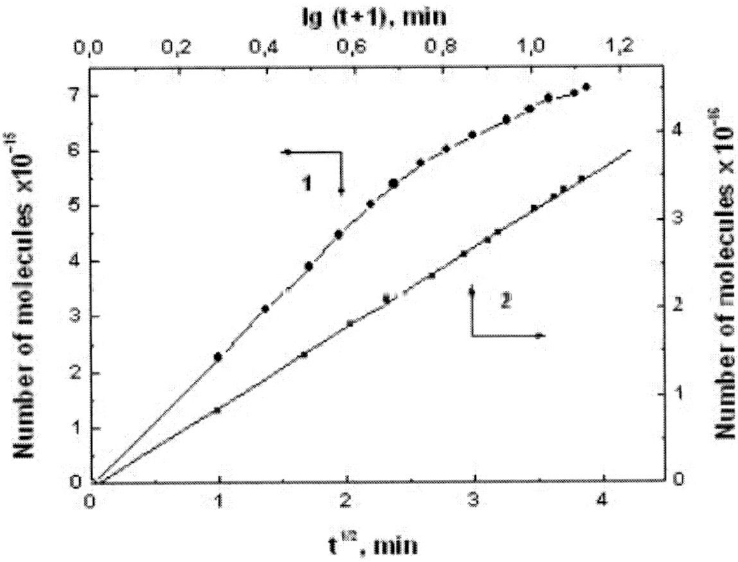

Figure 15. Kinetics of oxygen (1) and carbon monoxide (2) photoadsorption on titania $TiO_2(B)$.

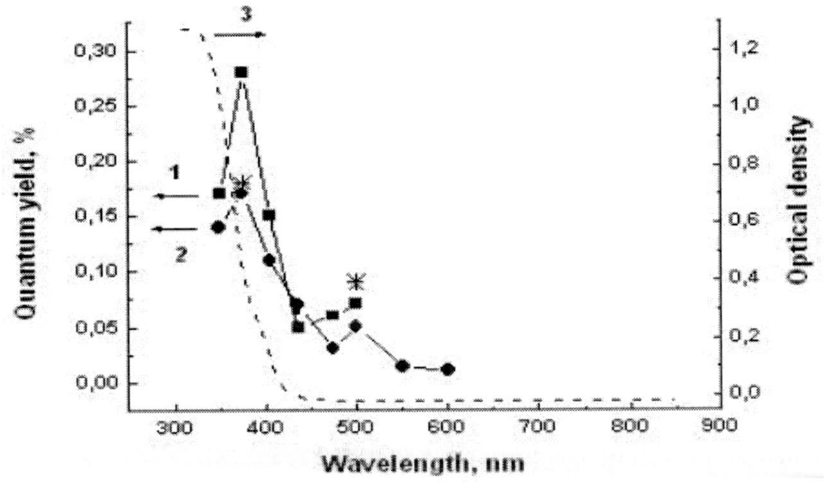

Figure 16. Spectral dependencies for titania $TiO_2(B)$: 1, 2 - quantum yields of oxygen photoadsorption after oxygen-vacuum treatment for 130 (1) and 300 (2) hrs; 2 – diffuse reflectance spectrum of $TiO_2(B)$. Starred points are quantum yield of oxygen photoadsorption after titania treatment in state (2) by water vapour.

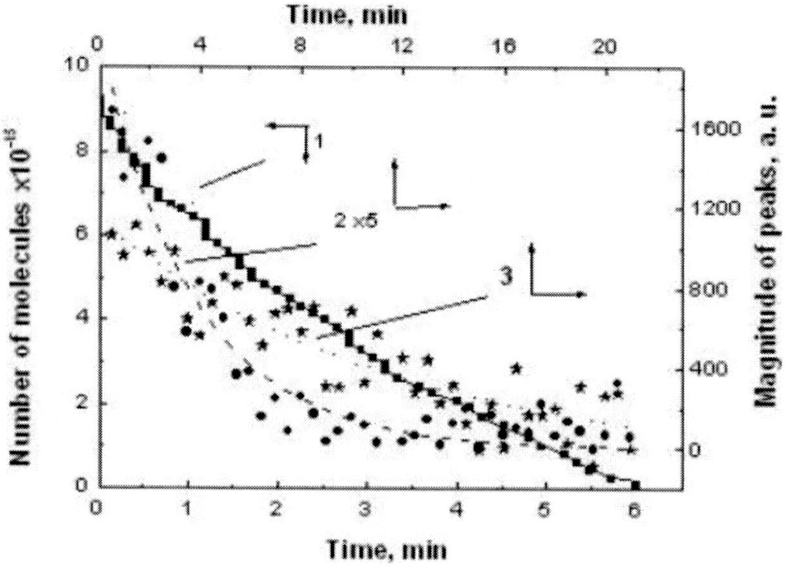

Figure 17. Kinetics of the photoinduced processes on TiO_2. 1 – photoadsorption of oxygen from dried air (manometric data); 2 – photoadsorption of oxygen from the mixture of dried air and Freon 134a (the alteration in magnitude of peak for *m/z* = 32 in the mass spectrum); 3 – decrease of amount of adsorbed CO_2 molecules under illumination of TiO_2 surface through the UV filter (the magnitude of peak for *m/z* 44 in the mass spectrum).

$TiO_2(C)$. In our experiments we observed oxygen photoadsorption on the TiO_2 surface from the air dried by passing through a trap cooled with liquid nitrogen. Photodesorption of carbon dioxide takes place simultaneously with the oxygen photoadsorption. The initial rate of oxygen photoadsorption exceeds that of the CO_2 photodesorption. So, the total pressure registered by a Pirani manometer decreases (photoadsorption). The kinetics of the total pressure decrease during titania illumination through a UV filter is shown in Figure 17, curve 1 (manometer data). The kinetics of the oxygen concentration decrease (oxygen photoadsorption) determined from changes in the intensity of peak with mass 32 (mass-spectrometer data) is shown in Figure 17, curve 2. This kinetics is described by a first-order equation with time constant $\tau_3 \approx 3$ min. This process ends only when all oxygen from the reaction volume is photoadsorbed. Even during the first illumination the amount of photoadsorbed oxygen exceeds that of photodesorbed CO_2 by a factor of 5. The kinetics of CO_2 photodesorption when oxygen is present in the gas phase is characterized by much longer time constant ($\tau_4 \approx 10$ min, Figure 17, curve 3) than those of the dark desorption (τ_1) and photodesorption of CO_2 (τ_2). According to the spectral dependencies of the quantum yield of oxygen photoadsorption from dried air, the photoadsorption was more efficient in the region of the intrinsic TiO_2 absorption (Figure 18, curves 2 and 3). The edge of the quantum yield spectrum of oxygen photoadsorption was shifted to longer wavelengths (550 nm) compared to that of the CO_2 photodesorption (500 nm, Figure 18).

So, it appears that the oxygen photoadsorption on the titania sample prepared by combustion of titanium particles in air does not result from the oxygen interaction with the surface electron sites induced by illumination. The oxygen photoadsorption (consumption of oxygen from the gas phase) occurs as a reaction of oxygen molecules with compounds adsorbed from the air earlier, i.e. after synthesis and storage of the titania sample.

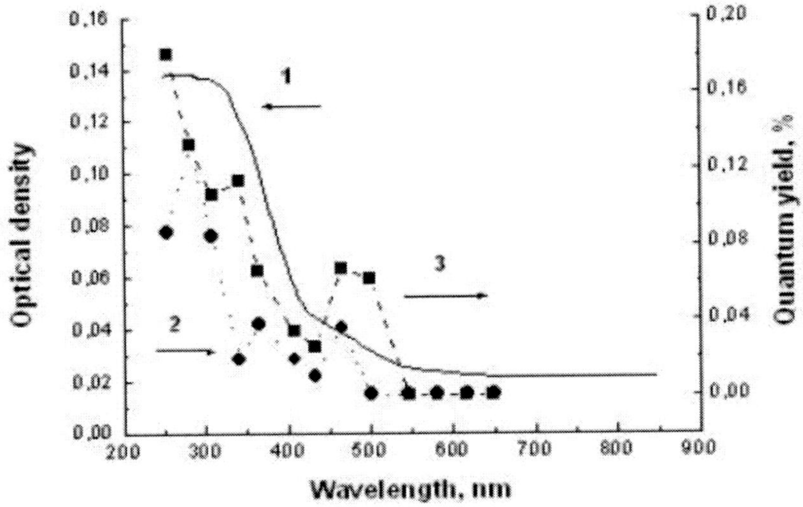

Figure 18. Spectral dependencies for titanium dioxide, prepared by burning of titanium particles under ambient air. 1 – diffuse reflection spectrum for TiO_2 which is the same as located on the wall of a reactor; 2 – photoadsorption of oxygen from the dried air; 3 – photoadsorption of oxygen from the mixture of dried air and Freon 134a.

The registered oxygen photoadsorption kinetics is close to the first order kinetics. Meanwhile, photoadsorption kinetics of simple gases on metal oxides after the thermal vacuum treatment of the oxides is more complex [61,63]. The reaction of oxygen with adsorbed compounds is also proven by the change of the time constant of CO_2 photodesorption in the presence of oxygen in the reaction volume and by the increase of the amount of CO_2 formed in this case. The consumption of oxygen from the gas phase under illumination for oxidation of organic substances adsorbed on TiO_2 was earlier suggested by Solonitsyn [64].

c). Photoadsorption of carbon monoxide, acetone, and freons

$TiO_2(A)$. No experimental evidences of the interaction of titanium dioxide with methane or halogen-containing methane derivative (Freon 22), with ethane or its halogen-containing derivatives (Freon 134a) were obtained.

Pphotocatalytic oxidation of the considered freons by surface oxygen or oxygen of gas phase does not occur, because peaks 33 (Freon 134a) or 51 (Freon 22) do not decrease under illumination both through the UV filter and without a filter (Figure 11, curve 2).

After saturation of the dark adsorption of acetone the irradiation of titania surface through a UV filter in the presence acetone the shade desorption of its is observed. This adsorption is detected by the increase of the M/e = 43 peak in the mass spectra of acetone (Figure 19, curve 1). The photodesorption of carbon dioxide is occurring under illumination in acetone as well as in freons (Figure 19, curve 2).

After light turning off the photodesorbed acetone is adsorbing on the titania surface. Under illumination the same processes proceed in the presence of mixture of acetone and dried air (or oxygen). But, in this case the oxygen consumption is detected from those mixtures (Figure 20, curves 1 and 2, correspondingly, for acetone and oxygen from air).

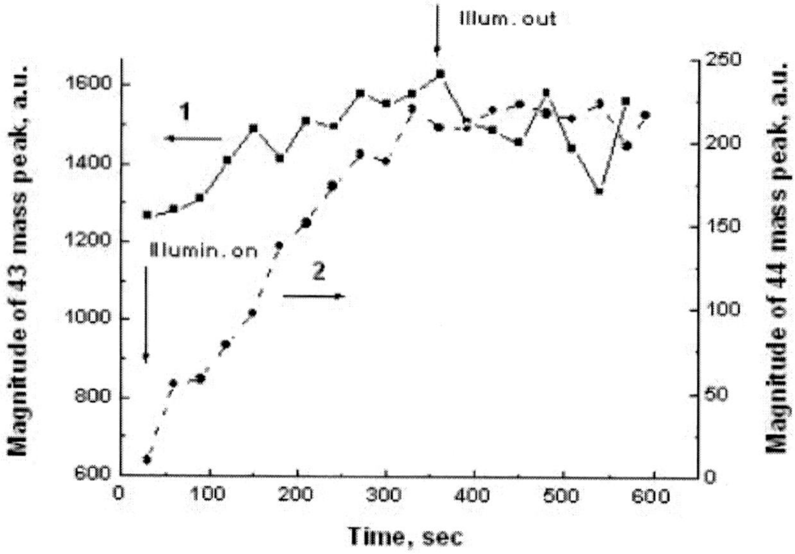

Figure 19. Kinetics of acetone (1) and CO_2 (2) photodesorption from titania surface under the unfiltered radiation of OSL-1 illuminator.

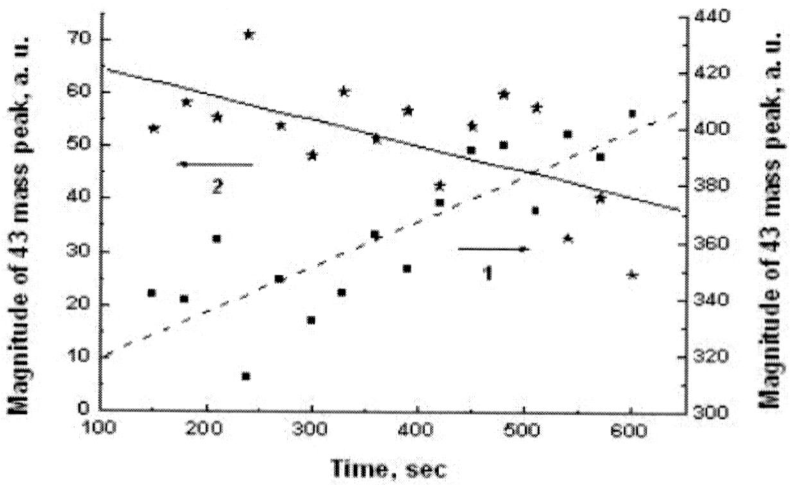

Figure 20. Kinetics of the alteration of 32 (1) and 43 (2) peaks in the mass spectra of oxygen and acetone, respectively, under illumination of the titania surface through UV filter in the presence of the mixture of acetone and oxygen.

According to Figure 19 and Figure 20 the alterations of the magnitude of mass peaks 43 (CH_3OCH_3), 32 (O_2), and 44 (CO_2) are equal, correspondingly, 70, 20, and 10 a. u. during 10 min. Thus, those processes are not connected with each other; i. e. the formation of carbon dioxide in the gas phase is not the result of photocatalytic acetone oxidation by oxygen [65].

The oxygen of reaction mixture containing acetone and oxygen is consumed with its photosorption on the titania surface. This process is more effective under illumination with the wavelength more 400 nm. Moreover, the photocatalytic oxidation of surface organic compounds proceeds with the formation of carbon dioxide in the gas phase. This process occurs under radiation with wavelength less 400 nm only. The effective quantum yields of oxygen consumption for those processes are given in Figure 21. The acetone concentration in gas phase is altered low. Consequently, the photocatalytic oxidation of acetone neither by gas phase oxygen nor by lattice oxygen do not proceed.

Thus, the prolonged illumination of titania surface in water vapor and in the presence of acetone leads to decreasing of photocatalytic activity in the carbon monoxide oxidation by oxygen or dried air.

$TiO_2(B)$. The rate of carbon monoxide photoadsorption was measured at pressure of 10^{-2} Torr. The intensity of monochromatic radiation was $1.9x10^{14}$ photon · s^{-1}. Kinetics curve of carbon monoxide photoadsorption on rutile is given in Figure 15, curve 2. The photoadsorption of CO at 290 K is not accompanied by carbon dioxide evolution, i. e. the product of interaction remain in the adsorbed state. The kinetics of carbon monoxide photoadsorption is described by a logarithmic equation.

In our experiments, the carbon monoxide photoadsorption was detected in the fundamental absorption region of titanium dioxide. Since in this range O^- ion-radical are formed, which interact with carbon monoxide at high rates, carbon monoxide photoadsorption corresponding to rutile photoreduction, occurs. The spectral dependency of the quantum yield of carbon monoxide photoadsorption is given in Figure 22, curve 1.

$TiO_2(C)$. No photoadsorption of Freon 22 (CHF_2Cl) and Freon 134a (CF_3CH_2F) was observed under illumination with unfiltered light of the DRSh-250 mercury lamp or illumination through a UV filter over titania prepared under ambient conditions and stored in air for a long time. No photoadsorption of these halocarbons was observed in the presence of dry air as well. According to the literature data no photocatalytic oxidation of methane is observed on TiO_2 [61, 65]. However, photoadsorption of methane and hydrogen was observed on rutile after high-temperature oxygen-vacuum treatment [62, 66, 67]. Photoadsorption of halogenated methane and ethane derivatives (Freon 22 and Freon 134a, respectively) was quite efficient on MgO and $CaCO_3$ exposed in air for long time [14, 68].

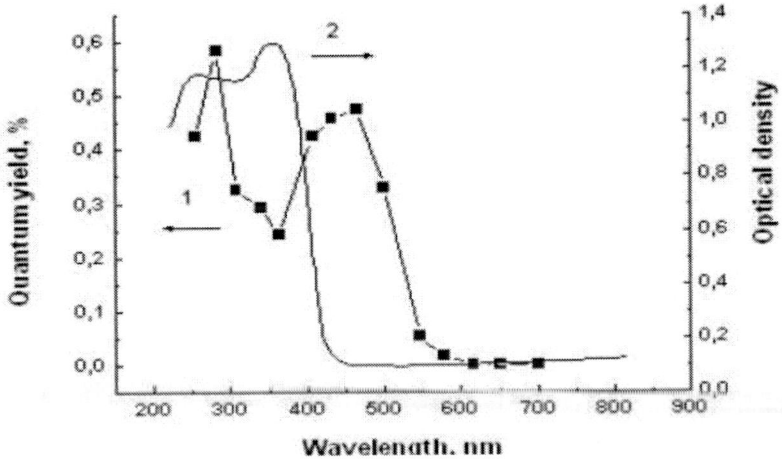

Figure 21. Spectral dependencies for titania TiO_2(A): 1 – quantum yield of oxygen photoadsorption in the presence of the mixture of acetone and oxygen; 2 – optical density of diffuse reflectance relative to MgO.

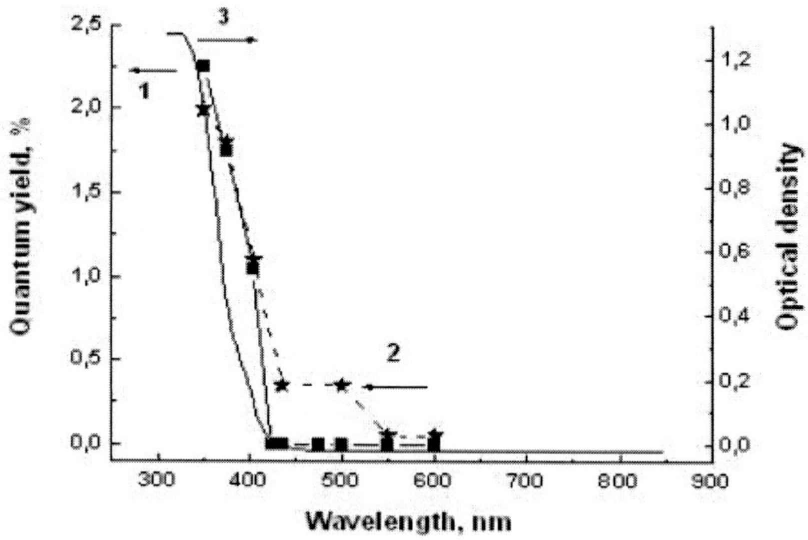

Figure 22. Spectral dependencies for titania TiO_2(B): 1 – quantum yield of carbon monoxide photocatalytic oxidation; 2 – optical density of diffuse reflectance; 3 – diffuse reflectance spectrum of TiO_2(B).

Photocatalytic oxidation of carbon monoxide [32] and some other gas-phase reaction, e.g. photocatalytic oxidation of C_{2+} alkanes [61,65, 69], aromatic hydrocarbons [70], photocatalytic destruction of chlorinated ethane derivatives [71, 72], etc., occur efficiently on the TiO_2 surface.

d). Photocatalytic oxidation of carbon monoxide

$TiO_2(A)$. It is discovered that in the spectrum of solar tropospheric radiation ($\lambda > 300$ nm), the photochemical activity of titanium dioxide produced through grinding the monocrystal under ambient air is significantly higher than of the compound produced by traditionally way, i.e. under conditions of preparing TiO_2 with a high specific surface.

It can be expected that the photochemical properties of other oxides produced from minerals (MgO from periclase, SiO_2 from quartz, Al_2O_3 from corundum, etc.) will significantly differ from the properties of compounds produced in laboratory or industrial conditions. The formation of oxide particles from minerals under natural conditions is accompanied by the formation of photochemical active solid tropospheric aerosol.

According to Ref. [15], the titanium dioxide is characterised by the absence of photocatalytic activity in the oxidation of methane, ethane, and their halogen derivatives like Freon 134a (CF_3–CH_2F) and Freon 22 (CHF_2Cl). However, this activity of TiO_2 relative to carbon oxide, unsaturated hydrocarbons, carbon acids, alcohols, and other organic compounds is very considerable [16].

The carbon dioxide formation is observed under the illumination of titania surface in vacuum, in the presence of halogenorganic compounds, acetone, and oxygen. The experiments with carbon monoxide enriched by ^{13}CO isotope show that the adsorption of carbon monoxide in darkness and under radiation does not lead to the formation of $^{13}CO_2$ in the gas phase. $^{3}CO_2$ is not generating also under irradiation of titania surface in the presence of oxygen-carbon oxide mixture, though the oxygen from mixture is consumed. At that, the photodesorption of ordinary CO is observed.

In the initial state before illumination the dark adsorption of oxygen on titania surface was absent accordingly by obtained manometer and mass-spectrometer data. It is probably a result of titania preparation by grinding of rutile crystal under ambient air, as since the titania surface is saturated with oxygen from ambient air.

Over a long period of illumination time by all radiation of OSL-1 illuminator in water vapor, in atmosphere of various gases (halogenorganic compounds, acetone) the titania surface is reducing as the result of lattice oxygen consumption for the carbon dioxide formation in the gas phase. At that, the formation of nonstoichiometric Ti^{3+} is detected by ESR method and by the absorption of light quanta with the wavelength more 400 nm and even by sight. Moreover, the darkness adsorption of oxygen is observed on reduced titania surface.

It is known that the phtocatalytic activity of metal oxides is decreasing in the course of photocatalytic reaction [65, 73]. Usually the initial activity can be restored after the oxygen treatment of metal oxides at high temperature (750-850 K). The cleaning of metal oxide surface from adsorbed compounds by high temperature oxygen vacuum treatment leads also to increasing photocatalytic activity [74]. The long time irradiation of metal oxide surface under ambient air cleans an oxide surface from adsorbed biographic organic compounds and leads to their oxidation into CO_2 and H_2O [15, 32].

We carried out the irradiation of partly reduced titania under the oxygen atmosphere for the restore of titania photocatalytic activity in the carbon monoxide oxidation. For that, the UV irradiation of the titania was performed with an OSL-1illuminator with DRSh-250 high pressure mercury lamp; water heat filter and UV filter in the presence of oxygen (at the initial pressure 10^{-1} Torr) in during 10 hrs.

After addition of carbon monoxide enriched by ^{13}CO isotope into reaction volume the illumination leads to the appearance in the gas phase not only CO_2, but also $^{13}CO_2$, i. e. to the photocatalytic carbon monoxide oxidation.

Thus, the high pressure of oxygen and the presence of solar radiation in the Troposphere favor in the long time course of the photocatalytic reactions which clean the atmosphere from dangerous compounds.

$TiO_2(B)$. The photocatalytic activity was measured on sample of steady activity towards photocatalytic carbon monoxide oxidation at 20 ^{O}C; the initial pressure of the stoichiometric mixture was 10^{-1} Torr. The reaction orders and the temperature dependence were determined under full-light irradiation with high-pressure mercury lamp supplied with a water filter. The spectral dependence was measured at a monochromatic radiation intensity of 1.9×10^{14} photon s^{-1}.

The spectral dependence of the quantum yield of photocatalytic carbon monoxide oxidation is given in Figure 22, curve 2.

Powdery titania is the most stable and most studied photocatalyst for both liquid- and gas-phase photocatalytic oxidation reactions. Note that for TiO_2 photocatalytic processes are observed solely and the photoadsorption mostly in the intrinsic absorption band. The quantum yield of these processes can be 1% and higher [75-77]. The maximum amount of photoadsorbed molecules achieved 0.01% of TiO_2 surface monolayer [78]. In the initial state (before high temperature oxygen-vacuum treatment) the adsorbed layer of TiO_2 contains both carbon dioxide and significant amount of water [79, 80]. Without oxygen-vacuum treatment and under irradiation in the oxygen atmosphere, oxygen consumption from the gas phase to oxidation of adsorbed organic compounds is observed [44]. The presence of such compounds on the TiO_2 surface with platinum leads to their dehydrogenation and formation of hydrogen (observed, for example, in Pt / TiO_2 suspended in 1N H_2SO_4 [56]).

TiO_2(C). The optical and chemical properties of titanium dioxide prepared by combustion of a pyrotechnic mixture are close to those of titania prepared by other methods. Absorption in the spectral region before the intrinsic absorption edge, which is related to changes in the titania stoichiometry, is also observed. Photodesorption of CO_2 and photoadsorption of oxygen are observed on TiO_2 in wide range of the solar troposphere irradiation spectrum. No reaction of the titania surface with methane and its halogenated derivatives as well as ethane and its halogenated derivatives was observed. Meanwhile, we observed photocatalytic oxidation of carbon monoxide.

It is natural to expect that titania prepared by combustion of the pyrotechnic mixture in air will be active in the same reactions where photocatalytic oxidation is observed. These are oxidation of ethylene and its halogenated derivatives, oxidation of organic acids, alcohols, etc. If such processes take place under conditions close to those observed in the combustion zone, they can contribute to cleaning of the environment from various gaseous pollutants.

e). On the mechanism of photocatalytic carbon monoxide oxidation

For occurring the photocatalytic reaction a necessary condition is the absorption of a light or UV radiation quanta with formation mobile, free electrons and holes in a conductivity band and valence band, accordingly, in a semiconductor crystal. And only in time of interaction of the adsorbed or surface forms of oxygen with free, mobile hole, changing reactionary ability of oxygen, to initiate the interaction with the carbon oxide or hydrocarbon happens it. Thus, this case is the initiation of a photocatalytic reaction by irradiation.

According to our data [14, 44], photocatalytic activity depends on the energy structure of metal oxides which is mainly determining by the electron configuration of the metal cation. For semiconductor oxides having the electron configuration of d^0- cation (TiO_2, Sc_2O_3, V_2O_5, and MoO_3) and for semiconductor oxides with the electron configuration of d^{10}- cation (ZnO, SnO_2, In_2O_3, and Nb_2O_5) the top of valence band is formed by *2p*-orbitals of oxygen. Completely occupied bands of above-mentioned metal oxides formed by *3 d*-orbitals of metal are placed below the top of valence band at a distance which is equal 9 eV [44]. For metal oxides, having the partly occupied atom *d*- shell of metal cation (Fe_2O_3, CoO, Co_3O_4, NiO, and CuO), the top of valence band is formed by *d*- states of cations. The band consisting of oxygen *2p*-orbitals is placed below overlapping with the *d* band of metal cation.

The metal oxides with the electron configurations of d^0 and d^{10} are active photocatalysts in the carbon monoxide oxidation. The metal oxides, having the metal cations with the partly occupied atom *d*- shell of metal cation, after the absorption of radiation from the fundamental absorption region of semiconductor metal oxide do not show photocatalytic activity in the photocatalytic oxidation of carbon monoxide [7, 32, 44].

The photocatalytic oxidation of carbon monoxide proceeds only under the action of light quanta from the fundamental absorption band of photocatalysts. Accordingly, the spectral dependencies of quantum yields the oxygen photoadsorption can not be the rate determining step and can not represented as the step of photocatalytic oxidation reaction. The oxygen photoadsorption is the process proceeding in parallel with photocatalytic oxidation under illumination of the unfiltered radiation.

The initial rate of photoreduction (photoadsorption carbon monoxide) is equal to the rate of the photocatalytic carbon monoxide oxidation. At experimental studies on the photocatalytic carbon monoxide oxidation are observed the first order for carbon monoxide and the zero order for oxygen. The activation energy of the photocatalytic oxidation is equal 7 kcal · mole^{-1} [44]. Since the amount of surface lattice oxygen is approximately a monolayer of surface, the mobile holes formed in the semiconductor oxide can effectively discharge surface lattice oxygen, which converts into the highly reactive form of oxygen. When the CO molecule to be oxidized meets the above surface oxygen form, an interaction of carbon monoxide with the activated lattice oxygen can occur. In other words, the chemical interaction happens at the contact of the oxidized molecule CO with the surface lattice oxygen discharging by mobile hole, as it is shown in Figure 23. The anion vacancies formed upon the interaction of lattice oxygen with carbon monoxide may be assumed to be filled due to the dissociative chemisorption of oxygen (redox mechanism). In agreement with the proposed a redox mechanism, the photocatalytic oxidation reaction does not require the presence an active surface center on the titania surface.

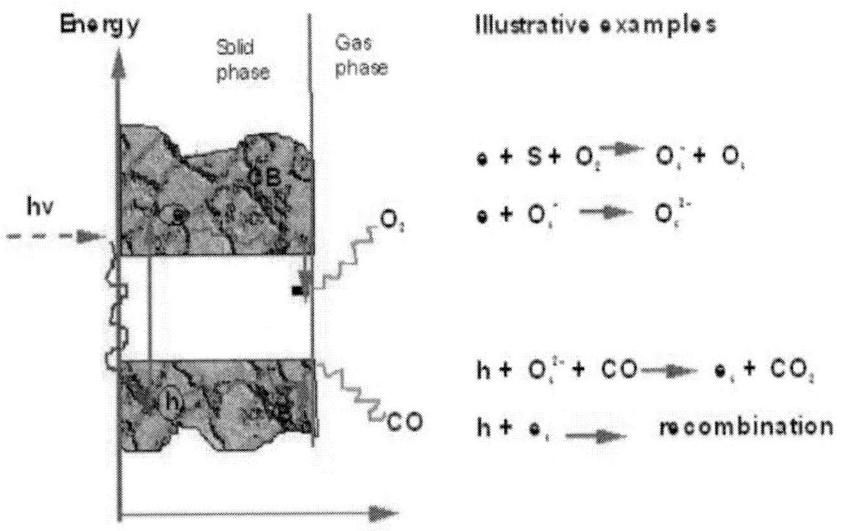

Figure 23. Scheme of the photocatalytic action of titanium dioxide for the carbon monoxide oxidation reaction.

In accordance with our experimental results, in the fundamental absorption region of semiconductor metal oxides (ZnO, SnO_2, TiO_2, Sc_2O_3, In_2O_3, V_2O_5, and MoO_3) the quantum yields of photocatalytic oxidation vary in a wide range from 30% to 0.01% under the same conditions of metal oxide pretreatment and quantum yield measurement (Table 2).

Such considerable distinction can not only be explained by different physical characteristic of used metal oxides (mobility of lighted charge carriers, recombination rate, state density of valence and conductivity bands, near surface electric fields and so on). Thus, the mechanism assumes that the quantum efficiency of photocatalytic reaction depends upon the bond energy of surface lattice oxygen.

Table 2. Quantum yield of carbon monoxide photocatalytic oxidation (fundamental absorption band), band gap (Eg), standard heat of formation (-ΔH0298), atomization energy at 0 K (D) and crystal structure of semiconductor metal oxides [81]

Metal oxide	Quantum yield, %	E_g, eV	$-\Delta H^0_{298}$, $kJ \cdot mol^{-1}$	D ,$kJ \cdot mol^{-1}$	Crystal system
ZnO	30	3.2	351	724	hexagonal
SnO_2	1	3.0	580	1370	tetragonal
TiO_2	1	3.7	944	1905	tetragonal
Sc_2O_3	0.5	2.7	1908	3151	cubic
In_2O_3	0.15	2.2	926	2152	cubic
V_2O_5	0.1	2.9	1552	3802	rhombic
MoO_3	0.01	5.4	745	2136	rhombic

Conclusion

The important effect of solar radiation in the troposphere is the initiation of photoadsorption and photocatalytic processes caused by absorption of radiation by titania particles. The realization of photoinduced processes on the titania particles, which are the phase chemical component of tropospheric solid aerosol, favors the cleaning of the gas phase of the troposphere from many polluting chemical compounds. And what is more, it is possibly to clean the atmosphere near a pollution source in the result of the production of titania aerosol by combustion of titanium microparticles in direct place of a contamination.

References

[1] Brimblecombe, P. *Air Composition and Chemistry*; Cambridge university: London, 1986; 351.

[2] Malakhov, V.V.; Vlasov, A.A; Boldireva, N.N; Dovlitova, L.S.; Pushkin, S.G. Khimiya v Interesakh Ustoichivogo Razvitiya (Chemistry for Sustainable Development). 1996, 3, 253-260.

[3] Pogova, V.P.; Kiselev, V.Ya.; Chursin, D.A.; Fedorova, N.V.; Skvortzov, V.A. Atmos. *Oceanic Opt.* 2002, 15, 555-557.

[4] Baryshev, V.P.; Bufetov, N.S.; Koutzenogii, K.P.; Makarov, V.I.; Smirnova, A.I. *Nucl. Instrum. and Meth. Phys. Res. A.* 1995, 359, 297-301.

[5] Kovalskaya, G.A. *Atmos. Oceanic Opt.* 2002, 15, 506-510.

[6] Kostov, I. *Mineralogy; Oliver and Boyd:* Edinburg and London, 1968; 584.

[7] Zakharenko, V. S. *Topics Catal.* 2005, 35, 231-236.
[8] Parmon, V. N.; Zakharenko, V. S. *CatTech.* 2001, 5, 96-115.
[9] Ausloos, P.; Rebbert, R. E.; Glasgov, L. J. *Res. NBS.* 1977, 82, 1-8.
[10] Isidorov, V. A.; Klokova, E. M.; Zgonnik, I. V.; Povarov, V.; Kolkova, S. *Catal. Today.* 1997, 30, 233-242.
[11] Wise, H.; Sancier K.M. *Cat. Lett.* 1991, 11, 277-284.
[12] Eapen, B. Z.; Hoffmann, V. K.; Schoenitz M.; Dreizin, E. L. *Combustion Science and Technology.* 2004, 176, 1055.
[13] Molodetsky, I.E.; Dreizin, E.L.; Vicenzi, E.P.; Law, C.K. *Combustion and Flame.* 1998, 112, 522.
[14] Zakharenko, V.S.; Parmon, V.N. Colloids and Surfaces A: Physicochem. *Eng. Aspects.* 1999, 151, 367-376.
[15] Hoffman, M.R.; Martin, S.T.; Choi W.; Bahneman D.W. *Chem. Rev.* 1995, 95, 69-96.
[16] Chen, X.; Mao, S.S. *Chem. Rev.* 2007, 107, 2891-2959.
[17] Lamb, B. In: *Handbook of Air Pollution Technology*; Calvert, S.; Englund, H.M.; John Wiley and Sons: New York, 1984; 125-167.
[18] Atkinson, R.; Baulch, D.L.; Cox R.A.J. *Phys. Chem. Ref. Data.* 1989, 18, 881 1096.
[19] Lloid, A. In: Handbook of Air Pollution Technology; Calvert, S.; Englund, H.M.; John Wiley and Sons: New York, 1984; 477-504.
[20] Cox, R.A.; Derwent, R.G.; Williams M.R. *Environ. Sci. Technol.* 1980, 14, 57-61.
[21] Calvert, J.G.; Su, F.; Bottenheim, J.W. *Atmos. Environ.* 1978, 12, 197-226.
[22] Isidorov, V.A. *Organic Chemistry of the Atmosphere*; Khimiya: St. Petersburg, 1992; 165.
[23] Leighton, P.A. *Photochemistry of Air Pollution*; New York: Academic Press, 1961; 320.
[24] Penkett, S.A. In: *The Chemistry of the Atmosphere: Its Impact on Global Change;* Calvert, J.G.; London: Blackwell Scientific Publication,1993; 327-344.
[25] Razumovskii, S.D. Zh. Fiz. Khimii. 1996, 70, 1325-1327.
[26] Gershenson, Yu.M.; Purmal, A.P. *Uspekhi khimii* (Advances in Chemistry). 1990, 59, 1729-1755.
[27] Boreskov, G.K. *Heterogeneous catalysis*; New York: Nova Science Publishers, 2003; 188.
[28] Zamaraev, K.I.; Khramov, M.I.; Parmon V.N. *Catal.Rev.-Sci.* 1994, 36, 617-644.
[29] Gretzel, M. *Energy Resources through Photochemistry and Catalysis*; Academic Press: New York, 1983; 520.
[30] Serpone, N.; Pelizzetti, E. *Photocatalysis. Fundamentals and Applications*; Wiley and Sons: New York, 1989; 490.
[31] Serpone, N.; Salinaro, A.; Emeline, A.; Ryabchuk, V. J. *Photochem. Photobiol. A: Chem.* 2000, 130, 83-91.
[32] Zakharenko, V.S. *Catal. Today*, 1997, 39, 243-249.
[33] Zakharenko, V.S.; Parmon, V.N.; Zamaraev, K.I. *Kinet. Catal.* 1997, 38, 140-144.
[34] Zakharenko, V.S.; Parmon, V.N. *Russian J. Phys. Chem.* 1999, 73, 124-129.
[35] Czochralski, Js. *Phys. Chem.* 1918, 92, 219-221.
[36] Khromova, S.A.; Karasev, V.V.; Onischuk, A.A.; Glotov, O.G.; Zarko V.E. In: *Nonequilibrium Processes*; Roy, G.; Frolov, S.; Starik, A.; Torus Press: Moskow, 2005; 225-234.
[37] Kochubey, D.I. EXAFS Spectroscopy of Catalysis; *Nauka*: Novosibirsk, 1992; 146.

[38] Klementev, K.V. J. *Phys. D: Appl. Phys*. 2001, 34, 209-217.
[39] Binsted, N.; Campbell, J.V.; Gurman, S.J.; Stephenson, P.C. EXCURV92 program code, SERC Daresbury Laboratory, UK,1991.
[40] Lisachenko, A.A.; Vilesov, F.I. *Uspekhi. Fotoniki* (Advances in Photonics). 1974, 4, 18-34.
[41] Volodin, A.M. *Russian J. Chem. Phys*. 1992, 11, 1054-1063.
[42] Basov, L.L.; Kotelnikov, V.A.; Lisachenko, A.A. *Uspekhi. Fotoniki* (Advances in Photonics). 1969, 1, 76-111.
[43] Parmon, V.N.; Zakharenko, V.S. *Khimiya v Interesakh Ustoichivogo Razvitiya* (Chemistry for Sustainable Development). 2001, 9, 461-483.
[44] Zakharenko, V.S. Thesis doct. sci.; *Boreskov Institute of Catalysis*: Novosibirsk, 1976; 30.
[45] Ivlev, L.S. *Khimicheskii sostav atmosfernykh aeroxolei* (Chemicai Composition and the Structure of Atmospheric Aerosols); Nauka: Leningrad, 1982; 480.
[46] Formenti, M.; Teichner, S.J. *Catalysis*. 1978, 2, 87-106.
[47] Bickley, R.J. *Catalysis*. 1982, 5, 308-332.
[48] Zakharenko, V.S.; Moseichuk, A.N. Atmos. *Oceanic Opt*. 2002,15, 495-500.
[49] Lyashenko, L.V.; Girokhovatskii, Ya.B. *Teoret. Eksperim. Khimiya* (Theoret. Eksperim. Chem.). 1974, 10, 186-192.
[50] Mozzanega, H. Thesis doct. sci.; Univ. Claude Bernard: Lyon, 1975; 110.
[51] Lyashenko, L.V.; Gorokhovatskii, Ya.B.; Stepanenko, V.I.; Yampolskaya, Ph.A. React. *Kinet. Catal. Lett*. 1979, 10, 13-17.
[52] Terenin, A.N. *Fotonika molecul krasitelei* (Photonics of dye molecules); Nauka: Leningrad, 1967; 616.
[53] Kuznetsov, V.N.; Lisachenko, A.A. *Russian J. Phys. Chem*. 1991, 65, 1568-1574.
[54] Zakharenko, V.S.; Moseichuk, A.N. *Atmos. Oceanic Opt*. 2005, 18, 454-458.
[55] Vanhieu, N.; Lichtman, D. *Surface Science*. 1981, 103, 535-541.
[56] Zakharenko, V. S.; Parmon, V.N. *Kinetics Catal*. 1996, 37, 427-430.
[57] Jackson, P.; Parfitt, G.D. *J. Chem. Soc. Faraday Trans*. I. 1972, 68, 896-905 .
[58] Zakharenko, V.S.; Khromova, S.A. *Russian J. Ecol. Chem*. 2006, 15, 226-234.
[59] Vavilov, V.S. *Effect of Radiation on Semiconductors*; Fizmatgiz, Moscow, 1963, 354.
[60] Gonzalez-Elipe, A.R.; Soria, J.J. *Chem. Soc. Faraday Trans*. I. 1986, 82, 739-743.
[61] Gravelle, P.C.; Juillet, F.; Meriaudeau, P.; Teichner, S.J. *Disc. Faraday Soc*. 1971, 52, 1862-1869.
[62] Bickley, R.J.; Stone, F.S. *J. Catalysis*. 1973, 31, 389-397.
[63] Lisebigler, A.; Lu, G.; Yates, J.T. *Chem. Rev*. 1995, 95, 735-758.
[64] Terenin, A.N.; Solonitsin, Yu.P. Disc. *Faraday Soc*. 1959, 28, 28-37.
[65] Mills, A.; Le Hunte, S.J. *Photochem. Photobiol*. A . 1997, 108, 1-25.
[66] Kaliaguine, S.L.; Shelimov, B.N.; Kazansky, V.B. *J. Catalysis*. 1978, 55, 384-393.
[67] Fujishima, A.; Rao, T.N.; Tryk, D.A. *Photochem. Photobiol*. C. 2000, 1, 1-21.
[68] Zakharenko, V.S.; Parmon, V.N. *Kinet. Catal*. 2000, 41, 834-837.
[69] Djeghri, N.; Formenti, M.; Juillet, F.; Teichner, S.J. *Faraday Disc. Chem. Soc*. 1974, 58, 185-193.
[70] Isidorov, V.A.; Klokova, E.M.; Zgonnik, I.V. Vestnik Leningrad University. 1990, 4, 61-65.

[71] Nimies, M.R.; Jacoby, W.A.; Blake, D.M.; Milne, T.A. *Environ. Sci. Technol.* 1993, 27, 732-739.

[72] Driessen, M.D.; Goodman, A.L.; Miller, T.M.; Zaharias. G.A.; Grassian, V.U. J. *Phys. Chem.* B. 1998, 102, 549-556.

[73] Zakharenko, V.S.; Cherkashin, A.E.; Keier, N.P.; React. Kinet. *Catal. Lett.* 1974, 1, 381-387.

[74] Muggli, D.S.; Falconer, J.L. J. *Catalysis*. 2000, 191, 318-325.

[75] Basov, L.L.; Solonitsin, Yu.P.; Terenin, A.N. Dokl. Acad. Nauk SSSR. 1965, 164, 122-124.

[76] Bickley, R.I.; Stone, F.S. J. *Catalysis*. 1973, 31, 398-407.

[77] Cherkashin, A.E.; Volodin, A.M.; Kosheev, C.V.; Zakharenko, V.S. *Uspekhi. Fotoniki* (Advances in Photonics). 1980, 7, 86-142.

[78] Solonitsin, Yu.P.; Kuzmin, G.N.; Shurigin, A.L.; Yurkin, V.M. *Kinet. Catal.* 1976, 17, 1267-1272.

[79] Dantshevskaya, M.N.; Panasyuk, G.N.; Figurovskaya, E.N. *Kinet. Catal.* 1969, 10, 930-934.

[80] Egashira, M.; Kavasumi, Sh.; Kagava, Sh.; Seiyama, T. *Bull. Chem. Soc.* Japan. 1978, 51, 3144-3152.

[81] Hand-book. *Physical-chemistry Properties of Oxides* (Russian); Samsonov, G.V.; Metalurgiya: Moscow, 1978, 456.

In: Rutile: Properties, Synthesis and Applications
Editor: It-Meng (Jim) Low

ISBN 978-1-61942-233-9

Chapter 6

RUTILE AS SUPERACID CATALYSTS: PREPARATION, MORPHOLOGY AND CATALYTIC ACTIVITY

Hussein A. Khalaf[1]
Chemistry Department
Faculty of Science, Omer Al-Mukhtar University
P.B.O. 919, El-Beida, Libya

ABSTRACT

The growth in population and the consequent demand for fuels and chemicals has had a major negative impact on the environment. It is now believed that catalysis can play a major role in environment protection and enable sustainable development by a number of ways. In green chemistry, catalysis can help through non-polluting processes that are atom efficient and produce negligible waste, end-of-pipe solutions, use of economically attractive alternate feedstocks, use of renewable feedstocks, producing biodegradable products, development of energy efficient processes and routes to alternate energy. Heterogeneous catalysts have begun to replace highly corrosive, hazardous and polluting homogenous catalysts. The replacement of liquid acids by solid superacids as catalysts is expected to ease their separation from reacting mixture, allowing continuous operation as well as regeneration and neutralization of the catalyst and lowering the cost of process installation and maintenance. Metal oxides such as SnO_2, TiO_2, VO_2 and ZrO_2 have been used as a precursor of solid superacids preparation. Among them, TiO_2 and TiO_2-based materials are of importance for sensors, pigments and photocatalysis. Titania has three polymorphs: anatase, brookite, and rutile which have different material properties including density, index of refraction, and catalytic properties. They have different applications in paint, optical coatings, beam splitters, and antireflection coatings as well as superacid catalysts because of their strong oxidizing power, high photostability and redox selectivity. Many methods e.g. impregnation, hydrothermal, sol-gel, co-precipitation and multi-gelation have been used for synthetic TiO_2-based catalysts. Different techniques have been used for characterization and studying the catalytic

[1] E-mail address: hkhalaf70@yahoo.com

activity of these rutile superacids. The obtained results reveal that these materials have higher acidity than concentrated H_2SO_4 and their acidity and morphology can vary depending on the synthetic methods. They can be used for several acidic reactions e.g. alkylation, isomerization, estrification, cracking and condensation.

Keywords: rutile, superacids, catalytic activity, green chemistry.

1. Introduction

Catalysis is of vital importance for the environment and for chemical industry, the number of catalysts applied in industry is very large and catalysts come in many different forms, from heterogeneous catalysts in the form of porous solids over homogeneous catalysts dissolved in the liquid reaction mixture to biological catalysts in the form of enzymes. Catalysis plays a very important role in the oil refining and petrochemical industry for the upgrading of hydrocarbons streams to modern transportation fuels.

Conventional acids such as H_2SO_4, HF, $AlCl_3$, BF_3, SbF_5, H_3PO_4 and HCl that used for organic reactions pose significant risks in handling, containment and disposal due to their toxic and corrosive nature. Thus, these acids have been replaced by heterogeneous solid acid catalysts. These have the advantages those they:

- Are relatively cheap
- Offer easy product separation
- Can reduce waste production
- Are easily handled and stored
- Can be tailored to suit the chemistry in which they are used.

Metal oxides have been observed to possess a strong acidity as well as a high catalytic activity. Modification of metal oxides with sulfate anion can generate a strong acidity, even stronger than 100% sulfuric acid, and hence they become superacid catalysts that useful in reactions like isomerizations, low temperature esterification, alkylation and cracking [1, 2]. Among various solid acid catalysts such as zeolites, clays and heteropolyacids, the sulfated metal oxides, based on zirconium, titanium, iron and tin oxides have emerged as powerful catalysts due to their superacidity, high reactivity at low temperatures and reusability [3, 4].

For sulfated metal oxides, the textural properties and catalytic activity can improve by addition of another metal component. Metal-promoted sulfated oxides are receiving increasing interest owing to their enhanced catalytic activity. The promoting effect of different metals like Mn, Fe, Ni, and Al on some metal oxides as Fe_2O_3 and ZrO_2 has been reported [5, 6]. Since metal oxide systems, containing titania and tin oxide, are widely used in oxidation reactions, it is expected that their sulfated analogues can show enhanced activity. Matsuhashi et al [7] reported that the addition of sulfate anion to SnO_2 improves its oxidation activity in the dehydrogenation of cyclohexanol. SnO_2-based catalysts exhibit good activity towards CO/O_2 and CO/NO reactions [8]. The activity and selectivity of SnO_2 catalysts can be substantially improved by incorporation of Heteroelements [9].

The catalytic properties of metal oxide-based catalysts depend on their morphology, size of crystallites and crystallographic structure. The rutile type structure is the most common

MX_2 structure and the electronic properties of this class of compounds vary widely from insulating to metallic behavior. The rutile type dioxides including the group of IVb metals is studied for understanding the structural and physical properties of it. Several oxides have the rutile structure e.g. GeO_2, SiO_2, TiO_2, SnO_2 and PbO_2 [10-12].

Titanium dioxide occurs primarily in three different forms: anatase, brookite, and rutile [10]. The most common are anatase and rutile, since brookite is rather unstable. The brookite type cannot be used in industries because of its instability at room temperature. The anatase type has the problems of poor light, heat resistance, and gradually decreasing in whiteness due to weather. It also has drawbacks for applications involving adsorption technology owing to its low surface energy. The rutile type has outdoor applicability because of its good light resistance and can be applied to surfaces by the use of adsorption technology without advanced skills or sophisticated equipment [13].

It is important to note that the catalytic properties of solid superacid catalysts strongly depend on the method of preparation, precipitating agent, sulfating agent, calcination temperature, pre-treatment, and storage. These diversities add to the disagreement of the acidity among various researchers. Even if the acidic properties of a solid acid could be accurately determined, it would still be difficult to relate the acidity to catalytic activity because of wide distribution of the type and strength of the surface acid sites and the uncertainty about which sites are catalytically active.

The topic of the present review is narrower: it deals with the synthesis and properties of solid superacids derived from metal oxides and their modified oxides for the goals of materials chemistry. By far the most studied among the solid superacids is TiO_2 and SnO_2 that having the rutile structure, which finds several applications in diverse areas and green chemistry. A number of methods is considered in this review, including hydrothermal, impregnation, precipitation, co-precipitation, sol-gel, multi-gelation and ultrasonically methods. The important part of this review considers various synthetic routes to rutile-based superacid catalysts and the properties of the obtained materials. Though this review is mostly devoted to study; preparative methods, morphology, physical properties, acidity and catalytic activity for TiO_2 and SnO_2-based superacid catalysts.

2. Solid Acids and Superacidity

Through 20th century, catalytic technologies have played a vital role in the economic development of the chemicals industry with a total contribution of 20% of world products Acid catalysis plays an important role in many important reactions of the chemical and petroleum industries. Conventional industrial acid catalysts, such as $AlCl_3$, H_2SO_4 and BF_3, have some considerable drawbacks because of their serve corrosives and high susceptibility to water. The search for environmentally benign heterogeneous catalysts has driven the worldwide carrying out research for new materials to substitute the current liquid acids and halogen-based solid acid [14]. Superacids catalysts displaying high thermostability, very strong acidity, and high catalytic activity, which have aroused increasing, interest [15-19]. The definition of solid superacids, according to Gillespie [20], is any solid acid having acidity stronger than that of 100% H_2SO_4, i.e. Ho < -12 [20-22], where H_o is the Hammett acidity function.

Recently, various kinds of solid superacids have been developed [15, 23-25]. Metal oxides and mixed oxides containing a small amount of sulfate ions, and those modified with platinum, are the first group. The second group is metal oxides, mixed oxides, graphite, metal salts, etc. treated or combined with antimony fluoride or aluminum chloride, and this group seem to be environmentally undesirable as catalysts. The third group is perfluorinated polymer sulfuric acid (Nafion-H). The fourth and fifth groups are H-ZSM-5 and a type of heteropolyacids, respectively. The last group is simply mixed oxides.

Solid superacids have been proved to be very effective from the viewpoint of activity, selectivity, reusability and in the preparation of various important chemicals which have a number of difficulties with homogeneous catalysts.

Homogenous liquid catalysts, for instance, are difficult to be separated from the product stream [26], besides that, large amounts of the catalyst are usually required, often leading to wasted catalyst. Furthermore, the liquid acids are corrosive to the reaction system and the liquid waste is an environmental hazard.

Solid superacids can be applied for different catalytic reactions [27-34] such as; estrification, isomerization, alkylation, polymerization, cracking and other acid catalyzed reactions (e.g., acylation, etherification, acetalization, nitration, disproportionation, etc.).

3. Measurement of Acidity and Surface Acid Sites

Acidity characterization is often performed to measure the density, strength, and type of acid active sites of solid catalysts. Such knowledge helps correlate observed catalyst activity with details from the catalyst preparation stage and provides valuable information for the elucidation of reaction mechanisms. One such acidity characteristic is the surface population distribution of Brönsted and Lewis acid sites. A variety of physiochemical techniques have been developed for the characterization of type, strength and numbers of acid sites on solid catalysts. One of the oldest techniques for measuring acidity is based on a proposal by Hammet [35] for ordering strengths of solid acids on the basis of amine titrations.

Unfortunately, this method is strictly applicable only to Brønsted acids. Since many oxide surfaces exhibit both Brønsted and Lewis acidity, both of which may interact with Hammet indicators, the nonspecifity of the indicator compromises the usefulness of the titration technique [36]. Indicators can be put to use to determine acidities of systems other than dilute aqueous solutions by basing an acidity scale on the concentrations of the indicator's acid and base forms. An example is the Hammett acidity function, which defines a scale of actual acidity using the spectroscopically measured exact concentrations of indicator molecules in their protonated (HIn) and deprotonated (In) states. If the reaction proceeds by means of proton transfer from the surface to the adsorbate, the acid strength is expressed by Hammett acidity function (H_o). The value of the Hammett acidity function [35] is then given by:

$$H_o = pKa + \log [B]/[BH^+] ,$$

where [B] and $[BH^+]$ are the concentrations of the neutral base (a basic indicator) and its conjugated acid, respectively, and pKa is $pKBH^+$. If the reaction proceeds by means of electron pair transfer from the adsorbate to the surface, H_o is expressed by:

$$H_o = pKa + \log [B]/[AB]$$

where, [AB] is the concentration of the neutral base that with the Lewis acid or electron pair acceptor [A]. The color of suitable indicators adsorbed on the surface can give a measure of acid strength; if the color is that of the acid form of the indicator, then the value of H_o is equal to, or lower than, the pKa of the conjugated acid of the indicator [37]. Lower value of H_o corresponds to greater acid strength.

Other physiochemical techniques used to characterize surface acidity and basicity include the adsorption of acidic and basic gas phase probe molecules combined with spectroscopic measurements (IR) and calorimetric, gravimetric, or thermal desorption measurements [38-46]. The use of these different techniques is closely related, and in combination can provide a characterization of type, number, and strength of acid/base sites on surfaces.

Infrared spectroscopy is an important technique in surface chemistry. It is an easy way to identify the presence of certain functional groups in a molecule. Also, one can use the unique collection of absorption bands to confirm the identity of a pure compound or to detect the presence of specific impurities. Sometimes the IR spectra of adsorbed probe molecules such as pyridine, CO and NO give valuable information on the adsorption sites that are present on a surface of the catalyst. A great advantage of IR spectroscopy is that the technique can be used to study catalysts in situ by the aid of specially designed cells.

IR spectroscopy is still remains the most widely used, and usually most effective, spectroscopic method for characterization of surface acid sites. Several studies [47-51] have been performed to investigate the interaction between surfaces and basic molecules by IR spectroscopy, in which it can give some information about the structure of adsorbed molecules, the mode of interaction between the surface and adsorbed species, surface structure and the nature of adsorption sites. The characterization of the acid sites by IR spectroscopy is based on the use of basic probe molecule; in general, pyridine (Py) is widely used as a probe molecule for estimating surface acidity of the catalysts. The acid strengths of sulfated zirconia and sulfated titania measured by the indicator methods are $H_o \leq -16.0$ [52] and $H_o \leq -14.5$ [53] respectively, being higher than that of 100% H_2SO_4.

Table 1. IR bands of pyridine on solid acid in the 1700-1400 cm^{-1} region*

Hydrogen bonded pyridine (cm^{-1})	Coordinately bonded pyridine (cm^{-1})	Pyridinium ion (cm^{-1})
1440 - 1447 (v.s.)	1447 – 1460 (v.s.)	
1485 – 1490 (w)	1488 – 1503 (v)	1485 – 1500 (v.s.)
		1540 (s)
1580 – 1600 (s)	- 1580 (s)	
	1600 – 1633 (s)	- 1620 (s)
		- 1640 (s)

Band intensities: v.s. = very strong; w. = weak; v. = variable.

In Table 1, the IR frequencies of the bands observed by adsorbing pyridine on solid acid are given. The characteristic bands of pyridine protonated by Brönsted acid site (pyridinium ions) observed at ~1540 and 1640 cm^{-1} while the other from pyridine coordinated to Lewis acid sites appear at ~1450 and 1620 cm^{-1} [54]. It is possible to calculate the number of

Brönsted and Lewis acid sites capable of retaining pyridine at certain desorption temperatures.

4. Rutile-Based Superacid Catalysts

Synthesis of these solid superacids is described first followed by the effects of the synthesis methods on the textural and surface properties. Structure of these catalysts and the catalytic performance will be discussed.

4.1. Preparative methods

The principal catalyst-preparation technique involves two stages. The first one is known as dispersion and is achieved by impregnation, adsorption from solution, co-precipitation, or deposition, while the second stage is variously called calcination or reduction [55]. It is brought about by a thermal treatment in either an inert atmosphere or an active atmosphere of either oxygen or hydrogen. When the active atmosphere is hydrogen the process is known as reduction. Although calcination/reduction can cause major problems in catalyst preparation on a large scale, it is a generalization to say that once the metal species have been bound to the support surface, its degree of dispersion and location will be retained during subsequent treatments.

Preparation of a given catalyst depend on some parameters e.g. temperature, pH, time of preparation, type of required catalyst and general characteristics of the products. And the process of preparation vary considerably from one catalyst to another, the processes by which the majority of industrial catalysts are manufactured can be grouped into three broad categories according to their products, as;

- Bulk catalysts and supports.
- Impregnated catalysts starting from performed supports.
- Mixed- agglomerated catalysts.

Bulk catalysts are comprised principally of active substances. Important examples include silica-alumina catalysts for cracking. The supports for such catalysts are made by similar processes and can be classed with bulk catalysts, as for example, aluminas, silica-aluminas, and molecular sieves [56]. In impregnated catalysts, the support gives the catalyst its texture, its mechanical resistance, its physical form and possibly certain activity, as in bifunctional catalysts. The category of mixed-agglomerated catalysts is less well defined and these catalysts are frequently classed in one or the other preceding categories. In principle, mixing the active substances with a powdered support or a precursor of the support, and then agglomerating the mixture to obtain these catalysts.

Impregnation methods [57] can be done by two ways, the first is in the case of no interaction between support and active catalytic species. In this way, if the support does not have its own catalytic activity, its role is to suitably present the catalytic agents; it gives the finished catalyst its form, its texture and its mechanical resistance. The second way is specific for the interaction between catalyst and support. The interaction may involve an ion exchange, an adsorption, or a chemical reaction.

Solid superacids are prepared by a technique similar to the original one [52, 58], namely impregnation methods by the introduction of a calculated amount of acid to the metal oxide or hydroxide. Impregnated catalysts prepared by impregnating a metal salt on a porous support. When liquid is slowly added to a porous solid powder, the liquid is first absorbed in the pores and the powder will flow as if it is dry. When the pores have been filled the outside of the grains rather suddenly become wet, the grains will tend to stick together and the powder will form lumps instead of flowing freely. The situation when the pores have been filled but the outside of the grains is dry is called *incipient wetness* and can easily be detected by shaking or stirring the powder.

Catalysts are prepared by impregnation by spraying a solution of a metal salt onto pellets of a porous support until incipient wetness. The pellets are then dried and calcined to transform the metal into insoluble form. The metal salt can be deposited homogeneously through the pellet or most of the metal may be deposited near the outside of the pellet. The distribution of the metal is controlled through the pH or through addition of chelating agents to the impregnation solution.

Recently, several techniques have been used to prepare the solid superacid catalysts including sol-gel, hydrothermal treatment, plasma sprayed coating and ultrasonic, this in addition to precipitation, co-precipitation and impregnation. Most of these methods are to be described below.

4.1.1. Pure Oxides

Several oxides having rutile structure, among them two important metal oxides, TiO_2 and SnO_2, have been studied here. Nanoparticles of TiO_2 have been synthesized using ultrasound technique. It has shown very rapid growth in its application to materials science due to its unique reaction effects [59]. Recently, use of cavitation in the synthesis process has yielded smaller particle size for a variety of oxides. Zhang and Reller [60] demonstrated that some properties of pure and doped titania are very sensitive to changes in crystallinity. Besides the crystalline structure, particle size and surface area are strongly influence catalytic activity. For example, the rutile phase is very efficient for the catalytic decomposition of H_2S; however, the anatase phase is more effective in many other catalytic processes [61]. Thermal stability of mesoporous structures is a critical parameter for application of catalysts. The mesoporous TiO_2, presents better performance as catalyst, due to easier accessibility of bigger molecules to the internal canals.

Home-prepared rutile TiO_2 has been prepared by Augugliaro et al [62] using slow addition of 20 mL of $TiCl_4$ to a 2 L beaker containing 1 L of doubly-distilled water under continuous mixing. After mixing for 10 min, the beaker containing the sol was sealed and kept in an oven for 2 days at 333 K. Then the resulting suspension was dried at 60 °C by means of the rotary evaporator, in order to obtain the final titania powdered catalyst.

Also Rutile TiO_2 nanoparticles were prepared by the sol-gel method [63] with 10ml titanium alkoxide, as the raw material, mixed with 40 ml 2-propanol in a dry atmosphere. This mixture was then added dropwise into another mixture consisting of 10ml water and 10ml 2-propanol. A yellowish transparent gel was formed after 1h stirring. The obtained gel was then dried at 105°C for several hours until it turned into a yellow block crystal. Calcination of the synthesized materials was carried out in air at 500°C for six hours. In general, the sol-gel method consists of the hydrolysis and condensation reactions which are catalysed in the presence of acid. The hydrolysis reaction leads to the formation of original nuclei of TiO_2 while the condensation leads to the growth of the original nuclei [64]. The reaction scheme is:

$$\mathrm{TiCl_4} + 4\mathrm{H_2O} \xleftrightarrow{H^+} Ti(OH)_4 + 4H^+ + 4Cl^- \quad \text{hydrolysis},$$

$$\mathrm{Ti(OH)_4} + 4\mathrm{H_2O} \xleftrightarrow{OH^-} TiO_2 + 6H_2O \quad \text{condensation}.$$

Zhang et al. [65] suggested that the equilibrium between the hydrolysis and the condensation reactions made the formation of TiO_2 particles in the rutile phase possible.

Rod-like and star-like nanocrystalline rutile titania were prepared by a sonochemical method in $TiCl_4$/HCl solution [59]. In a typical procedure, 2 ml of pure $TiCl_4$ was added dropwise to 100 ml of hydrochloric acid solution. Ultrasound irradiation was accomplished with a high intensity ultrasonic probe immersed directly in the reaction solution under ambient air for 50 min. White precipitates were obtained at the end of reaction. The precipitates were centrifuged, washed with distilled water, and dried in air at 60 °C.

The multi-gelation method has a good performance for controlling pore structures, i.e., pore size distributions as shown in Figure 1 [66, 67]. Multi-gelation method is the procedure to produce the inorganic oxides gel by swinging of pH of solution dexterously several times. This method can be employed in preparation of hydrous titanium.

It can control particle size of TiO_2 uniformly with a designed pore size. Scheme 1 shows the preparation flow diagram of the multi-gelation method to synthesize TiO_2. The raw materials are acidic $TiCl_4$ and basic ammonia solutions. Hydroxyl gel of TiO_2 was synthesized by swinging pH from $TiCl_4$ and ammonia solutions.

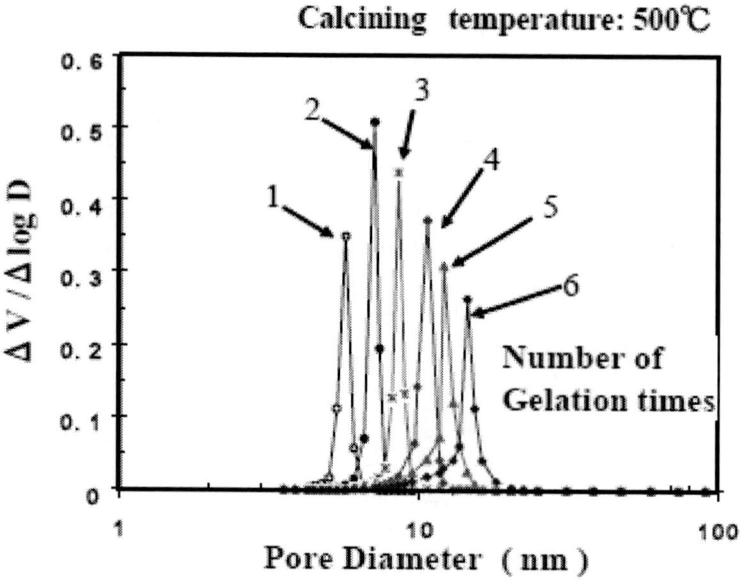

Figure 1. Pore Distributions of TiO_2 Controlled by Multi-Gelation Method [66, 67].

$TiCl_4$ and ammonia solutions were supplied to the gelation vessel alternately, and hydrous titanium was synthesized. Particles of hydrous titanium were controlled in desirable particle size by alternately supplying $TiCl_4$ or ammonia solution. Hydrous titanium was washed by water to remove ammonium chloride. After filtration, TiO_2 as a catalyst or a catalyst carrier was molded to cylindrical shapes by the extruder, for example. And it was dried at 120 °C and calcined at 500 °C in the standard procedure.

Khalaf et al. [68] prepared tin oxide gel and used it as a precursor of sulfated and supported oxide, by a slow dropwise addition of a 1:1 ammonium hydroxide solution to a warm (60-80 °C) 0.3 M aqueous solution of tin(IV) chloride ($SnCl_4.5H_2O$), with a continuous stirring till pH = 8 is reached. The white precipitate was left over-night before being filtered and washed thoroughly with 2% CH_3COONH_4 solution until all chloride was eliminated (silver nitrate test), and then dried at 110 °C till constant weight. The dried material was ground to 250 mesh size and kept dry over P_2O_5 desiccators. The dry gel thus obtained was calcined at 600 °C for three hours to produce SnO_2 (rutile structure).

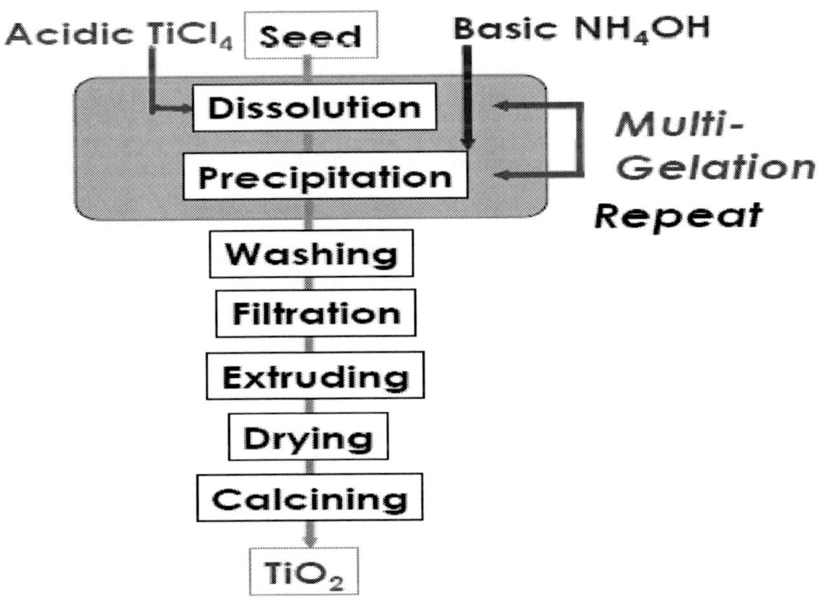

Scheme 1. Preparation Flow Diagram of TiO_2 by Multi-Gelation Method.

4.1.2. Modified Oxides

To control the properties of the metal oxide catalysts, some additives including metals, metal oxides and/or anions can lead to catalysts that are active and more resistant against deactivation. Therefore, several metals such as platinum, palladium, aluminum, nickel, cobalt and manganese or oxides such as alumina, silica, copper oxide, zirconia and vanadia or sulfates and phosphates have been added to metal oxides, producing active and selective catalysts. All these categories will be discussed below. The presence of dopants modifies strongly the anatase–rutile transition, changes the photo-reactivity of nanosized TiO_2 particles, and enhances the catalytic properties of supported metals or oxides [69].

a) Mixed Oxides

Certain metal oxides, when mixed with each others, show high activity and selectivity superior to those exhibited by the individual oxides. The enhancement in activity has been attributed to various causes [70-72]. Thus, in most cases the activity has been studied in relation to only one of the properties of the system like the nature of added oxide, coordination number of the added cation, preparation methods, and composition. Not many attempts have been made to correlate as many of theses aspects as possible with catalytic activity for a particular system.

Different methods have been used for the preparation of titania mixed with other oxides. The most widely accepted model for the generation of acid sites on the mixed oxides is that the surface acidity results from the charge imbalance based on the generation of Ti-O-Zr bonding [73-74]. Tanabe's model predicts Lewis acidity for zirconia rich and Brønsted acidity for titania rich chemically mixed TiO_2/ZrO_2. Tanabe's hypotheses were tested by studying the adsorption of pyridine. It was found that in TiO_2/ZrO_2 binary oxides the number of Lewis sites per square meter was equivalent to that in the pure oxides [75]. Sol gel method is widely employed due to its capability in controlling the textural and surface properties of the resulting oxides. The preparation of titania-zirconia mixed oxides with various ZrO_2 content in TiO_2 by the sol gel method was reported by Manriquez et al. [76] The number of acid sites is dramatically increased. Their activity in the isopropanol decomposition showed a good correlation between the acid-basic properties and the selectivity to propane, acetone, and isopropyl ether [77].

Kristiani et al. [78] prepared a series of TiO_2-ZrO_2 mixed oxides with various titanium contents from 0 to 100 wt%, following their procedure; appropriate amounts of the titanium and zirconium alkoxides were simultaneously added dropwisely by drop to the n-Butanol solution at 0 °C and the pH was adjusted at 3.0 with HNO_3. The mixture was added to H_2SO_4 then put in reflux at 70 °C for 5h.Water for the hydrolysis step was added drop wise. The amounts for the two alkoxides used for the different synthesis are reported in Table 2. After gelling, all the catalyst samples were immersed overnight at room temperature in cetyltrimethyl ammonium chloride solution, decanted, and then dried overnight at 70 °C. Finally, the catalyst was calcined at 500 °C for 2h in air.

Table 2. Experimental ratio used in the synthesis of the TiO_2-ZrO_2 mixed oxides

TiO_2-ZrO_2 (weight ratio) (wt%)	*Titanium butoxide (mol)*	*Zirconia butoxide (mol)*
10-90	0.05	0.29
50-50	0.25	0.17
90-10	0.45	0.03

Also TiO_2-ZrO_2 binary oxide (1:1) was prepared [79] by a homogeneous co-precipitation method with in situ generated ammonium hydroxide by decomposition of urea at 368 K. For this purpose, an aqueous solution containing the requisite quantities of $TiCl_4$, $ZrOCl_2$, and urea were heated together to 95 ^{o}C with vigorous stirring. In about 6 h of heating, as decomposition of urea progressed to a certain extent, the pH of the solution increased from 2 to more than 7 and the formation of precipitate gradually occurred. The precipitate was heated for 6 h more to facilitate aging. The resulting precipitate was filtered off, washed, dried and calcined at 800 °C for 5 h in air atmosphere and stored in dry nitrogen atmosphere.

In the preparation of V_2O_5/TiO_2 system [80] obtained by the sol–gel method, the rutile phase was found to predominate in pure TiO_2 calcined at 450 °C as a result of the decrease of phase transition temperature promoted by the sol–gel method under acid conditions. The anatase phase became predominant at 450 °C as the amount of vanadium increased from 6 to 9 wt%. A structural change in the TiO_2 phase from predominantly anatase to totally rutile with increased calcinations temperature was observed in 6 wt% samples.

Vanadium supported on titania is commonly used as a catalyst for a number of industrially important reactions. From a scientific point of view, this catalytic system is an interesting example of a strong interaction between the support (TiO_2) and the active phase (V_2O_5). In particular the spreading of vanadium over the TiO_2 support leads to a modification of the chemical–physical peculiarities of the former and to an enhancement of its catalytic properties [81-84]. The structural characteristics and the activity and selectivity of vanadium/titania systems depend mainly on the method and conditions of preparation, such as crystallographic titania structures, vanadium load and calcination temperature [81]. Several procedures have been proposed to prepare V_2O_5/TiO_2 catalytic systems: impregnation of V_2O_5 on TiO_2 aerogels [85, 86], co-gelling of vanadia and titania [87, 88], and preparation of aerogels by two-step procedure [89].

The molybdate and tungstate promoted TiO_2-ZrO_2 catalysts, containing 10 wt.% MoO_3 or WO_3 were prepared by a suspension impregnation method [79]. To incorporate molybdenum- or tungsten-oxide, the required quantities of ammonium heptamolybdate or ammonium metatungstate were dissolved in excess water and the finely powdered oven dried hydrous titania–zirconia support was added to this solution and refluxed at 110 °C for 2 h. The excess water was then evaporated on a water-bath with continuous stirring. The resulting sample was oven dried at 100 °C for 12 h and calcined at 800 °C for 5 h in air atmosphere and stored in dry nitrogen atmosphere.

b) Sulfated Oxides

Recently sulfated metal oxides have been used as catalysts in several organic transformations due to its high efficiency [90], non-pollutant and non-corrosive aspects, lower cost and greater stability. It has the strongest acidity on the surface. It can be easily recovered from the reaction mixture by simple filtration and re-used after activation. Thus the sulfated tin oxide (SO_4^{-2}/SnO_2) is superior to other known acids [91]. Sulfated tin oxide has a higher surface area as compared to aluminosilicates or zeolites. Thus it is widely used in hydro cracking of paraffin's, dehydration of alcohol, esterification, and alkylation of olefins, photochemical catalysis and protection of aldehyde, ketones and alcohols [92] and olefin polymerization [93]. Sulfated tin oxide samples were prepared by impregnation of $SnO_2.XH_2O$ gels with the appropriate amount of 1 M H_2SO_4 solution to obtain, 6, 12 and 18 wt.% sulfates [94]. These gels were dried at 110 °C for 24 h, followed by calcination at 600 °C for 3 h. These rutile-catalysts show good results for phenol hydrolysis. Mesoporous sulfated titania xerogels with certain structural and catalytic properties were prepared by the modified sol-gel method [95] using surfactant as templates. The sulfate loading of these catalyst materials were varied in the range of 0 to 50 wt% SO_4^{2-}.

The properties of catalysts strongly depend on the preparation procedures. The high porosity and the large specific surface area of materials prepared by the sol-gel method make them very attractive from a catalytic point of view [96]. The sol-gel method not only allows good control of the characteristics of the support, but also offers the possibility of preparing oxide-supported metal catalysts from a homogenous solution containing both the catalytic metal precursor and the support precursor [97]. A metal alkoxide undergoes hydrolysis and subsequent condensation in an alcohol solvent, forming a polymer oxide network entraining the alcohol. For catalytic uses, the liquid solvent must be removed from the alcogel. In conventional drying, a liquid-vapor interface forms in the pores and the corresponding surface tension collapses the oxide network, thereby reducing the surface area [26]. To prevent the collapsing of the oxide network, most work usually apply the removal of solvents by supercritical drying [26, 97]. Another different approach during the removal of solvents is by using surfactants as a template to fabricate porous materials and to improve the alcogel properties [97, 98]. By immersing the wet alcogels in surfactant solution before drying, the oxide network will be prevented from collapsing. Metal oxides, especially ZrO_2, TiO_2, Fe_2O_3, and mixed oxides modified by sulfate ions have been found to be promising catalysts for low temperature esterification, isomerization, alkylation, and cracking reactions [78]. Hino and Arata [99], and Arata and Hino [100] further reported that the solid superacids could be synthesized by supporting WO_3 or MoO_3 on ZrO_2 or TiO_2 under certain preparation conditions.

It has been known that sulfate addition is widely used to enhance the surface acidity of the solid catalysts, known as solid superacid catalysts. Some sulfated metal oxides and mixed metal oxides, such as SO_4^{2-}/ZrO_2, SO_4^{2-}/SnO_2, SO_4^{2-}/TiO_2, SO_4^{2-}/Fe_2O_3, SO_4^{2-}/SiO_2-TiO_2, and SO_4^{2-}/SiO_2-ZrO_2 possessed superacidity properties [101-104]. Sulfated metal oxides have also been impregnated onto silica, alumina, and porous materials [105-107]. The strong acid properties are related to the sulfate ions. The types of acid sites and some structural model have been presented and reported as either Lewis or Bronsted acids. These solid superacids have been applied as acid catalysts in the reactions such as isomerization, cracking, alkylation, acylation, and esterification [108-111].

c) Metal Promotion

The addition of some metals to these sulfated oxides can lead to catalysts that are more active and more resistant to deactivation [6, 112]. Therefore, several metals such as platinum, palladium, aluminum, nickel, cobalt, and manganese have been added to sulfated metal oxides, producing active and selective catalysts [113]. Recently, Gao et al. [114, 115] found that the addition of aluminum to sulfated zirconium oxide significantly enhances the activity and stability of the catalysts for n-butane isomerization. They also found that the addition of Al^{3+} to pure oxides (zirconium, titanium, and iron oxides) before sulfation increases the acidity, activity, and stability of the catalyst for n-butane isomerization. The promoting effect of transition metals is generally recognized to be related to a bifunctional mechanism associated with the isomerization reaction, where the presence of transition metals accelerates the dehydrogenation of the reactants [114, 116].

The metals Au, Pt, Pd, Ir, Rh, etc. set on oxides are extensively used as heterogeneous catalysts; their catalytic activity, however, decreases when they suffer poisoning, sintering, and volatilization [117], being sintering the most common process of deactivation. The rate and extent of sintering depend on the oxide support, the supported metal, the catalyst promoter, and the reaction temperature as well as the reaction atmosphere [118]. The gold particles of these catalysts are more reactive when the support is a reducible metal oxide such as TiO_2 [119, 120]. Also, the gold particles are highly reactive when their dimensions are between 2 and 5 nm [120-122]. Due to the low melting point of gold particles with these low dimensions, they sinter themselves easily at low temperatures limiting the practical applications of the corresponding catalysts [123-125].

It is well known that a strong metal–support interaction will occur on TiO_2-supported group VIII metals after high-temperature reduction [126], which plays important roles as follows: (1) metals dispersed on TiO_2 support were significantly more stable than on other supports [127]; (2) a special interface between metal and partially reduced TiO_2 support was formed that includes a metal atom, a Ti^{3+} ion, and an anion vacancy [128, 129]; (3) electron migration from the support to metal occurred and the extra electrons saturated the unfilled d orbital of the metal atoms [130]; (4) partially reduced TiO_2 migrated to the surface of metals [131, 132]. Fluorine shows high electronegative and small atomic radius. Modification of solid acid supports (Al_2O_3, SiO_2, SiO_2–Al_2O_3, TiO_2, etc.) using fluoride has received much attention [133 135]. The modification of the supports can change the interaction with the active phase, the surface acidity and consequently the catalytic activity. Some studies suggested that fluoride modification of support could partly promote the catalytic activity such as hydrodesulfurizaiton (HDS), hydrodenitrogenation (HDN) and hydrocracking. The previous studies [136] show that Co–Mo catalysts supported on fluorine-modified alumina and fluorine-modified titania performed higher HDS activity than those unmodified catalysts. They concluded that the addition of fluoride to the catalyst support results in higher dispersion of the molybdenum sulfide phase so that lead to more active HDS activity.

Tin oxide modified with copper was prepared [68] by co-precipitation methods with different loading levels of Cu (5, 10, 15 and 20%wt). This method depends on dropwise addition of a 1:1 ammonium hydroxide solution to a warm (60-80 °C) 0.3 M aqueous solution of tin(IV) chloride ($SnCl_4.5H_2O$) mixed with the calculated amount of copper acetate $Cu(CH_3COO)_2.H_2O$ with a continuous stirring till pH = 8 is reached. The precipitate was left over-night, filtered, washed, and then dried at 110 °C till constant weight. The products are calcined at 600 °C for 3 h and by investigation using XRD, it is found that they have rutile structure.

4.2 Morphology and physical properties

Titanium dioxide presents three crystalline phases: anatase (tetragonal), rutile (tetragonal) and brookite (orthorhombic). Submitting TiO_2 to high thermal treatments provokes the transformation of anatase or brookite into the more stable rutile phase. The physicochemical properties of TiO_2 are different for different polymorphs and are also influenced by their structure. The control of the crystalline structure is thus very important for TiO_2 to be applied. Rutile is the only stable phase in three modifications in all the temperature range and the other two modifications can exist only metastable phase. Each structure shows different physical properties, such as refractive index and chemical and photochemical reactivity. Each application usually requires specific crystalline structure and size of the primary particle.

Rutile is tetragonal and fully described by three parameters: a, c and, u. The parameters a and c are shown in Figure 2, u defines the fractional coordinate of the O atom [137].

The three polymorphic forms of titanium dioxide are shown in Figure 3. Crystallographic data on the three oxide modifications are summarized in Table 3. In all three oxide modifications, each titanium atom is coordinated to six almost equidistant oxygen atoms, and each oxygen atom to three titanium atoms [138]. In the case of anatase, the TiO_6 octahedron is slightly distorted, with two Ti-O bonds slightly greater than the other four, and with some of the O-Ti-O bond angles deviating from 90°. The distortion is greater in anatase than rutile. The structures of anatase and rutile crystals have been described frequently in terms of chains of TiO_6 octahedra having common edges. Two or four edges are shared in rutile and anatase, respectively. The third form of titanium dioxide, brookite, the inter-atomic distances and the O-Ti-O bond angles are similar to those of rutile and anatase. The essential difference is that there are six different Ti-O bonds ranging from 1.87 to 2.04 Å.

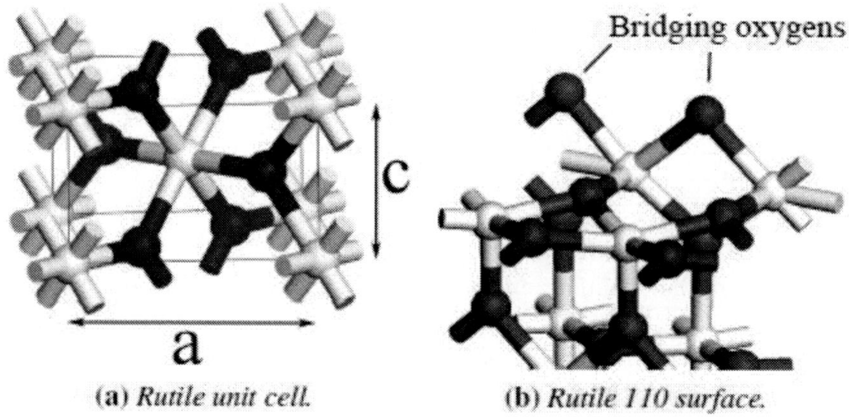

Figure 2. Unit cell of rutile TiO2 and the stoichiometric [110] surface. Titanium atoms are white, oxygen atoms are black. The oxygen atoms at the top of (b) are bridging oxygen atoms. When there is no oxygen atom present the site is an oxygen vacancy [137].

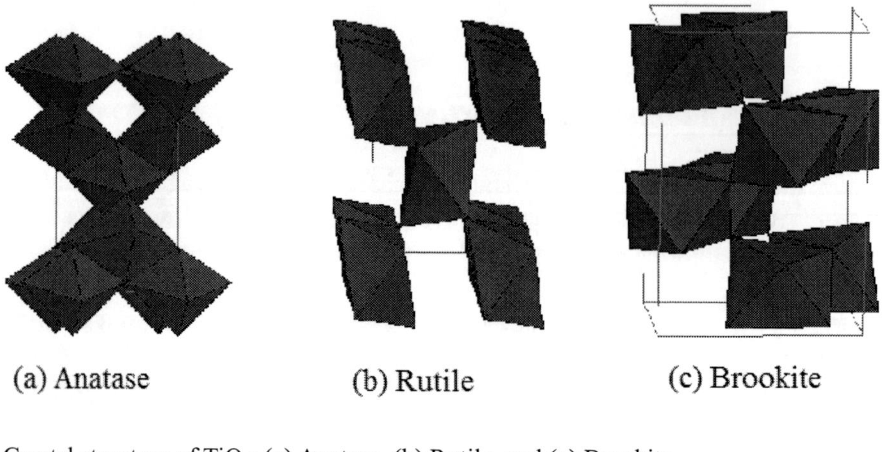

Figure 3. Crystal structure of TiO_2; (a) Anatase, (b) Rutile, and (c) Brookite.

Table 3. X-ray data on TiO_2 modification [139]

	Space group	Z	Cell parameters (Å) a	b	c	Ti-O (Å)*
Anatase	$C_{4h}^{19} = C4/amc$	8	5.36		9.53	1.91 (2)
						1.95 (4)
Brookite	$D_{2h}^{15} = Pbca$	8	9.15	5.44	5.14	1.84-2.03
Rutile	$D_{4h}^{14} = P4_2/mnm$	2	4.954		2.959	1.944 (4)
						1.988 (2)

* The number in parentheses refers to the number of equivalent oxygen atoms at the stated distances from a titanium atom.

Accordingly, there are 12 different O-Ti-O bond angles ranging from 77° to 105°. Brookite is formed by joining together the distorted TiO_6 octahedral sharing three edges. All three oxide modifications are birefringent; anatase is uniaxial negative, brookite is biaxial positive and rutile is uniaxial positive. Further data are given in Table 3.

To elucidate the accurate temperature of anatase-rutile transformation, tests of calcinations were carried out at 600, 700, 800, 870, 940 and 970°C for 3 hours and at 980°C for 4 hours [139]. The results show that no anatase-rutile transformation was detected for 3 h of calcinations at 970°C (Figure 4). The Figure 5 shows the XRD patterns obtained for calcinations at 980°C, considering different periods. The XRD patterns show that the rutile phase starts to form only at 980°C. After 4 hours, rutile is the unique phase present.

Most heterogeneous catalysts are porous solids. The porosity arises from the preparation methods of the solids. For example the precipitation from a solution originates precursor particles that agglomerate and form a porous structure.

Crystallization in hydrothermal conditions can produce zeolites, in which the peculiar disposition of the 'building units' generates intra-crystalline cavities of molecular size. As mentioned above, during the thermal treatment the elimination of the volatile material (burning, evaporation) produces cavities that represent both the result of the solid rearrangement and the exit way of the removed material. Different shaping procedures, employed to obtain the catalysts suitable for industrial reactors (tableting, extrusion, spray drying, etc.), give rise to stable aggregates of particles that contain a porous structure. Then a typical catalyst contains one or more group of pores whose size and volume depend on preparation method. The pores are classified in different classes depending on their size: micropores ($\varphi < 2$ nm), mesopores ($2 < \varphi < 50$ nm) and macropores ($\varphi > 50$ nm).

The knowledge of the morphological parameters is very important to understand the catalyst evolution during the preparation procedure and to give a feedback useful for modifying the method and to obtain the desired results. It is a matter of fact that a better comprehension of the catalytic behaviour requires the knowledge of the morphological characteristics. In fact the catalytic process takes place at the surface. To reach the surface sites the molecules of the reagents must diffuse through the porous system, while the reaction products follow the opposite path. Mass transfer process inside the granules depends on pore size (bulk diffusion in macropores, Knudsen diffusion in mesopores and molecular diffusion in micropores).

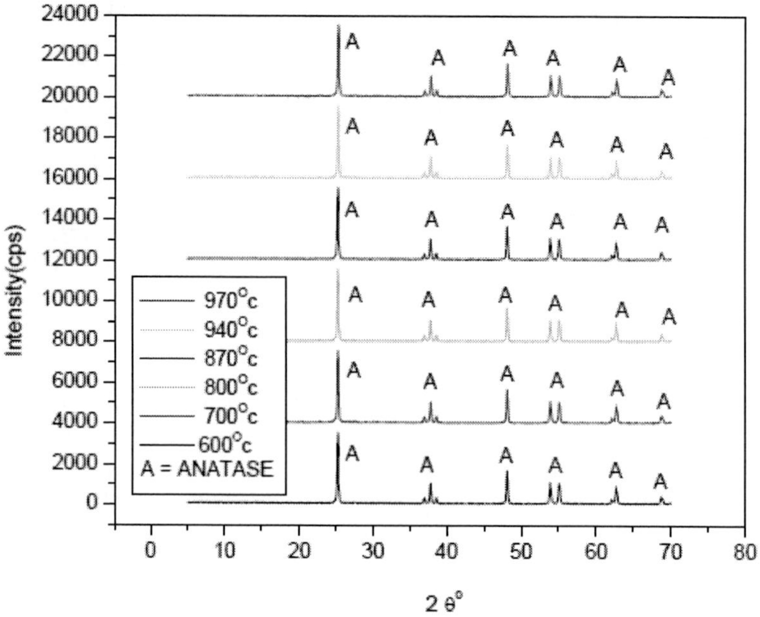

Figure 4. XRD Patterns of TiO_2, calcined for 3h at 600, 700, 870, 940 and 970 °C [139].

Also deactivation phenomena (i.e. organic material deposition) are greatly influenced by pore size; deposition of carbonaceous material with blockage of micropore mouth and coverage of the whole wall in meso and macropores are typical examples. The morphological characteristics of interest are the specific surface area, the specific pore volume, and the distribution of the area and of the pore size. The methods for pore determination are numerous; the right choice depends on type of pore and, as will be clarified later, on the use we want to do of such data. The most utilized technique tat used for this purpose suggested by BET [140]. The porous structure enables the solid to have a total surface much higher than that corresponding to the external one. Most common catalysts have a specific surface area between 1 and 1000 $m^2\ g^{-1}$, while their external specific surface area is in the range 0.01-10 $m^2\ g^{-1}$. Titania and modified Titania have different specific surface areas depending on the source of it, the methods used for preparation and the type and the amount of modifiers. The specific surface areas of the pure titania and titania mixed with zirconia prepared by sol-gel [78] are reported in Table 4. For TiO_2 and ZrO_2 the corresponding specific surface areas were 56.5 $m^2\ g^{-1}$ and 51.9 $m^2\ g^{-1}$, respectively. In mixed oxides, the BET specific surface area varies depending of the ZrO_2 content, which were 110.2 (90:10), 182 (50:50) and 79.3 $m^2\ g^{-1}$ (10:90).

On the other hand, the type of modifier has an effect on the structural data as shown in Table 5 [141]. For the Pt modified titania catalyst, the surface area, the pore volume and porosity were slightly higher than that of the pure TiO_2. For the binary metal oxide catalysts, the addition of a second metal oxide increased the specific surface area of the catalyst. Especially, the surface area of the SiO_2–TiO_2 catalyst was remarkably increased up to 229.8 $m^2\ g^{-1}$. However, the effects of the additives on the pore volume and porosity were different. SiO_2–TiO_2 and ZrO_2–TiO_2 catalysts gave higher pore volume and porosity, while MoO_3–TiO_2 catalyst presented the lowest pore volume and porosity, which were 0.095 $cm^3\ g^{-1}$ and 27%, respectively.

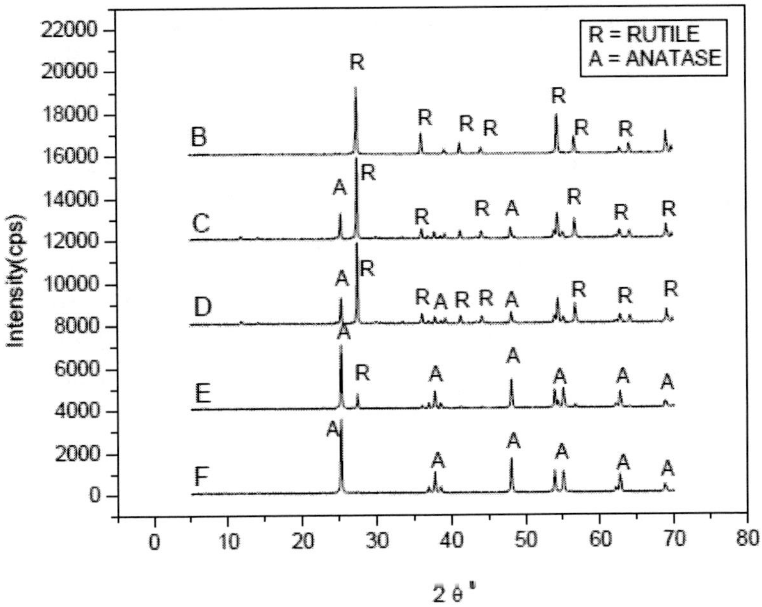

Figure 5. XRD Patterns of TiO_2, calcined at 980 °C for 4h (B), 2h (D), 1h (E) and 0.5 h (F) at 600, 700, 870, 940 and 970 °C [139].

Table 4. Specific surface area of pure and mixed titania-zirconia [78]

Catalyst	**S_{BET} ($m^2.g^{-1}$)**
ZrO_2	51.9
TiO_2-ZrO_2 (10:90)	79.3
TiO_2-ZrO_2 (50:50)	182
TiO_2-ZrO_2 (90:10)	110.2
TiO_2	56.5

Table 5. Textural data of pure and modified titania [141]

Sample	SBET ($m^2.g^{-1}$)	VP ($cm^3.g^{-1}$)	Porosity (%)	Crystallite size (nm)
TiO_2	72.1	0.131	33.8	8.5
Pt/TiO_2	80.4	0.140	35.3	9.4
SiO_2-TiO_2	229.8	0.156	37.8	6.2
ZrO_2-TiO_2	132.7	0.137	34.8	6.2
WO_3-TiO_2	129.6	0.128	33.3	7.8
MoO_3-TiO_2	139.7	0.095	27.0	6.7

For sulfated titania, it is found that the addition of sulfate has great effect on the porosity and surface area of the samples [95]. Transmission electron micrographs of the prepared sulfated titania xerogels are shown in Figure 6. The addition of sulfate increased the surface area and total pore volume of the samples significantly. Adding 5 mol% SO_4^{2-} to the gel increased the surface area to about 70.3 $m^2\ g^{-1}$. Further sulfate addition to 10 and 25 mol% increased the surface area to 95.6 and 85 $m^2\ g^{-1}$ respectively. The highest surface area was observed from sample TS10. The increase of sulfate loading also increased the total pore volume of the samples. The initial pore volume of titania xerogel was 0.0019 $cm^3\ g^{-1}$. The total pore volume after sulfate loading of 5, 10 and 25 mol% was 0.028, 0.039 and 0.06 $cm^3\ g^{-1}$ respectively. Sample TS0, which is pure titania oxide without addition of sulfate ions, exhibit reflections arising from both rutile ($2\theta = 27.5^\circ$, 36° and 55.6°) and anatase ($2\theta = 25.4^\circ$ and 47.9°). Increasing the sulfate loading above 5wt%, however, caused a sharp increase in the intensity of the anatase phase and a significant loss in the rutile phase. The addition of sulfate loading also gave rise to the crystalline titania sulfate.

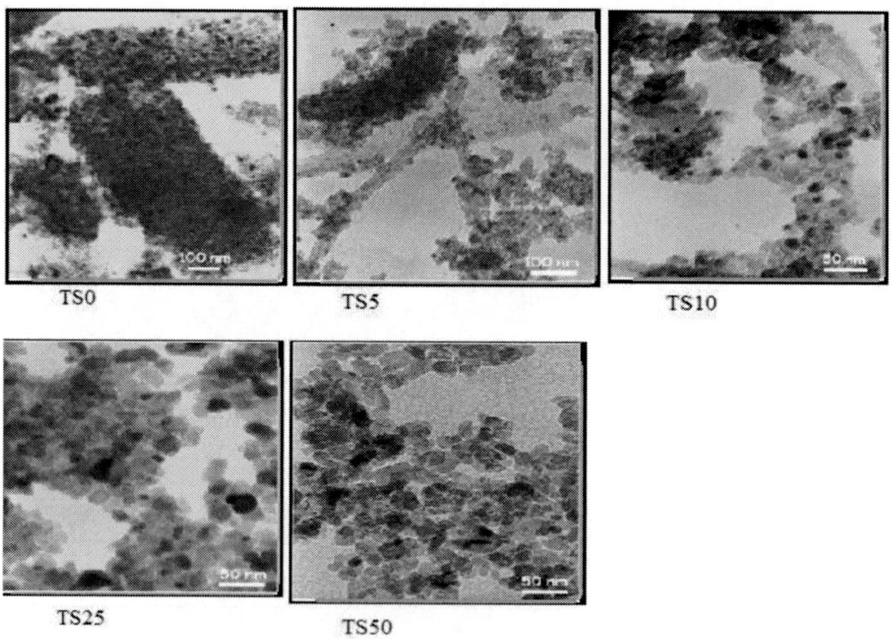

Figure 6. Transmission electron micrographs of the modified sulfated Titania xerogels at various sulfate concentrations [95].

Table 6. textural data for sulfated Titania [95]

Sulfated titania Xerogels samples	*Sulfate (wt %)*	*SBET ($m^2.g^{-1}$)*	Vp (cm^3g^{-1})
TS0	0	4.71	0.0019
TS5	5	70.27	0.028
TS10	10	95.64	0.039
TS25	25	85.06	0.06

Table 7. Textural data for tin oxide-based catalysts [68, 94, 135, 142]

Sample*	S_{BET} m^2g^{-1}	Vp_{cum}	Vp_{tot}	Crystallite Size (Å)
		cm^3g^{-1}		
TO	11	0.016	0.0325	185
5CuTO	11	0.006	0.0093	307
10CuTO	12	0.007	0.0108	177
20CuTO	14	0.008	0.0124	156
6STO	35	0.036	0.0387	88
12STO	60	0.035	0.048	50
18STO	88	0.066	0.1036	31
5CuSTO	11	0.035	0.0510	185
10CuSTO	13	0.021	0.041	160
20CuSTO	15	0.013	0.0201	145

* [TO= Tin oxide; STO= Sulfated Tin Oxide; CuTO = Tin Oxide modified with Copper; CuSTO= Tin Oxide sulfated with 6%SO42- modified with Copper and all the numbers refer to the amount of modifier].

For the preparation of tin (IV) oxide modified with copper ions followed by sulfation [68, 94, 135, 142], it is found that the addition of Cu or sulfate ions to tin oxide has no effect on the crystalline structure (rutile phase) but there is a slightly higher degree of crystallization after modifying with 5%Cu and by increasing the loading level of Cu there is a decrease in the degree of crystallization and, consequently, increase the specific surface area. The surface texture study showed, also, that the addition of sulfate resulted in modified surface area and porosity (Table 7).

4.3. Acidity and Structure

It has been found [143] that superacid sites are formed during sulfation and phosphation of the oxides, regardless of the source of both sulfur and phosphorus, i.e., H_2S, H_2SO_4, $(NH_4)_2SO_4$, or SO_2 for sulfur and H_3PO_4, $(NH_4)_2HPO_4$, $(NH_4)H_2PO_4$, or P_2O_5 for phosphorus, provided that the optimal calcination temperature for each one is used.

Several authors [111, 144-148] reported that the incorporation of some anions like sulfate or phosphate to metal oxide can increase the acidity of these catalysts and have been classified as superacids. The acid strength for the sulfated metal oxides is high, stronger than 100% sulfuric acid [144, 145]. It has been reported that for SO_4^{2-}/TiO_2, the Hammet acidity is, $H_0 \leq -14.52$, as compared with that of H_2SO_4, $H_0 \leq -12$ [149]. The sulfate ion can be introduced from H_2SO_4, $(NH_4)_2SO_4$, SO_2, SO_3, and H_2S. It was reported that the existence of covalent S-O bonds in sulfur complexes formed on metal oxides are responsible for the generation of acidity [150-152].

The superacid properties of sulfated titania were considered to be responsible for the high catalytic activity in various acid-catalyzed reactions such as the alkylation of derivatives of benzene, cracking of paraffins, and dimerization of ethylene [153-157].

Saur et al. [158] suggested the utility of sulfated Titania for NO reduction and CO oxidation because of its non effectiveness to sulfur poisoning. The nature of acid sites may be defined by the presence of surface protons leading to the Brønsted sites or cationic centers due to unsaturation in coordination, which explain the Lewis acidity. The surface acidity of the sulfated samples is determined by the irreversible adsorption of pyridine. Pyridine adsorbs only on the strong acid sites [159]. The results from the qualitative measurements of the surface acidity of the samples are shown in Table 8. The highest sulfate acidity, 3.61 mmol g^{-1} was obtained for sample TS10 which also shows having the highest specific surface area, as measured by BET method, Table 8. The specific surface area correlates with the acidity of the samples. High surface areas may increase the possibility to adsorb more base molecules which in result gives a high acidity.

The type of acidity sites of the samples was defined by pyridine IR spectroscopy, which is shown in Figure 7 for the sulfated titania catalyst at various sulfate loadings. Prior to characterization, the samples were calcined at 500°C for 4 hours. Indicative of Lewis acidity in the band range of 1392 to 1394 cm^{-1} also appeared for all samples. From the sharpness of the bands, it is indicated for all samples that the Lewis acid site is stronger than the Brønsted acid site.

This shows that all samples posses both the Brønsted acid and Lewis acid site on their surface. Upon various sulfate loading, the results show that the Lewis acid bands sharpened only due to the addition of sulfate from 5wt% to 10wt%, shown by the spectra of sample TS10 compared to TS5.

Addition of sulfate loading to 25wt% and 50wt% in sample TS25 and TS50, the intensity reduced. On the other hand, the Brønsted acid bands remain constant and show no increase of also show the increase in sharpness, although not as significant as the Lewis acid peak. In the sulfated metallic oxides there may be as many Lewis acid sites as Brønsted acid sites.

The superacidity of these materials is attributed to the Brønsted acid sites, created or already existing, whose acidity is increased by the presence of neighboring strong Lewis acid sites. The strength of these Lewis acid sites is due to an inductive effect exercised by sulfate on the metallic cation, which becomes more deficient in electrons [160], as seen in the Scheme 2. Total number of acid sites and their relative strength, for the catalysts can be measured by means of a potentiometric titration with 0.1N n-butylamine. To explain the obtained results, it was suggested that the initial electrode potentional (Ei) indicates the maximum acid strength of the sites.

Table 8. Surface acidity of sulfated titania catalysts [95]

Sulfated Titania Xerogels samples	Sulfate (wt %)	Surface acidity using Py adsorption ($mmol.g^{-1}$)
TS0	0	1.22
TS5	5	3.61
TS10	10	2.44
TS25	25	2.42

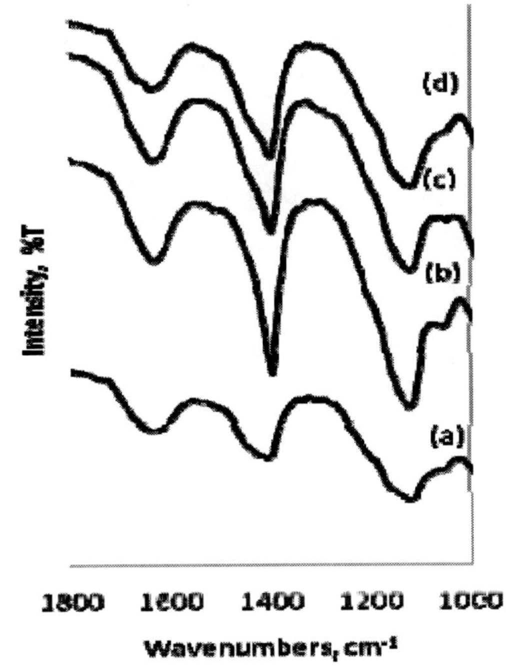

Figure 7. FTIR spectra of sulfated titania catalyst samples at various sulfate concentrations after pyridine adsorption (a)TS5 (b)TS10 (c) TS25 and (d) TS50 [95]

The value of meq amine/g solid, where the plateau is reached in titration curves, indicates the total number of acid sites [161-163]. On the other hand, the acid strength of these sites can be classified according to the following scale [163, 164]: Ei > 100 mV (very strong sites), 0 < Ei < 100 mV (strong sites), 100 < Ei < 0 (weak sites) and Ei < 100 mV (very weak sites).

Data cited in Table 9 and Figures 7-9 show the potentiometric titration results for tin(IV) oxide based catalysts [68, 94, 135, 142]. From these results one can concluded that the TO sample has weak acid sites and its maximum strength is equal -65 mV. The addition of Cu into pure tin oxide causes an increase in its acidic strength, but still in weak range, to become -31, -12 and +2 mV for the samples 5CuTO, 10CuTO and 20CuTO [135]. As to sulfated tin oxide samples [94] (6STO, 12STO and 18STO), it is found that the addition of sulfate into tin oxide can increase the acidic strength gradually and create very strong acid sites on the surface of tin oxide. The acidic strength is +47, +112 and +144 mV for 6STO, 12STO and 18STO, respectively.

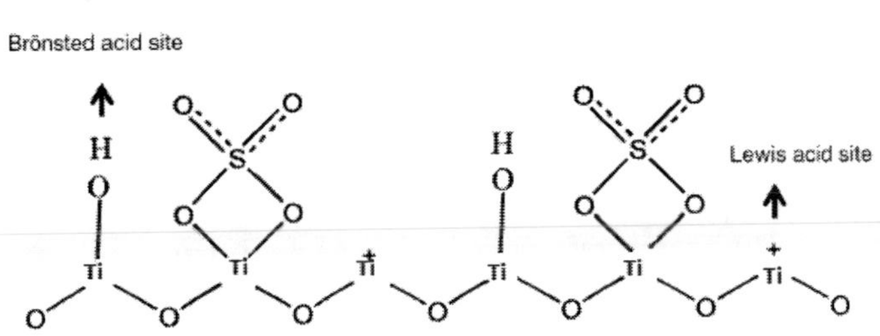

Scheme 2. Bronsted and Lewis acid sites in sulfated titania.

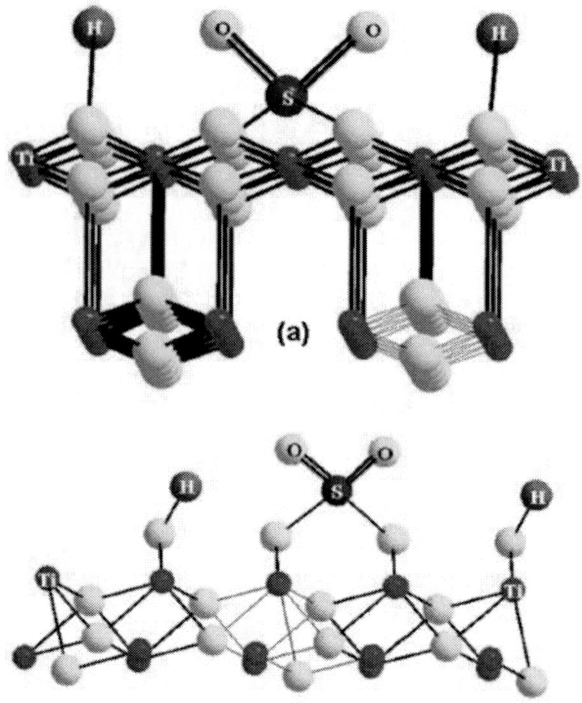

(b)

Scheme 3. Surface structures for the sulfate formation on hydrated titania: (a) chelating bidentate; (b) bridged bidentate.

On the other hand, for copper supported sulfated tin oxide catalysts [68] (5CuSTO, 10CuSTO and 20CuSTO) it is clearly that these samples have higher acidic strength in comparison with sulfated tin oxide (6STO) and have higher acidic strength in comparison with copper supported tin oxide (xCuTO). Their acidic strength is +50, +58 and +62 mV. Thus the role of copper in acidic strength has positive effect when added to sulfated tin oxide, i.e. 6STO catalyst has +47 mV and become +50, +58 and 62 mV after addition of 5, 10 and 20%Cu respectively.

Table 9. Potentiometric titration data for the catalysts [68, 94, 135, 142]

Sample	Ei (mV) =maximum acid strength	No of acid sites (meq g^{-1})
TO	- 65	0.11
6STO	+ 47	0.30
12STO	+ 112	0.40
18STO	+ 144	0.40
5CuTO	- 31	0.15
10CuTO	- 12	0.30
20CuTO	+ 2	0.40
5CuSTO	+ 50	0.50
10CuSTO	+ 58	0.50
20CuSTO	+ 62	0.70

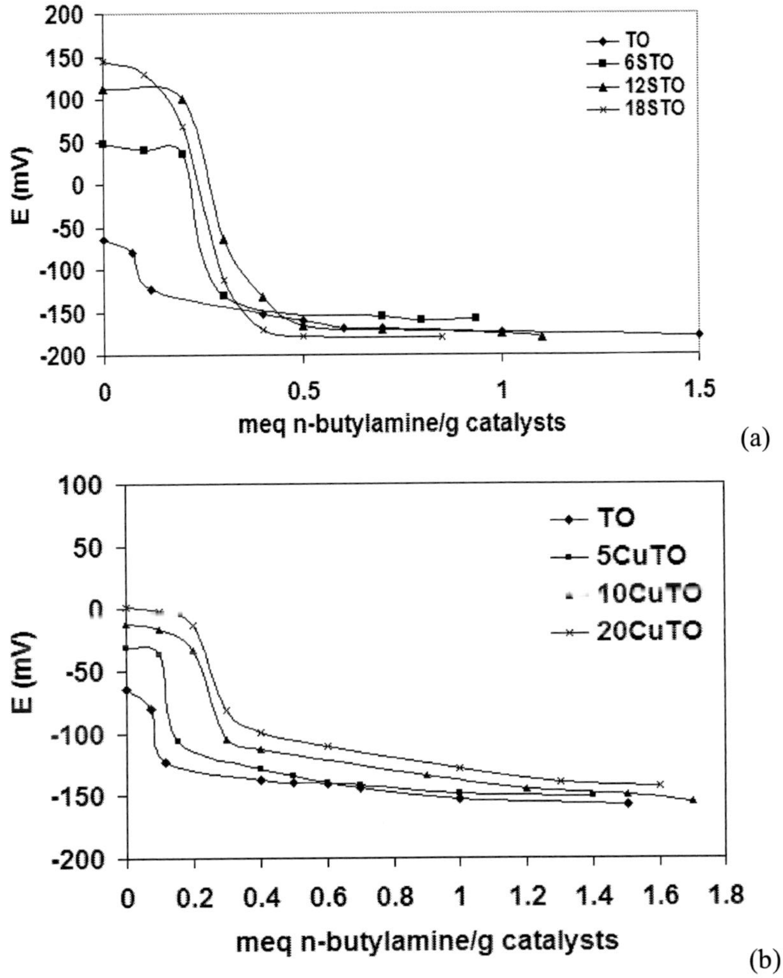

Figure 8. Potentiometric titration curves for (a) sulfated tin oxide catalysts [94], and (b) Cu supported tin oxide catalysts [135].

4.4 Catalytic Activity

Acid catalyzed reactions have played a significant role in the synthesis of a variety of compounds. While most such reactions were initially run with simple protic or Lewis acids, the use of such materials now presents serious environmental problems concerning the disposal of the unused acids and their salts. To remedy this situation, solid acids have been developed to promote a number of different reactions.

Generally, these solid acids are easy to handle and are non-toxic, non-volatile and non-corrosive [164]. Superacid catalysts can be applied for different catalytic reactions [165-167] such as; esterification, isomerization, alkylation, polymerization, cracking, other acid catalyzed reactions (acylation, etherification, acetalization, nitration, disproportionation, etc.). The isomerization reaction of n-butane is not catalyzed even by 100% H_2SO_4 at room temperature. However, SO_4^{2-}/ZrO_2 [19] SO_4^{2-}/TiO_2 and SO_4^{2-}/Fe_2O_3 [168] were found to catalyze the isomerization reaction at room temperature, the main product being 2-methylpropane.

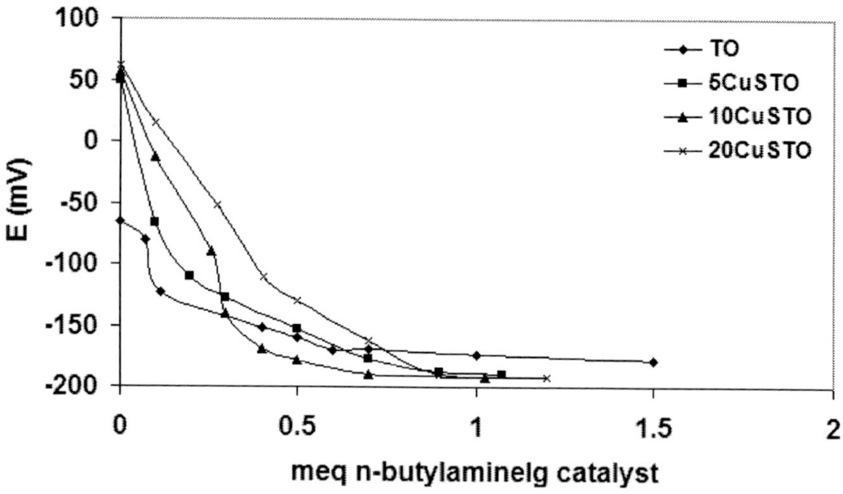

Figure 9. Potentiometric titration curves for Cu supported sulfated tin oxide catalysts [68].

The corresponding catalytic properties of the sulfated titania samples towards citronellal isomerization was explored [95]. Isomerization of citronellal was conducted over 4 types of sulfated titania xerogels (TS5, TS10, TS25 and TS50) at various sulfate concentration (5 wt%, 10 wt%, 25 wt% and 50 wt%, respectively). Fresh titania oxide without sulfate, TS0 was also used as catalyst for comparison. Besides varying the catalyst at different sulfate loadings, the effect of solvent in the isomerization reaction was also studied.

Table 10. Activity and selectivity towards (-)-isopulegol without solvent

Catalysts	Sulfate Wt%	Conversion of Citronellal %	Selectivity to (-)-Isopulegol
TS0	0	52	10.2
TS5	5	99	0
TS10	10	99	1
TS25	25	99	2
TS50	50	98	9

The solvent used was toluene, which is a non-polar solvent. The effect on solvent towards the catalyst activity and selectivity was compared to those without solvent.

Scheme 4 show the conversion of (+)-citronellal to isopulegol isomers. From Table 10 where the reaction is conducted without any solvent, it is shown that the sulfate concentration has an effect on the selectivity to (-)-isopulegol, the higher the sulfate concentration the higher the yield of isopulegol. Isomerization over strong acid catalyst have a result in nearly 100% conversion of citronellal, however the selectivity to isopulegol is poor, still below 10%. Strong acids are active but have low selectivity due to side reactions such as cracking, dehydration and etherification [169]. These side reactions have higher reaction rate compared to isomerization, therefore the superacid catalyst tend accelerate the formation of side products of isopulegol from dehydration, cracking and esterification. As the reaction time increases, the yield of side products also increases, which causes the selectivity of the reaction to isopulegol to decrease. Moreover, besides the side reactions, the isomerization reaction itself also generates other isomers from isopulegol such as neo-isopulegol, iso-isopulegol and neo-isopulegol (Scheme 4). This contributes to the low selectivity of catalyst towards (-)-isopulegol. However, it is also shown that in the reaction without any sulfate loading (TS0) and solvent, the selectivity is much higher as 10.2%. This may be explained that because there is no solvent loading, the acidity is mild, and less side reaction has occurred, therefore the selectivity is much higher. However, the activity of catalyst TS0 in no solvent condition is much lower given the conversion of citronellal of only 52%.

In Table 11, it is shown that the activity and selectivity of catalyst towards isopulegol very much increased. The conversion of citronellal is mostly 100%, except for catalyst TS0 which only gives 27% of conversion. The highest selectivity for (-) isopulegol was attained for catalyst TS5 at 54.6% after 30 minutes of reaction. At higher sulfate concentrations the selectivity tends to decrease. From the pyridine IR spectra (Figure 7) it is observed that TS5 has both Brønsted and Lewis acid site at equal strength. This shows that good activity and selectivity for isomerization of citronellal are obtained for catalyst samples having both Brønsted and Lewis acidity at equal strength. Samples having stronger Brønsted acid sites decrease the selectivity, although the activity is still good. The result of this work was similar to the ones conducted by Yadav and Nair [112], whereas they used sulfated zirconia as catalyst. Given the same solvent and reaction time, their result was 91% citronellal conversion and the selectivity was 65%. Chuah and coworkers [169] also has conducted the same reaction using the same catalyst, sulfated zirconia. After 10 minutes of reaction, the conversion reached 100%, with selectivity to (-) isopulegol of 61%. They have concluded that strong acid catalyst, which have strong Bronsted site, are not favorable for the reaction since it generates more side products and decreases the selectivity.

Table 11. Activity and selectivity towards (-)-isopulegol with toluene as solvent

Catalysts	Sulfate Wt%	Conversion of Citronellal %	Selectivity to (-)-Isopulegol
TS0	0	27	20.1
TS5	5	100	54.6
TS10	10	100	27.6
TS25	25	100	48.9
TS50	50	100	45.4

The addition of toluene as solvent has given a significant effect both on the activity and selectivity of the catalyst (Table 10 and 11). The overall conversion of citronellal was generally >98%, except for the reactions using fresh titania oxide as catalyst. The selectivity was much higher for reactions in toluene than without solvents. In toluene, the selectivity has increased from below 10% to above 25% of (-)isopulegol. In other works [170], it is more favorable to use a polar solvent such as acetonitrile and 2-propanol.

Hydroxylation of phenols give different three compounds according to Scheme 5, these products are catechol (CAT), hydroquinone (HQ) and resorcinol (RC). Hydroquinone is a reducing agent that can reduces unexposed silver ion to metallic silver and in turn is oxidized to benzoquinone (BQ) as in Scheme 5. Similar interconversions of HQ and BQ are important in certain biological oxidation-reduction reaction.

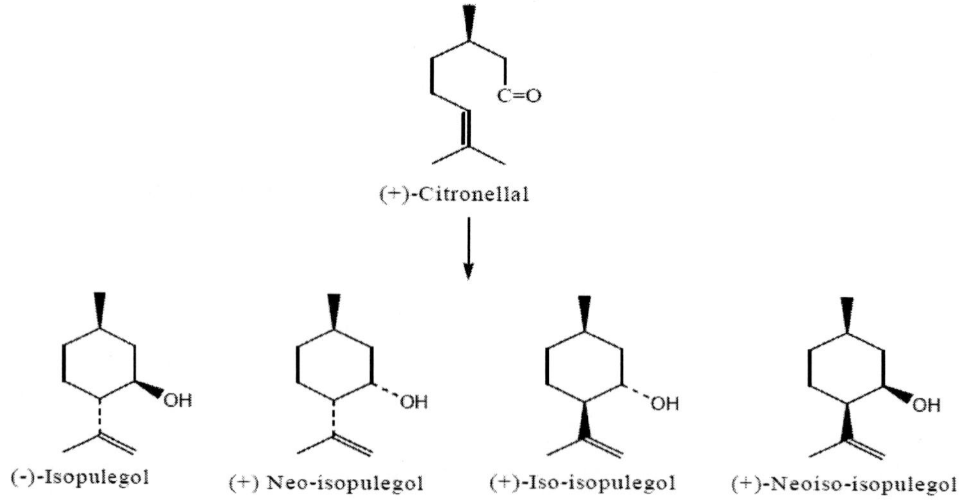

Scheme 4. Cyclisation of (+)-citronellal to isopulegol isomers.

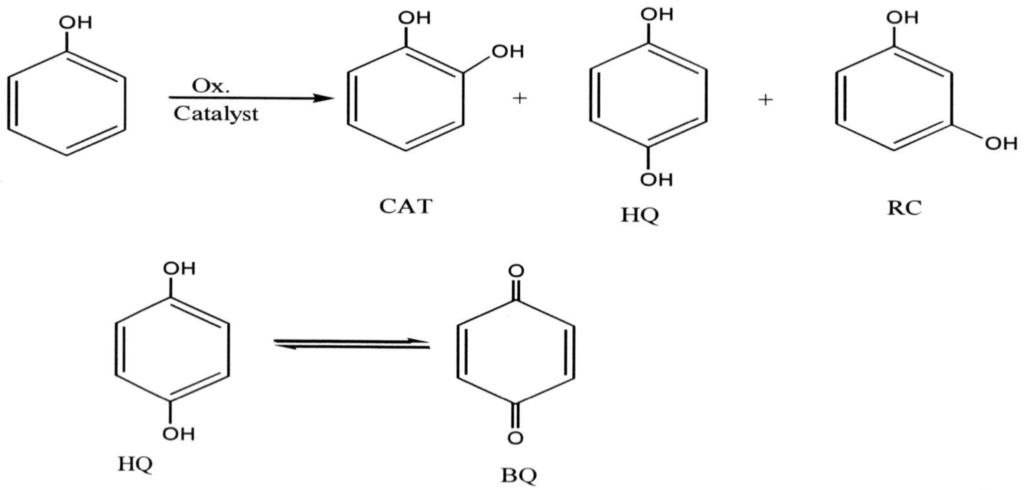

Scheme 5. Hydroxylation of phenol.

Catechol and hydroquinone are two high value phenolic derivatives widely used as photography chemicals, antioxidants, polymerization inhibitors and also in pesticides, flavoring agents, and medicine [171]. As other oxidation reactions, phenol hydroxylation has been usually carried out using stoichiometric oxidants. Because of the problems associated with updated methodologies, one of the major goals of chemistry research is the replacement of wasteful and toxic stoichiometric oxidants by ecologically friendly technologies. Therefore, phenol direct hydroxylation with H_2O_2 is of great interest to chemists nowadays. Hydroxylation of phenol with H_2O_2 using microporous materials and its commercial utilization by Enichem workers is a major advance in this area [172]. In the search for other transition metals, it has been reported [173] that copper sites are active in phenol hydroxylation due to their ability to activate H_2O_2 and to form free radicals. Copper containing porous materials appear to be promising catalysts for phenol hydroxylation with H_2O_2 [174].

The results of the catalytic activities using the SnO_2-based catalysts in methanol solvent at 60 °C were listed in Table 12. From these data it is clear that there is no products are detected when using TO and 18STO samples as catalysts, whereas the Cu supported samples show a high catalytic activity for hydroxylation of phenol. For 20CuTO catalyst, it is found that the phenol conversion is equal 14.8% and the reaction tends to give HQ as the main product. By comparing the catalytic performance of Cu supported sulfated tin oxide 20CuSTO with 20CuTO, the phenol conversion on 20CuSTO is 7% higher than that on 20CuTO. Also, the HQ selectivity in case of 20CuSTO (76.4%) is higher than that of 20CuTO sample (52.9%). This revealing that 20CuSTO is a good catalyst for the hydroxylation of phenol. These results are in agreement with the previous obtained data [175, 176]. It is found that the product selectivity is affected by the nature of the solvent [176]. Within protic solvents such as methanol, hydroquinone prevails over catechol, or while within an aprotic solvent such as acetone, catechol prevails over hydroquinone.

Table 12. Conversion % of phenol and selectivities % on some selected catalysts at 60 °C

Sample	Phenol Conversion (%)	Selectivity (%)			
		CAT	HQ	BQ	RC
TO	0.0	-	-	-	-
18STO	0.0	-	-	-	-
20CuTO	14.8	45.3	52.9	Trace	Trace
20CuSTO	22	21.8	76.4	Trace	Trace

Step 1

$$H_2O_2 + Cu^{2+}STO \longrightarrow HO_2^- + H^+ + Cu^{1+}STO$$

$$H_2O_2 + Cu^{1+}STO \longrightarrow {}^{\bullet}OH + OH^- + Cu^{2+}STO$$

Step 2

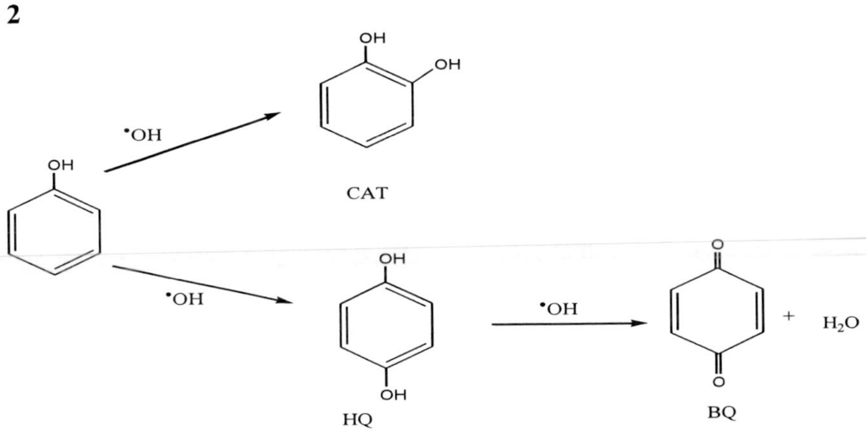

Step 3

$$HO_2^- + Cu^{2+}STO \longrightarrow O_2 + H^+ + Cu^{1+}STO$$

$$H^+ + OH^- \longrightarrow H_2O$$

Scheme 6.

Germain et al. [177] demonstrated that the acidity of the catalyst presents an important effect upon the induction period of the reaction and upon the autocatalysis phenomenon. In the same way, the acidity also intervenes in the over-oxidation of phenol.

It has been observed that the nature of the solvent affects the product distribution. The hypothesis is that in the presence of acetone, the interaction between the Bronsted sites of the surface catalyst and the phenol is favored.

The reaction mechanism for the oxidation of aromatic compounds employing transition metals has been previously studied [178]. The proposed reaction path involves following steps shown in Scheme 6. In step 1, the interaction of CuTO or CuSTO catalysts and hydrogen peroxide yields HO• and HO_2• species, via a redox mechanism. In step 2, hydroquinone and catechol are subsequently obtained in parallel processes, by the attack of HO• radicals to the aromatic ring, benzoquinone can be formed by the consecutive oxidation of hydroquinone. In addition, a little tar was formed during the procedure of phenol hydroxylation due to over-oxidation of BQ. In step 3, oxygen and water are formed in side reactions by the decomposition of H_2O_2.

CONCLUSIONS

In conclusion, solid superacids have been shown to be very effective from the viewpoint of activity, selectivity, reusability and in the preparation of various important chemicals which have a number of difficulties with homogeneous catalysts. To control the properties of these catalysts, some additives including metals, metal oxides and/or anions have been added. Therefore, several metals such as platinum, palladium, aluminum, nickel, cobalt and manganese or oxides such as alumina, silica, copper oxide, zirconia and vanadia or sulfates and phosphates have been added to metal oxides, producing active and selective catalysts. Oxides have the rutile structure e.g. GeO_2, SiO_2, TiO_2, SnO_2 and PbO_2 have been used as a precursor of solid superacids preparation.

Among them, TiO_2 and SnO_2 and their based materials are of importance for preparation of solid superacid catalysts because of their strong oxidizing power, high photo-stability and redox selectivity.

Various methods e.g. impregnation, hydrothermal, sol-gel, co-precipitation and multi-gelation have been used for synthetic TiO_2 and SnO_2 and their based catalysts. The obtained results reveal that these materials have higher acidity than concentrated H_2SO_4 and their acidity and morphology can vary depending on the synthetic methods, the type and amount of additives. They can be used for several acidic reactions e.g. alkylation, isomerization, estrification, cracking and condensation.

REFERENCES

[1] T.M. Jyothi, K. Sreekumar, M.B. Talawar, S.P. Mirajkar, B.S. Rao1, S. Sugunan Polish. *J. Chem.*,74, 801, 2000.
[2] A. Hess, E. Kemnitz, *Appl. Catal.* A, 149, 373, 1997.
[3] K. Arata, *Adv. Catal.*. 37, 165, 1990.
[4] K. Arata, *Appl. Catal. A: Gen.*, 146, 3, 1996.
[5] X. Song, K.R. Reddy, A. Sayari, *J. Catal.*, 161, 206, 1996.
[6] H. A. Khalaf, *Monatsh. Chem.,* 140, 669, 2009.
[7] H. Matsuhashi, M. Hino, K. *Arata, Chem. Lett.*, 1027, 1998.
[8] L. Wang, A. Kong, B. Chen, H. Ding, Y. Shan, M. He, J. *Mol. Catal A: Chem.*, 230, 143, 2005.
[9] L.G. Bruck, S.N. Gorodskii, A.V. Zeigarnik, *J. Mol. Catal. A: Chem.*, 131, 29, 1998.
[10] J.F. Banfield, D.R. Veblen, *Am. Mineral*, 77, 545, 1992.
[11] A.M. Hofmeister, A. *Mineralogist*, 75, 1238, 1990.
[12] J. Mimaki, T. Tsuchiya, T. Yamanaka, Z. K*ristallog*r., 215, 419, 2000.
[13] M. Kim, J.G. Chung, J. *Coll. Interf. Sci.*, 233, 31, 2001.
[14] K. Tanabe, M. Misono, Y. Ono and H. Hattori, *New solid acids and bases*, Kodansha, Tokyo, Elsevier, Amesterdam, 1989.
[15] G.A.H. Mekhemer, A.K.H. Nohman, N.E. Fouad, H.A. Khalaf, *Coll. Surf. A: Physic. Eng. Asp.,* 161,439, 2000.
[16] F. Garin, D. Adriamasinoro, A. Abdulsamad, J. Sommer, *J. Catal.,* 131, 199, 1991.
[17] B. H. Davis, R. A. Keogh, R. Sirinivasan *Catal. Today*, 20, 219, 1994.
[18] Z. Gao, J. M. Chen, Y. Tang, Acta. *Chim. Sin.*, 52, 36, 1994.
[19] C. Morterra, G. Cerrato, F. Pinna, M. Signoretto, G. Strukul, J. Catal., 149, 181, 1994.
[20] R.G. Gillespie, *Acc. Chem. Res.*, 1, 202, 1968.
[21] K. Tanabe, Proc. *Int. Congr. Catal.* 9th*, Catgory*, 5, 1988.
[22] R.G. Gillespie, *Adv. Phys. Org. Chem.*, 9, 1, 1972.
[23] T. Yamaguchi, *Appl. Catal.* A, 61, 1, 1990.
[24] G.A. Olah, G.K.S. Parkash, J. Sommer, *Superacids,* Wiley, New York, 1985.
[25] K. Tanabe, H. Hattori, T. Yamaguchi, Critical Rev. in *Surf. Chem.*, 1, 1, 1990.
[26] D.A. Ward, E. I. Ko, *J. Catal.*, 150, 18, 1994.
[27] P.Y. Chen, K.C. Wu, W.C. Lin, C.H. Hwang, Poster Presentation, International Chemical Congress of Pacific Basin Societies, 1995.
[28] K. Matsuzawa, *Prepr. Am. Chem. Soc. Div. Pet. Chem.*, 42, 734, 1997.
[29] T.S. Thorat, V.M. Yadav, G.D. Yadav, *Appl. Catal.* A., 90, 73, 1992.
[30] K. Hatakeyama, T. Suzuka and M. Yamane, Sekiyu Gakkaishi, 34, 267, 1991.
[31] G.D. Yadav and T.S. *Thorat, Ind. Eng. Chem. Res.*, 35, 721, 1996.
[32] K. Murata, *JETI,* 45, 54, 1997.
[33] M.E. Quiroga, N.S. Figoli and U.A. Sedran, *Chemical Engineering Journal*, 67, 199, 1997.
[34] K.M. Parida, P. K. Pattnayak, *Catal. Lett.*, 47, 255, 1997.
[35] L.P. Hammet, *Chem. Rev.*, 16, 67, 1935.
[36] J. Kijenski, A. Baiker, *Catal. Today*, 5, 1, 1989.

[37] J.A. Dumesic, J.J. Fripiat (Eds). Acidity in Aluminan., Amorphous and Crystalline Silico-Aluminas, *Topics in Catalysis*, 4, 1997.

[38] K. Tanabe, *Solid Acids and Bases*, Academic Press, New York, 1970.

[39] H.A. Khalaf, MSc. Thesis, Minia University, El-Minia, Egypt, 1999.

[40] J.R. Anderson, M. Boudart, *Catalysis: Science and Technology*, Vol. 2, Ed. (Springer-Verlag, Berlin) p. 231, 1981.

[41] G. Busca, V. Lorenzelli, *Materials Chemistry*, 7, 89, 1982.

[42] M.C. Kung, H.H. Kung, *Catal Rev.-Sci. Eng.*, 27, 425, 1985.

[43] H.A. Benesi, B. H. C. Winquist, *Adv. Catal.*, 27, 97, 1978.

[44] H. Knozinger, *Adv. Catal.*, 25, 184, 1985.

[45] N. Cardona-Martinez and J. Dumesic, *Adv. Catal.*, 38, 149, 1992.

[46] A. Auroux and A. Gervasini, *J. Phys. Chem.*, 94, 6371, 1994

[47] E. A. Paukshtis and E. N. Yurchenko, *Russ. Chem. Rev.* (Eng. Trans), 52, 242, 1983.

[48] H. Knozinger, in: Elementary reaction steps in heterogeneous catalysis, R. W. Joyner et al. (eds.) Kluwer Academic publ., 267, 1993.

[49] M.L. Hair, *Infrared spectroscopy in surface chemistry*, Marcel Dekker, New York, 1962.

[50] L.H. Little, *Infrared Spectroscopy of Adsorbed Species*, Academic Press, New York, 1966.

[51] R.P. Eischens, S.A. Francis, W.A. Pliskin, *J. Chem. Phys.*, 22, 1786, 1954.

[52] M. Hino and K. Arata, *J. Chem. Soc.*, *Chem. Commun.*, 851, 1980.

[53] M. Hino and K. Arata, *Chem. Lett.,* 1259, 1979.

[54] J.E. Mapes, R.R. Eischens, *J. Phys. Chem.*, 58, 809, 1954.

[55] B. Delmon, P. Grange, P.A. Jacobs, G. Poncelet, ed., '*Preparation of Catalysts 11'*, Elsevier, Amsterdam, 1979.

[56] J.M. Thomas, W.J. Thomas, *Principles and practice of heterogeneous catalyst*, New York, VCH, 1996.

[57] J.F. Le Page, *Applied heterogeneous catalysis*, Edltious Technip, Paris, 1978.

[58] H.A. Khalaf, Ph.D. thesis, Minia university, El-Minia, Eygpt, 2005.

[59] K. Yang, J. Zhu, J. Zhu, S. Huang, X. Zhu, G. Ma, Materials Letters, 57, 4639, 2003.

[60] Y.H. Zhang, A. Reller, *Materials Letters*, 54, 375, 2002.

[61] Y.H. Zhang, A. Reller, *Materials Letters*, 57, 4108, 2003.

[62] V. Augugliaro, V. Loddo, M.J. Lopez-Munoz, C. Marquez- Alvarez,G. Palmisano, L. Palmisano, S.Yurdakal, Photochem. *Photobiol. Sci.,* 8, 663, 2009.

[63] A. Karami, J. *Iran. Chem. Soc.*, 7, S154, 2010.

[64] S. R. Kumar, C. Suresh, K.A. Vasudevan, N.R. Suja, P. Mukundan, K.G.K. Warrier, *Mater. Lett.*, 38, 161, 1999.

[65] Q. Zhang, L. Gao, J. Guo, *Appl. Catalysis B: Environmental*, 26, 207, 2000.

[66] S. Inoue, H. Kudou, A. Muto, T. Ono, The 255th ACS National Meeting in Fuel Symposia New Orleans, LA, March 23-27, 2003.

[67] S. Inoue, A. Muto, H. Kudou,T. Ono, *Am. Chem. Soc.*, *Div. Fuel Chem.*, 48(2), 496, 2003.

[68] H.A. Khalaf, S.E. Mansour E.A. El-Madani, *Monatsh. Chem.*, 141(7), 729, 2010.

[69] P. Baltazar, V.H. Lara, G. Cordoba, Arroyo, J. *Sol-Gel Sci. Techn.*, 37, 129, 2006.

[70] S.K. Bhattacharya, K.S. De, S.N. Pandao, G.V. Chandrasekhar, Proc. III Int. *Congr. Catal.*, 1.474, 1964.

[71] D. Gutschich, J. Scheva, G. Rienaecher, Z. *Onorg. Allg. Chem.*, 367, 34, 1969.
[72] I.V. Nicolescu, M. Spinzi, M. Gruia, A. Papia, V. Dumitrescu, H.G. Viewing, S. Nowak, F.G. Koennecke, *Revue roum. Chim.*, 11, 363, 1966.
[73] D. Das, H.K. Mishra, K.M. Parida, A.K. Dalai, J. *Mol. Catal. A: Chemical*, 189, 271, 2002.
[74] Y. Saih, K. Segawa, *Catal. Today*, 86, 61, 2003.
[75] M.A. Ecormier, K. Wilson, A.F. Lee, *J. Catal.*, 215, 57, 2004.
[76] M.E. Manriquez, T. Lopez, R. Gomez, J. Navarrete, *J. Mol. Catal.* A, 220, 229, 2004.
[77] N. Ravasio, N. Poli, M. Saba, F. Zaccheria, *Topics in Catalysis*, 13, 195, 2000.
[78] A. Kristiani, S N. Aisyiyah Jenie, J A. Laksmono, S. Tursiloadi, *Inter. J. Adv. Eng. Sci. and Tech.*, 1, 038, 2010.
[79] B.M. Reddy, P.M. Sreekanth, Y.Yamada, Q. Xu, T. Kobayashi, *Applied Catalysis A: General*, 228, 269, 2002.
[80] C.B. Rodella, V.R. Mastelaro, *J. Phys. and Chem. of Solids*, 64, 833, 2003.
[81] B. Grzybowska-Swierkosz, *Appl. Catal. A*, 157, 263, 1997.
[82] I.E. Wachs, B.M. Weckhuysen, *Appl. Catal. A,* 157, 67, 1997.
[83] P. Courtine, E. Bordes, *Appl. Catal. A,* 157, 45, 1997.
[84] G.C. Bond, *Appl. Catal. A*, 157, 91, 1997.
[85] O. Zegaoui, C. Hoang-Van, M. Karroua, *Appl. Catal. B*, 9, 211, 1996.
[86] M.A. Reiche, E. Ortelli, A. Baiker, *Appl. Catal. B: Envi*ron., 23, 187, 1999.
[87] M.A. Cauqui, J.M. Rodrı´guez-Izquierdo, J. *Non-Cryst. Solids*, 147, 724, 1992.
[88] M. Schneider, M. Maciejewski, S. Tschudin, A. Wokaun, A. Baiker, *J. Catal.,* 149, 326, 1994.
[89] C. Hoang-Van, O. Zegaoui, P. Richat, J. *Non-Cryst. Solids*, 225, 157, 1998.
[90] E. Igiesia, S.L. Sale, G.M. Kramer, *J. Catal.*, 144, 238, 1993.
[91] H. Matsuhshi, H. Miyazaki, K. Arata, *Chem. Lett.*, 442, 483, 2001.
[92] C. Lin, D. Shawn, Y.Yang, P. Tsung, *Catal. Lett.*, 23, 24, 2001.
[93] C.P. Nicholas, T. Marks, J. *Nano Letters*, 4, 1557, 2004.
[94] H.A. Khalaf, S.E. Mansour, E.A. El-Madani, *JAAUBAS,* 10, 2011.
[95] S.N. Aisyiyah Jenie, D.S. Kusuma, A. Kristiani, J.A. Laksmono, S. Tursiloadi, *IJBAS-IJENS*, 10, 5, 2010.
[96] K.R. Sunajadevi, S. Sugunan, *React. Kine. Catal. Lett.*, 82, 11, 2004.
[97] S. Tursiloadi, H. Imai, H. Hirashima, J. *non-Crystalline Solids,* 350, 271, 2004.
[98] J.G. Weissman, E.I. Ko, S. Kaytal, *Appl. Catal.* A., 94, 45, 1993.
[99] M. Hino, K. Arata, *J. Chem. Soc. Chem. Commun.*, 1259, 1998.
[100] K. Arata, M. Hino, *Chem. Lett.*, 971, 1989.
[101] H. Yang, R. Lu, J. Zhao, X. Yang, L. Shen, Z. Wang, Mater. *Chemistry and Physics*, 80, 68, 2003.
[102] S.M. Jung, O. Dupont, P. Grange, *Appl. Catal. A: General*, 208, 393, 2001.
[103] Y. Huang, B. Zhao, Y. Xie, *Appl. Catal. A: General*, 171, 75, 1998.
[104] C.X. Miao, Z. Gao, *Mater. Chemistry and Physics*, 50, 15, 1997.
[105] T. Lei, J.S. Xu, Y. Tang, W.M. Hua, Z. Gao, *Appl. Catal. A: General*, 192, 181, 2000.
[106] J.R. Sohn, D.H. Seo, *Catal. Today*, 87, 219, 2003.
[107] K. Arata, M. Hino, *Appl. Catal.*, 59, 197, 1990.
[108] W. Hua, A. Goeppert, J. Sommer, *J. Catal.,* 197, 406, 2001.
[109] Q.H. Xia, K. Hidajat, S. Kawi, *Chem. Commun.*, 2229, 2000.

[110] S.K. Samantaray, K. Parida, *J Mol Cat A Chemical*, 176, 151, 2001.
[111] H.A. Khalaf, G.A.H. Mekhemer, A.K.H. Nohman, S.A.A. Mansour, *Monatsh Chem* 138, 641, 2007.
[112] G.D. Yadav, J.J. Nair, *Micropor Mesopor Mater*, 33, 1, 1999.
[113] A.L.C. Pereira, S.G. Marchetti, A. Albornoz, P. Reyes, M. Oportus, M.D.C. Range *Appl. Catal. A Gen.*, 334, 187, 2008.
[114] W. Hua, Y. Xia, Y. Yue, Z. Gao, *J. Catal.*, 196, 104, 2000.
[115] Z. Gao, Y.D. Xia, W.M. Hua, C.X. Miao, *Top. Catal.*, 6, 101, 1998.
[116] X. Song, A. Sayari, *Catal. Rev. Sci. Eng.*, 38, 329, 1996.
[117] C.H. Bartholomew, *Appl. Catal. A*, 107, 1, 1993.
[118] M.T. Bore, H.N. Pham, E.E. Switzer, T.L. Ward, A. Fukuoka, A.K. Datye, *J. Phys. Chem. B*, 109, 2873, 2005.
[119] M. Haruta, S. Tsubota, T. Kobayashi, H. Kageyama, M.J. Genet, B. Delmon, *J. Catal.*, 144, 175, 1993.
[120] G.C. Bond, D.T. Thompson, *Catal. Rev. Sci. Eng.*, 41, 319, 1999.
[121] R. Zanella, S. Giorgio, C.H. Shin, C.R. Henry, C. Louis, *J. Catal.,* 222, 357, 2004.
[122] T.V.W. Janssens, B.S. Clausen, B. Hvolbaek, H. Falsig, C.H. Christensen, T. Bligaard, J.K. Norskov, *Top. Catal.*, 44, 15, 2007.
[123] G. Pattrick, E. van-der-Lingen, C.W. Corti, R. J. Holliday, D.T. Thompson, *Top. Catal.,* 30, 273, 2004.
[124] M. Haruta, M. Daté, *Appl. Catal. A,* 222, 427, 2001.
[125] W.F. Yan, S.M. Mahurin, S.H. Overbury, S. Dai, *Top. Catal.*, 39, 199, 2006.
[126] S.J. Tauster, S.C. Fung, R.T.K. Baker, J.A. Horsley, *Science*, 211, 1121, 1981.
[127] A.J. Simoens, R.T.K. Baker, D.J. Dwyer, C.R.F. Lund, R.J. Madon, *J. Catal.*, 86, 359, 1984.
[128] R. Burch, A.R. Flambard, *J. Catal.*, 85, 8, 1984.
[129] J.D. Bracey, R. Burch, *J. Catal.*, 86, 384, 1984.
[130] J.M. Herrmann, *J. Catal.*, 89, 404, 1984.
[131] J. Santos, J. Philips, J.A. Dumesic, *J. Catal.*, 81, 147, 1983.
[132] D.E. Resasco, G.L. Haller, *J. Catal.*, 82, 279, 1983.
[133] J. Ramirez, R. Cuevas, A. Lopez-Agudo, S. Mendioroz, J.L. Fierro, *Appl. Catal.*, 87, 233, 1989.
[134] C. Papadopoulos, A. Lycourghiotis, P. Grange, B. Delmon, *Appl. Catal.*, 38, 255, 1988.
[135] H. A. Khalaf, E. A. El-Madani, S. E. Mansour, *Global Journal of Inorganic Chemistry*, 2, 102, 2011.
[136] D. Wang, W. Li, M. Zhang, K. Tao, *Appl. Catal. A Gen.*, 317, 105, 2007.
[137] S.S. Lee, S.H. Byeon, *Bull. Korean Chem. Soc.*, 25, 1051, 2004.
[138] R.J.H. Clark, The Chemistry of Titanium and Vanadium. Amsterdam: Elsevier, 1968.
[139] J.G. dos Santos, T. Ogasawara R.A. Corrêa, *Brazilian Journal of Chemical Engineering*, 26, 555, 2009.
[140] K.S.W. Sing, in J.M. Thomas and R.M. Lambert (Editors), *Characterisation of Catalysts,* Wiley, Chichester, p. 12, 1980.
[141] Y. Zhang, G. Xiong, N. Yao, W. Yang, X. Fu, *Catalysis Today*, 68, 89, 2001.
[142] E. A. El-Madani, MSc.Thesis, omer Al-Mukhtar University, El-Beida, Libya, 2009.
[143] H. Yanga, R. Lua, L. Shena, L. Songa, J. Zhaoa, Z. Wanga and Li Wangb, *Mater. Letters,* 57, 2572, 2003.

[144] F. Lonyi, J. Valyon, J. Engelhardt, F. Mizukami, *J. Catal.*, 160, 279, 1996.
[145] A.K.H. Nohman, G.A. Mekhemer, N.E. Fouad, and H.A. Khalaf, *Ads. Sci. & Tech.*, 17, 8,1999.
[146] A. Corma, A. Martinez, C. Martinez, *Appl. Catal.*, 144, 249, 1996.
[147] T. Lopez, P. Bosch, F. Tzompantzi, R. Gomez, J. Navarrete, E. Lopez-Salinas, M.E. Llanos, *Appl. Catal. A*, 197, 107, 2000.
[148] K.T. Wan, C.B. Khouw, M.E. Davis, *J. Catal.,* 158, 311, 1996.
[149] J.P. Chen, R.T. Yang, *J. Catal.*, 139, 277, 1993.
[150] A. Kayo, T. Yamaguchi, K. Tanabe, *J. Catal.*, 83, 99, 1983.
[151] T. Yamaguchi, T. Jin, K. Tanabe, *J. Phys. Chem.*, 90, 3148, 1986.
[152] T. Jin, T. Yamaguchi, K. Tanabe, *J. Phys. Chem.*, 90, 4794, 1989.
[153] M. Hino, S. Kobayashi, K. Arata, *J. Am. Chem. Soc.*, 101, 6439, 1979.
[154] M.S. Scurrell, M. S. *Appl. Catal.*, 34, 109, 1987.
[155] M. Hino, K. Arata, K. J. Chem. Soc., *Chem. Commun.*, 112, 1985.
[156] J.C. Yori, J.C. Luy, J.M. Parera, *Catal.* Today, 5, 493, 1989.
[157] J.R. Sohn, H.J. Kim, *J. Catal.*, 101, 428, 1986.
[158] O. Saur, M. Bensitel, A.B. M. Saad, J.C. Lavalley, C.P. Tripp, B.A. Morrow, B. A. *J. Catal.*, 99, 104, 1986.
[159] X. Bokhimi, A. Morales, E. Ortiz, T. Lopez, R.Gomez and J. Navarrete, J. *Sol,-Gel. Sci. and Tech.,* 29, 31, 2004.
[160] L.K. Noda, R.M. Almeida, L.F.D. Probst, N.S. Goncalves, *J. Mol. Catal.* A: Chemical, 225, 39, 2005.
[161] E .A. El-Sharkawy, A.S. Khder and A.I. Ahmed, Micropor. *Mesopor. Mater.*, 102, 128, 2007.
[162] R. Cid and G. Pecci, *J. Appl. Catal. A: Gen.*, 14, 15, 1985.
[163] P. Sharma, S. Vyas and A. Patel, *J. Mol. Catal.*, 214, 281, 2004.
[164] M. L. Ferreira and E. H. Rueda, *J. Mol. Catal.* A, 178, 147, 2002.
[165] W. Pryzystajko, R. Fiedorow and I. G. Dalla Lana, *Appl. Catal.*, 15, 265, 1985.
[166] T.S. Yang, T.H. Chang C. T. Yandeh, *J. Mol. Catal.*, 115, 339, 1997.
[167] L.K. Noda, R. Monteiro de Almeidaa, N.S. Gonçalvesa, L.F. Dias Probsta and O. Salab, *Catal. Today*, 85, 69, 2003.
[168] D.A. Ward and E.I. Ko, *J. Catal.*, 157, 321, 1995.
[169] G.K. Chuah, L.S. Jaenicke, L.J. Harrison, *J. Catal.*, 200, 352-359, 2001.
[170] Z. Yongzhong, N. Yuntong, S. Jaenicke, G.K. Chuah, *J. Catal.*, 229, 404, 2005.
[171] S. B. Tendulkar, S.S. Tambe, I. Chandra, P.V. Rao, R.V. Naik and B.D. Kulkarni, Ind. *Eng. Chem. Res.,* 37, 2081, 1998.
[172] Q. Yang, C. Li. S. Yuan, J. Li. P. Ying, Q. Xin and W. Shi, *J. Catal.*, 183, 128, 1999.
[173] J. Sun, X. Meng, Y. Shi. R. Wang, S. Feng, D. Jiang, R. Xu and F. S. Xiao, *J. Catal.*, 193, 199, 2000.
[174] S. Valange, Z. Gabelica, M. Abdellaoui, J.M. Clancens and J. Barrault, *Micropor. Mesopor. Mater.,* 30, 177, 1999.
[175] A. Thangaraj, R. Kumar and P. Ratnasamy, *J. Catal.*, 131, 294, 1991.
[176] A. Tuel, S. Moussa-Khouzami, Y. Ben Yaarit and C. M. Naccache, *J. Mol. Catal.*, 68, 45, 1991.
[177] A. Germain M. Allian and E. Figueras, Proceedings 5th European workshop meeting on selective oxidation by heterogeneous catalysis. Berlin: Dechema e.V, 1995.

[178] Y.H. Yue, A. Gedeon, J.L. Bonardet, N. Melosh, J.B.D Espinose, J. Fraissard, *Chem. Commun.,* 1967, 1999.

In: Rutile: Properties, Synthesis and Applications
Editor: It-Meng (Jim) Low

ISBN 978-1-61942-233-9

Chapter 7

THE COORDINATION STATE OF TITANIUM ATOMS AT THE SURFACE OF INDUSTRIAL TiO_2 SPECIMENS ACCORDING TO THE DATA OF ELECTRONIC SPECTROSCOPY OF DIFFUSION REFLECTION

Eugene A. Sosnov
Saint-Petersburg State Institute of Technology
St. Petersburg, Russia

ABSTRACT

Given the essential role of the surface in the physico-chemical processes, it is important to identify the structure of the surface layer within the thickness of several nanometers. One of the effective physico-chemical techniques to evaluate the electronic state of the solid materials surface is the Electronic Spectroscopy of Diffusion Reflection (ESDR) in the UV-Vis-spectrum. On the example of titanium dioxide (rutile) of various origins and two-phase pyrogenic titanium dioxide P25 (Degussa) was demonstrated the possibility of mathematical processing of ESDR-spectra to determine the coordination state of polyhedron with different spatial environment of titanium atoms. The results of mathematical processing of experimental spectra are compared with other methods of ESDR-spectra presentations (definition of the absorption edge by the method of tangents, the representation in coordinates of Gurevich-Kubelka-Munk (F_R), $(\alpha \ h\nu)^m \propto (h\nu - E_g)$, $(F_R \ h\nu)^m \propto (h\nu - E_g)$) and hardware differentiation of the spectra. It is shown that the technological prehistory and chemical modification of surface of industrial dispersed TiO_2 having a significant impact on the coordination state of titanium atoms on the titanium dioxide surface.

INTRODUCTION

One of the important factors determining the sorption, catalytic and other properties of solids is their phase composition. A typical example of above is titanium dioxide, which is due to chemical stability in acidic and neutral liquids, electrical conductivity, photocatalytic activity is widely used in the manufacture of fillers for composite polymer materials, in dielectric ceramics, pigments, electronic devices, photoelectrochemical cells, photocatalysis, chemical sensors and solar cells, etc. [1].

In this case, depending on the crystalline modification (Anatase, Rutile and Brookite) titanium dioxide shows different physical and chemical properties. Since anatase modification of TiO_2 is an active catalyst for dehydration, and rutile - for dehydrogenation. And at simultaneous presence of amorphous, anatase and rutile component there was increasing of photocatalytic activity of titanium-oxide material compared to the material consisting of only one crystallographic phase [2, 3].

Given the essential role of the surface in these physico-chemical processes, it is highly important to identify not only the structure of the bulk phase, but also the structure of the superficial layer within the thickness of several nanometers.

Traditionally, however, the information about the phase composition generally considers to the volume of solids and contains no data on the structure of their surface layers. One of the effective physico-chemical techniques to evaluate the electronic state of the surface of solids is the Electron Spectroscopy of Diffuse Reflectance (ESDR) in the UV- and Visible spectrum. Considering the depth of radiation penetration into solids in electronic spectra, the method can be used to identify the phase composition of the surface layer ~5 nm thick [4].

1. Theoretical Foundations of ESDR

Consider the electronic structure of solid oxide and manifestations of its features in the optical electronic spectra. Figure 1 shows the total energy scheme, reflecting the specificity of electronic interactions in oxides.

In this scheme the valence *ns-*, *np-* and *(n-1)d*-orbitals of the element forming the oxide, and *2s-*, *2p*-orbitals of oxygen take into account. The collective energy states of the solid particle that form energy layer are represented in the central part of scheme. The legitimacy of using molecular orbital theory [5] is valid provided that the solid particle could be regarded as a “giant macromolecule” (supramolecular structure) [6], which may possess a periodic structure.

In the interaction between valence orbitals of the metal and oxygen atoms in a solid occur quasi-continuous band states responsible for the formation of metal-oxygen bonds $MO^{B}{}_{M\text{–}O}$ (the zone of binding electronic states).

Higher energy is a narrow zone of non-bonding orbitals (Available from participating in chemical bonds, electrons of oxygen are localized on the oxygen atoms, and in the scheme - at non-bonding pseudo-atomic orbitals $MO^{NB}{}_{O}$). Note that the gap between $MO^{NB}{}_{O}$ and $MO^{B}{}_{M\text{–}O}$ showed at the scheme may absent (wide band of binding states). In this case, the whole continuum is defined as the valence band.

Next is a zone formed by *(n-1)d*-valence orbitals of the metal. Typically, this zone is narrow; its electrons filling degree can be easily determined from the valence configuration of the atom at the formal oxidation state (for Ti^{4+} - $4s^0 3d^0$). Must take into account that the d-orbitals of the metal split in the crystal field of different symmetry (Figure 2).

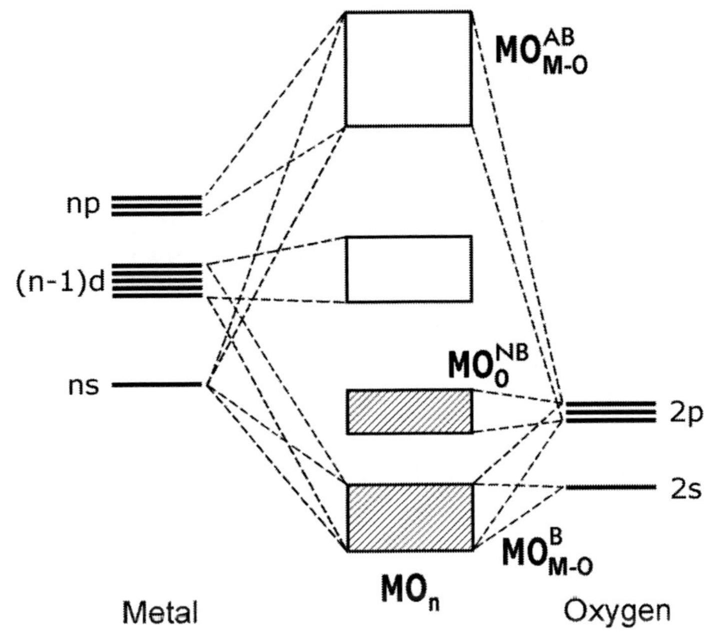

Figure 1. Scheme of the electronic structure of oxides.

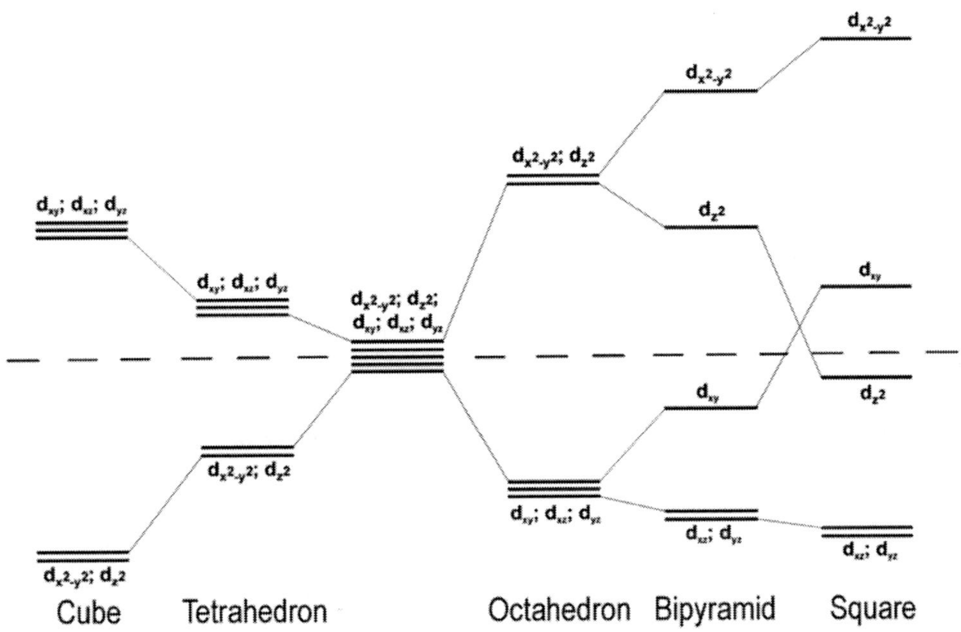

Figure 2. d-Orbital splitting in crystal field of different symmetries.

But in solids the splitted orbitals can overlap without the formation of the gap between the subbands. This process leads to a change in the energy gap between the non-bonding orbitals of oxygen and free d-orbitals of the metal, which depends on the coordination environment of metal atom.

At the top of the scheme is a zone of free antibonding states $MO^{AB}{}_{M\text{–}O}$. Their number corresponds to the number of employed bonding orbitals $MO^{B}{}_{M\text{–}O}$.

Short-wavelength region of the optical electronic spectra formed, usually, as a result of electron transitions from nonbonding orbitals of oxygen $MO^{NB}{}_{O}$ to *(n-l)d*-states of metal, which manifest themselves as broad and intense absorption bands at the edge of Visible and near-UV region. The absorption bands edge corresponds to the electrons transition from the highest filled orbitals of the valence band to the lowest unoccupied orbital of the conduction band. The corresponding energy gap is determined in solid-state physics as “forbidden band width”, and it’s a fundamental characteristic of the solid.

2. Representation of the ESDR-Spectra

Despite the individual manifestations of the absorption band for various polymorphous modifications of titanium dioxide [7, 8], in the case of the simultaneous presence of several different types of coordination structures on the TiO_2 surface, the allocation of their contribution to the spectral transitions are quite difficult [9-12]. The possibilities of various representations of ESDR-spectra will be considered at the example of model two-phase system – pyrogenic titanium dioxide P25 Degussa. One of the frequently used methods for the estimation of the coordination state of atoms on the surface from electronic diffuse reflectance spectra is the determination of the long-wave fundamental absorption edge by the method of tangents [13, 14] (Figure 3a).

The results obtained this way are compared with the optical forbidden bandwidths for material polymorphs, which allow the structure of the surface of the sample to be qualitatively estimated [8, 15]. In spite of the apparent simplicity and reliability of the method, it cannot be used to obtain reliable data on the structure of the surface of materials for two reasons.

1. The position of the absorption band edge determined by the method of tangents substantially depends on the selection of the curve region used for linear extrapolation [11]. For instance, the approximation of the spectral curve of P25 (Figure 3a) over the range 25 500–24 500 cm^{-1} (approximation validity coefficient R^2 = 0.991) places the absorption band edge at 24 250 cm^{-1} (412.4 nm), whereas spectrum linearization over the range 26 000–24 500 cm^{-1} (R^2 = 0.994) shifts the position of the absorption band edge to longer waves, 24 190 cm^{-1} (413.4 nm).

2. If several structures with different first coordination spheres are present on the surface, the method of tangents usually allows the absorption band edge for only one type of structures to be determined, most often, for structures with the smallest optical forbidden band width [11]. For our sample, the absorption band edge is situated at 24 190 cm^{-1} (413.4 nm). This gives a 3.00 eV optical forbidden bandwidth, as is characteristic of the rutile TiO_2 polymorph [8], whereas, according to the X-ray data, the P25 sample is two-phase and anatase content in the batch of material is not constant [16] and is within 60-90%. In addition, the band gap, determined by the method of tangent depends on what kind of representation of

the spectrum used for the analysis: as a reflection spectrum (R_D - ν) or transmittance - (T - ν) [7, 8].

Another method for estimating surface structures is the representation of diffuse reflectance spectra in the Gurevich-Kubelka-Munk coordinates (F_R) [17]:

$$F_R = \frac{(1 - R_D)^2}{2R_D} = \frac{\alpha}{S} \approx c\,, \tag{1}$$

where R_D is the diffuse reflectance and α and S are the effective absorption and scattering coefficient, respectively.

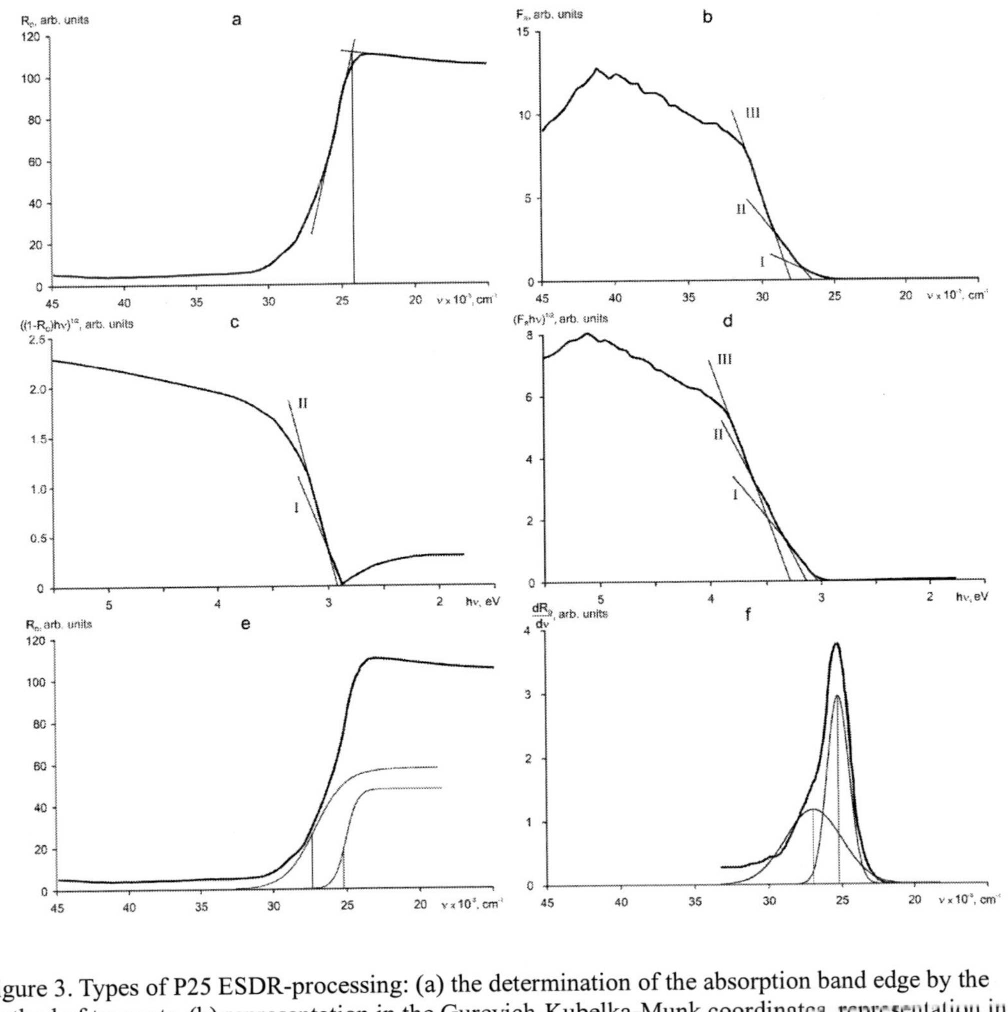

Figure 3. Types of P25 ESDR-processing: (a) the determination of the absorption band edge by the method of tangents, (b) representation in the Gurevich-Kubelka-Munk coordinates, representation in the (c) $((1-R_D)\ h\nu)^{1/2} - h\nu$ and (d) $(F_R\ \ h\nu)^{1/2} - h\nu$ coordinates, (e) spectrum separation into components described by the Fermi–Dirac distribution, and (f) processing of the results of the instrumental differentiation of ESDR.

The representation of diffuse reflectance spectra in these coordinates (Figure 3b) also has serious shortcomings. The disproportion of the sizes of linearized regions that appears because of spectrum restructuring in the Gurevich-Kubelka-Munk coordinates prevents us from the determination of quantitative ratios between surface structures with different coordination [9, 11, 18]. In addition, the position of the approximating straight line as previously depends on the selection of the region of experimental data. In particular, the shape of the spectrum in this representation allows several curve regions linearizable with approximation validity coefficients $R^2 > 0.99$ to be identified (Figure 3b, regions I, II, and III), and the selection of regions for approximation can be fairly arbitrary. At the same time, if electronic diffuse reflectance spectra in the Gurevich-Kubelka-Munk coordinates are processed according to [19], where a descending spectrum region (region III) over the range 30-70% of the maximum F_R value is used for approximation, the characteristics of only one of the coordination structures are determined (in our case, the structure with the forbidden band width E_g of 3.48 eV, which can be interpreted as the coordination state of titanium in the anatase polymorph.

The forbidden bandwidth of semiconducting materials can be determined using the representation of optical spectra in the coordinates [20]

$$(\alpha h\nu)^m \propto (h\nu - E_g), \tag{2}$$

where α is the coefficient of radiation absorption by the sample, cm^{-1}; $h\nu$ is the quantum energy, eV; and m is the coefficient depending on the type of electron transitions from the valence to conduction band. For allowed (the summed momentum of electron spins does not change) direct transitions, $m = 2$, and, for indirect transitions, $m = 1/2$.

If α cannot be determined directly, its normalized calculated values can be used [21],

$$\alpha = 1 - (T + R), \tag{3}$$

where T is the normalized thin film transmission and R is the normalized thin film reflectance. With electronic diffuse reflectance spectra, representation (3) becomes

$$\alpha = 1 - R_D, \tag{4}$$

which introduces determination error E_g because the transmission spectrum of the material cannot be measured.

Since the Gurevich-Kubelka-Munk representation is proportional to the absorption coefficient, it was suggested in [19, 22, 23] that α in (2) could be replaced with the spectrum representation in the Gurevich-Kubelka-Munk coordinates, $(F_R\, h\nu)^m \propto (h\nu - E_g)$. Because the band gap in TiO_2 was determined by indirect electron transitions in Figure 3c & d shows the spectra of titanium dioxide P25 in coordinates $((1-R_D)h\nu)^{1/2} - h\nu$ and $(F_R\, h\nu)^{1/2} - h\nu$, respectively. At the same time, these methods for spectral data representing have the same shortcomings as the representation of spectra in the Gurevich-Kubelka-Munk coordinates. In the spectrum also can be arbitrarily allocate some regions for a linear approximation with a reliability coefficient of $R^2 > 0.995$. It follows that these procedures also cannot be used to correctly describe the structure and determine the anatase/rutile quantitative ratio on the surface of the P25 sample.

3. Methods of Separation the ESDR-Spectrum into Components

Earlier studies [10-12], we were invited to consider the edge of fundamental absorption spectra not from the optics standpoint (i.e., radiation interaction with an infinitely thick scattering layer of closely packed disperse medium particles [24]), as well as a quantum phenomenon – the probability finding an electron with energy E_i close to the quantum transition point (E_0). The probability of the transition is then described by the Fermi–Dirac distribution of particles with half-integral spins [25] (Figure 4),

$$n_i = \frac{n_0}{1+\exp\left(\left(E_i - E_0\right)/kT\right)}, \quad (5)$$

where n_i the number of carriers have overcome the band gap; n_0 is the number of mobile carriers; k is the Boltzmann constant; and T is the temperature, K.

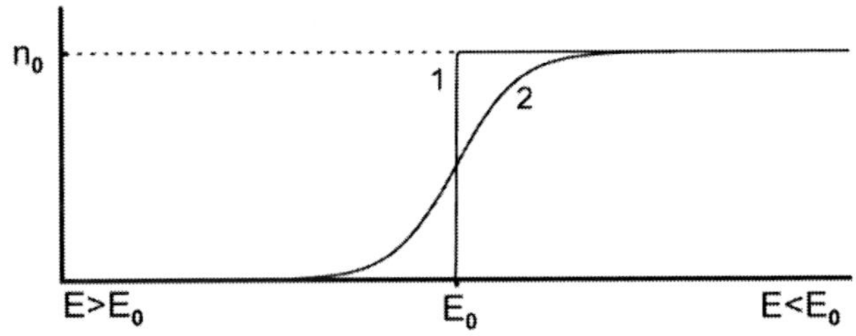

Figure 4. Fermi–Dirac distribution curve at T = (1) 0 K and (2) > 0 K.

At T = 0 K, we then observe a jumptype quantum transition at the $E_i = E_0$ point (Figure 4, curve 1). At T > 0 K, the quantum transition becomes smeared because of electron thermal energy, and it acquires the form of an S-shaped curve (Figure 4, curve 2).

We can apply such a representation to the nature of electronic diffuse reflectance spectra and, in the presence of several (j) types of surface structures with different coordination states on the TiO_2 surface, describe each transition by the Fermi-Dirac distribution in the form

$$I_{ji} = \frac{I_{0j}}{1+\exp\left(\left(E_i - E_{0j}\right)/\beta_j\right)}, \quad (6)$$

where I_{0j} is the transition intensity for the j-th type of surface structures, I_{ij} is the contribution of j-th type structures at the i-th point of the spectrum, E_{0j} is the quantum transition energy for surface structures of the j-th type, and β_j is the coefficient characterizing smearing of the transition band.

The absorption band edge in electronic diffuse reflectance spectra can be represented in the form

$$I_i = \sum_j I_{ji} = \sum_j \frac{I_{0j}}{1+\exp\left(\left(E_i - E_{0j}\right)/\beta_j\right)}, \tag{7}$$

and mathematical processing can be applied to separate such overlapping spectral transitions (Figure 3e).

As distinct from the ideal model structure, the real titanium-oxide complexes on the TiO2 surface, though they have similar coordination structures, can be characterized by somewhat different lengths and angles of valence bonds. This can be caused by the differences in the electron density on oxygen atoms situated in the volume and at the boundary with air. Such distortions of the polyhedron structure should influence the shape of the distribution curve and increase smearing of the quantum transition region above temperature-induced broadening (β > kT, where kT is ~0.025 eV at T = 293 K). Actually, the β coefficient determined from the Fermi–Dirac distribution characterizes the degree of temperature and configurational disordering of surface titanium-oxide structures.

It should be noted that the literature clearly separate "electronic" and "optical" band gap of semiconductors [20, 26]. The proposed method of ESDR-spectra separation into components described by Fermi-Dirac distribution, gives the energy characteristics of surface structures, comparable to the "electronic" band gap of semiconductors.

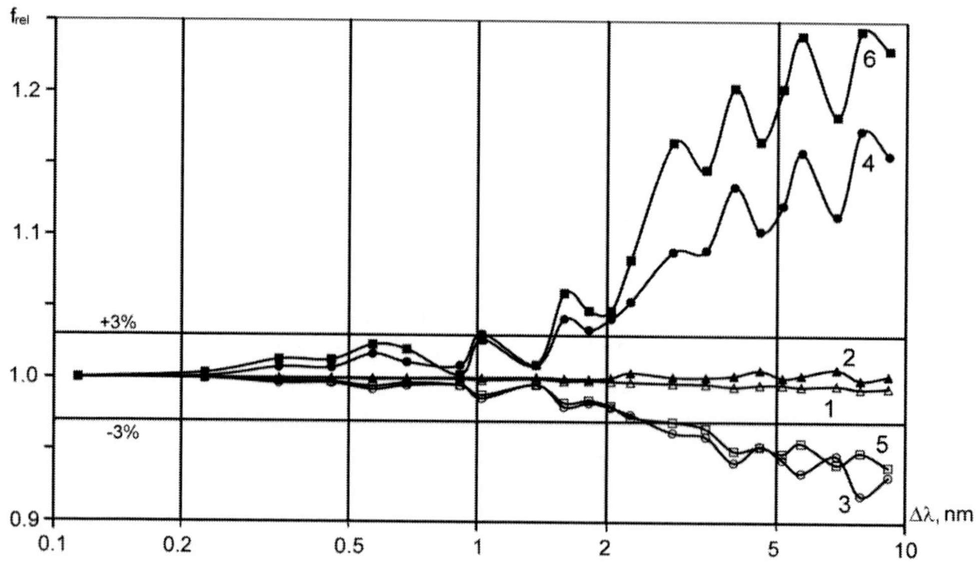

Figure 5. Influence of spectrum scanning step on the relative (1, 2) E_0, (3, 4) I_0, and (5, 6) β values of the distinguished spectral bands. The curves characterize bands (1, 3, 5) 1 and (2, 4, 6) 2 of P25.

To estimate the influence of scanning conditions on the accuracy of the determination of the characteristics of the spectral bands of interest, a series of electronic diffuse reflectance spectra were obtained for P25 sample using different scan steps (from 0.1 to 10 nm). The dependence of calculation results on the scan step is shown in Figure 5. As follows from these data, scanning step changes from 0.1 to 2 nm do not cause significant changes in the determination of band characteristics, and deviations of the I_0 and β values do not exceed 3%. An increase in the scanning step to 10 nm considerably increases the errors in these values, which can be as large as 20–25%. The error in the position of the middle of the electronic transition (E_0) is substantially smaller and does not exceed 1% over the whole range of scanning step values under consideration.

In recent years, the exact spectral curve peak positions have been determined using the first derivatives of spectral functions [27, 28]. Overlap of bands related to different coordination structures is, however, also possible for differential curves (Figure 3f), which requires an additional mathematical processing for the separation of peaks. We used the nonlinear regression method to decompose the curve of the instrumental differentiation of the spectrum of P25 into Gauss components,

$$\left(\frac{dI}{d\nu}\right)_i = \sum_j \frac{A_j}{w_j\sqrt{\pi/2}} \exp\left(-2\left(\frac{E_i - E_{0j}}{w_j}\right)^2\right). \tag{8}$$

The results are given in Table 1.

Table 1. Optical forbidden bandwidths and anatase and rutile contents (x_A and x_R) in the surface layer of P25 pyrogenic titanium dioxide obtained as a result of electronic diffuse reflectance spectrum processing by various methods

Method	Anatase		Rutile	
	x_A, %	E_0, eV	x_R, %	E_0, eV
According to processing by the method of tangents from the absorption band edge	–	–	–	3.00
Gurevich-Kubelka-Munk	–	3.48	–	–
$((1-R_D)h\nu)^{1/2}$	–	–	–	2.87
$(F_R h\nu)^{1/2}$	–	3.29	–	–
Spectrum separation into components according to Eq. (7)	54.8 *	3.39	45.2 *	3.13
Instrumental differentiation curve separation into Gauss components according to Eq. (8)	53.3 **	3.34	46.7 **	3.13

* Determined from the intensity (I, %) of the transition.

** Determined from the area bounded by the Gauss curve.

A comparison of the results of P25 electronic diffuse reflectance spectrum processing obtained using various procedures (Table 1) shows that the results obtained by graphic methods (the determination of the absorption band edge and linearization of spectrum regions in the Gurevich-Kubelka-Munk, $((1-R_D)h\nu)^{1/2}$ or $(F_R h\nu)^{1/2}$ representation) cannot be used to correctly describe not only the quantitative but also qualitative composition of surface structures. At the same time, the separation of spectra into components described by the Fermi–Dirac distribution and the separation of the instrumental differentiation curve into components described by the Gauss distribution allow not only the structure of coordination polyhedra but also the anatase/rutile ratio on the surface of the material to be determined. The difference between the data obtained by these methods is small (within measurement errors of 1–2%).

4. THE COORDINATION STATE OF TITANIUM ATOMS IN TiO_2

Study of coordination state of titanium atoms on the surface of its dioxide was performed on a number of samples available, according to X-ray diffraction data, both rutile and anatase crystal structure. Characteristics of the samples and their chemical composition are presented in Table 2.

It should be noted that not only rutile of different origin (samples 5-8), but also industrial pigment rutile-based materials from different manufacturers (samples 9-13), whose surface contains an ultra-thin protective coatings based on SiO_2 and Al_2O_3 were studied.

The electronic diffuse reflectance spectra of the titanium dioxide samples were obtained on spectrophotometers equipped with an integrating sphere with respect to a MgO optical reference (differential procedure) over the frequency range 50 000–15 000 cm^{-1} (250–700 nm). The resolving power was 1 nm (160 – 20 cm^{-1}). Simultaneously with spectrum recording, its instrumental differentiation was performed. Samples of the obtained spectra are shown in Figure 6.

To obtain quantitative data characterizing the titanium coordination station the surface of its dioxide, we isolated the region of the long-wave absorption band edge from the electronic diffuse reflectance spectra (from the minimum to maximum R_D value). We assumed that the superposition of quantum transitions described by distribution (7) was observed in this region. The spectral regions were therefore divided into components by the gradient descent method; the results are given in Table 3.

Note that, if the sample contains coordination polyhedra of only one type (with similar configuration), mathematical processing reduces the intensity of the second band to zero. Simultaneously using nonlinear regression was conducted decomposition curves of hardware differentiation of the spectra of the samples in the Gaussian components.

When considering the results presented in Table 3, it should be borne in mind that, if λ or E_0 are fundamental in character and characterize the configuration state of the titanium atom, the absolute band intensities (I_0) obtained as a result of mathematical processing depend on the reference sample. For this reason, only the relative intensities $\left(I_j = I_{0j} \Big/ \sum_j I_{0j}\right)$ are characteristic. The determination of quantitative characteristics from the data obtained in the instrumental differentiation of the spectra requires calculations of the areas bounded by Gauss curves.

Table 2. Characteristics of TiO_2 samples

Sample		XRD, A/R, %	S_{SP}, m^2/g	d, mkm	Content, %		
№	Description				TiO_2	SiO_2	Al_2O_3
1	Anatase Aldrich # 248576	100 / 0	11.0	0.14	≥ 99.0	–	–
2	Anatase A-1 (GOST 9808-84) JSC Sumykhimprom (Ukraine)	100 / 0	17.6	0.09	≥ 98.0	–	–
3	Anatase plasmochemical, produced from $TiCl_4$	100 / 0	133	0.012	≥ 99.0	–	–
4	AEROXIDE TiO_2 P25 (Degussa) Evonik Degussa Corporation	80 / 20	51.7	0.03	≥ 99.5	≤ 0.2	≤ 0.3
5	Rutile P-1 (GOST 9808-84) CJSC "Crimea TITAN" (Ukraine)	10 / 90	8.0	0.18	≥ 98.0	–	–
6	Rutile P-02 (GOST 9808-84) JSC Sumykhimprom (Ukraine)	3 / 97	6.5	0.22	≥ 93.0	–	–
7	Rutile P-1 (GOST 9808-84) Pigment Holding Company (Russia)	0 / 100	11.6	0.12	≥ 98.0	–	–
8	Rutile OSCh 7-3 (4N5) (TU 6-09-3811-79) PROMCHIM Group (Russia)	0 / 100	1.4	1.0	99.995	–	–
9	DuPont Ti-Pure R-706	0 / 100	12.4	0.12	94.3	3.0	2.5
10	DuPont Ti-Pure R-902	0 / 100	14.6	0.10	91.0	1.4	4.3
11	RL-68 Thann et Mulhouse (France)	0 / 100	16.5	0.09	95.0	+ *	+ *
12	Kemira 402 KEMIRA Pigments OY (Finland)	0 / 100	11.2	0.13	96.0	–	1.3
13	Kronos 2230 Cristal Global (France)	0 / 100	16.7	0.09	96.0	+ *	+ *

* Quantitative characteristics are not reported.

Comparing the results obtained by separation of ESDR-spectrum into components, described by Fermi-Dirac distribution, and results of separation of hardware differentiation spectra into the Gaussian curves, it should be noted sufficiently close quantitative characteristics obtained by both methods. At the same time, the expansion curves of hardware differentiation have a less sensitivity (the method could not distinguish the blurred or non-intensive transitions in the spectrum of samples 6 and 7). This shortcoming is caused, apparently, an error of fixing the actual values of the first derivative due to the oscillation of luminous flux devices. Therefore, in further consideration of the titanium coordination state on the TiO_2 surface will be based on the separation of initial spectra.

Analysis of the results presented in Table 3, shows that anatase (samples 1 - 3) is usually characterized by only one band in the spectrum with E_0 = 3.3 - 3.4 eV. However, in the spectrum of sample 1 (anatase Aldrich) observed the appearance of the transition at ~ 400 nm (E_0 = 3.09 eV), which is characteristic of rutile modification of TiO_2.

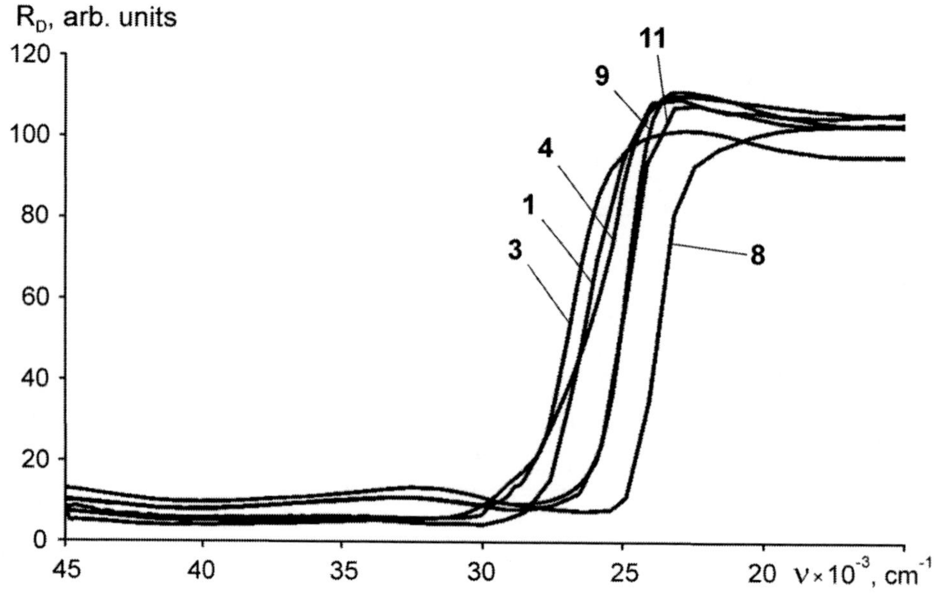

Figure 6. ESDR-spectra of titanium dioxide samples. Numbers on the figure are correspond to sample numbers in Table 2.

For samples of rutile (5 - 8) are also commonly observed one transition band in the optical spectrum - in the range 405 - 420 nm (E_0= 2.95 - 3.1 eV). Technological peculiarities of the dispersed rutile have significant influence on the distortion degree of coordination polyhedron (and therefore on the value E_0 in ESDR-spectrum). Thus, for rutile samples obtained by liquid-phase hydrolysis of $TiCl_4$ (7, 8) followed by heat treatment, the values of E_0 are 2.95 - 2.96 eV, whereas for the samples obtained on sulfate technology (5, 6), the titanium coordination polyhedra on surfaces have a less perfect structure. This is reflected in lower values of E_0 (3.0 – 3.1 eV), specified for the fundamental absorption bands of rutile.

Since the titanium atoms in anatase and rutile have an octahedral coordination environment of oxygen, it should be noted that as the alignment of Ti-O lengths and bond angles ∠ O-Ti-O (and, consequently, increasing the symmetry of coordination polyhedron of titanium atoms), the decrease of width band gap of material (from 3.3 – 3.4 to 2.9 – 3.1 eV) observed.

In the spectrum of sample 7 except for the aforementioned band at 420 nm observed a broad diffuse transition in the region of 470 nm (E_0 ~2.7 eV). Industrial TiO_2 samples to stabilize their properties (regulation photocatalytic activity, resistance to degradation, etc.) usually have modifiers, such as iron salts, promoting a partial recovery of the titanium ions on the rutile surface to Ti^{3+}, which gives the TiO_2 faint bluish tinge. The appearance of the ESDR-spectrum of sample of the second absorption band with E_0 ~2.7 eV may be associated with the formation on the surface of TiO_2 Ti^{3+} ions with titanium coordination state, which is typical for Ti_2O_3 polyhedra [29]. For the rutile of high purity (sample 8), when the total amount of impurities in the material does not exceed 0.01%, only one type of coordination complexes on the surface observed.

Table 3. Results of ESDR-spectra processing of titanium dioxide samples

Sample	Band 1						Band 2						Method of tangents from the absorption band edge	
	Instrumental differentiation curve separation into Gauss components			Spectrum separation into components according to Eq. (7)			Instrumental differentiation curve separation into Gauss components			Spectrum separation into components according to Eq. (7)				
	λ, nm	E_0, eV	I, %	λ, nm	E_0, eV	I, %	λ, nm	E_0, eV	I, %	λ, nm	E_0, eV	I, %	λ, nm	E_0, eV
1	365	3.39	81.3	377	3.29	82.7	401	3.09	18.7	400	3.10	17.2	404	3.07
2	371	3.34	100	373	3.32	100	–	–	–	–	–	–	393	3.16
3	370	3.35	100	371	3.34	100	–	–	–	–	–	–	396	3.13
4	371	3.34	53.3	366	3.39	54.8	396	3.13	46.7	396	3.13	45.2	413	3.00
5	401	3.09	100	404	3.07	100	–	–	–	–	–	–	425	2.92
6	–	–	–	375	3.30	1.8	405	3.07	100	405	3.06	98.2	421	2.95
7	421	2.95	100	419	2.96	88.0	–	–	–	462	2.68	12.0	439	2.82
8	421	2.95	100	420	2.96	100	–	–	–	–	–	–	435	2.85
9	385	3.21	24.2	387	3.20	20.1	401	3.09	75.8	403	3.07	79.9	418	2.97
10	394	3.15	65.4	396	3.13	63.0	405	3.06	34.6	409	3.03	37.0	420	2.95
11	375	3.30	9.5	378	3.29	7.0	401	3.09	90.5	402	3.08	93.0	417	2.97
12	387	3.20	23.9	383	3.24	21.5	404	3.07	76.1	405	3.06	78.5	418	2.97
13	378	3.28	19.1	383	3.24	17.7	405	3.06	80.1	406	3.06	82.3	421	2.95

For sample 4 (P25 Degussa), having a two-phase structure, as a result of mathematical processing of the ESDR-spectrum enable to allocate two bands that characterize the simultaneous presence of titanium-oxygen polyhedra with the coordination environment characteristic to anatase ($E_0 = 3.39$ eV) and rutile ($E_0 = 3.13$ eV).

However relationship between the anatase and rutile modifications on the sample surface is 54 : 46, which differs from the ratio of titanium oxide phases, which are marked according to X-ray diffraction (80 : 20). This feature of the material - the simultaneous presence at surface structures, which accelerate different catalytic processes (e.g., dehydration and dehydrogenation), apparently, explains the high catalytic characteristics of P25, far surpassing the mechanical mixtures of anatase & rutile [8].

In contrast to P25 for the rutile samples containing according to X-ray diffraction data the impurity of anatase phase (5, 6) the appearance of transitions characteristic of the anatase coordination in the spectra don't observed. Such discrepancy between the structural and spectroscopic analysis results can be called by multistage obtaining of rutile. In the process of rutile obtaining first formed anatase, which then transforms into rutile [30, 31]. Since the phase transition from anatase to rutile starts at the surface, in the particle's center can be saved an anatase core, not affecting to titanium coordination state on the particle surface. The same process (stopped early) explains the appearance in the spectrum of sample 1, the bands characteristic of the titanium coordination state in rutile.

Of particular it needs to note results obtained for industrial pigment materials (samples 9 - 13). These materials, according to passports, are highly dispersed rutile, which surface is protected by thin layer of SiO_2 and Al_2O_3. The appearance at the TiO_2/SiO_2 interface chemical bonds Ti-O-Si leads to change in the structure of the coordination polyhedra of titanium atoms. This can be expressed in the elongation of the Ti-O bonds, appear on the boundary TiO_2/SiO_2 titanium atoms in an octahedral environment with a different degree of distortion and splitting of the rutile band in ESDR-spectrum into two components with different energies. It must be emphasized that the method of tangents from the absorption band edge allows us to characterize the coordination state of titanium in only one of coordination polyhedra present at the surface (Table 3). The values of E_0, obtained by this method are lower than those obtained by mathematical treatment of spectra.

CONCLUSION

Two methods have been proposed for handling the complex Electronic Spectrum of Diffuse Reflection, which allow distinguishing the contribution of surface structures with different coordination state of titanium atom. It is shown that the technological prehistory and chemical modification of the surface of industrial titanium dioxide have a significant impact on the coordination state of titanium atoms on the material's surface.

REFERENCES

[1] Shabanova, N. A., Popov, V. V., & Sarkisov, P. D. *Chemistry and Technology of Nanodispersive Oxides*; Akademkniga: Moscow, 2006; 309 p. [in Russian].

[2] Artem'ev, Y. M., Ryabchuk, V. K. *Introduction to Heterogeneous Photocatalysis*; St. Peterburg State Univ.: St. Petersburg, 1999; 304 p. [in Russian].
[3] Ambrus, Z., Mogyorósi, K., Szalai, Á., Alapi, T., Demeter, K., Dombi, A., & Sipos, P. *Appl Catal A*. 2008, *vol 340, no 2*, 153-161.
[4] Kiselev, V. F., Krylov, O. V. *Adsorption and Catalysis on Transition Metals and Their Oxides*; Springer-Verlag: Berlin & New York, NY, 1989; 445 p.
[5] Atkins, P.; Friedman, R. *Molecular Quantum Mechanics;* Oxford Univ. Press: Oxford, 2005; 573 p.
[6] Aleskovskii, V. B. *Chemistry of Supramolecular Compounds*; St. Petersburg State Univ.: St. Petersburg, 1996; 256 p. [in Russian].
[7] Marfunin, A. S. *Physics of Minerals and Inorganic Materials: An Introduction*; Springer-Verlag: Berlin & New York, NY, 1979; 340 p.
[8] Pak, V. N., Ventov, N.G. *Zh Fiz Khim*. 1975. *vol 49, no 10*, 2535-2537.
[9] Gao, X., Bare, S. R., Fierro, J. L. G., Banares, M. A., & Wachs, I. E. *J Phys Chem B*. 1998, *vol 102, no 29*, 5653-5666.
[10] Sosnov, E. A., Malkov, A. A., & Malygin, A. A. *Integrated Ferroelectrics*. 2009, *vol 103, no 1,* 41-51.
[11] Sosnov, E. A., Malkov, A. A., & Malygin, A. A. *Rus J Phys Chem A*. 2009, *vol 83, no 4*, 642-648.
[12] Sosnov, E. A., Vasil'eva, K. L., & Malkov, A. A. *Rus J Phys Chem A*. 2010, *vol 84, no 6*, 1028-1032.
[13] Pak, V. N. *Zh Fiz Khim*. 1976, *vol 50, no 5*, 1266-1268.
[14] Pak, V. N., Tsvetkov, V. K. In *Workshop on the Chemistry of Solids*; Kol'tsov, S. I., Korsakov, V. G., & Smirnov, V. M.; Ed.; Leningrad State Univ.: Leningrad, 1985; pp 160-174. [in Russian].
[15] Kanna, M., Wongnawa, S. *Mater Chem & Phys*. 2008, *vol 110, no 1*, 166-175.
[16] Ohtani, B.; Prieto-Mahaney, O. O.; Li, D.; & Abe, R. *J Photochem & Photobiology A*. 2010, *vol 216, no 2-3*, 179-182.
[17] Ivanov, A. P. *Optics of Scattering Media*; Nauka i Tekhnika: Minsk, 1969; 592 p. [in Russian].
[18] Klaas, J., Schulz-Ekloff, G., & Jaeger N. I. *J Phys Chem B*. 1997, *vol 101, no 8*, 1305-1311.
[19] Gao, X., Bare, S. R., Weckhuysen, B. M., & Wachs, I. E. *J Phys Chem B*. 1998, *vol 102, no 52*, 10842-10852.
[20] Pankove, J. I. *Optical Processes in Semiconductors;* Dover Publications: New York, NY, 1971; 422 p.
[21] Radecka, M. *Materials Science-Poland*. 2006, *vol 24, no 1*, 285-290.
[22] Olthof, B., Khodakov, A., Bell, A. T., & Iglesia E. *J Phys Chem B*. 2000, *vol 104, no 7*, 1516-1528.
[23] Augugliaro, V., Kisch, H., Loddo, V., Lopez-Munoz, M. J., Marquez-Alvarez, C., Palmisano, G., Palmisano, L., Parrino, F., & Yurdakal, S. *Appl Catal A*. 2008, *vol 349, no 1-2*, 189-197.
[24] Ivanov, A. P., Loiko, V. A., & Dik, V. P. *Light Propagation through Close-Packed Disperse Media*; Nauka i Tekhnika: Minsk, 1988; 191 p. [in Russian].
[25] Yavorskii, B. M., Detlaf, A. A., & Lebedev A. K. *Handbook on Physics*; Oniks: Moscow, 2006; 1056 p. [in Russian].
[26] Savel'ev, I. V. *Course of General Physics*; Lan': St.Petersburg, 2008; Vol. 3, 320 p. [in Russian].
[27] Janus, M., Morawski, A. W. *Appl Catal B*. 2007, *vol 75, no 1-2*, 118-123.
[28] Chiarello, G. L., Selli, E., & Forni, L. *Appl Catal B*. 2008, *vol 84, no 1-2*, 332–339.

[29] Gosztola, D. J., Saponjic, Z., Chen, L., & Rajh, R. In *5th Int. Conf. on Synchrotron Radiation in Materials Science (SRMS-5)*; Chicago, July 30-Aug. 2, 2006; pp 107.
[30] Khazin, L. G. *Titanium Dioxide*; Khimia: Leningrad, 1970; 176 p. [in Russian].
[31] Dobrovol'skii, I. P., Lenev, L. M., Antipov I. V., & Nazarov Yu. N. *Some Physico-Chemical Basis of the Synthesis of the Chloride Titanium Dioxide*; South-Ural Publ. House: Chelyabinsk, 1974; 179 p. [in Russian].

In: Rutile: Properties, Synthesis and Applications
Editor: It-Meng (Jim) Low

ISBN 978-1-61942-233-9

Chapter 8

FERROMAGNETISM IN TRANSITION METAL DOPED RUTILE TiO_2

A. Mudarra Navarro[1], L. A. Errico[1,2], C. Rodriguez Torres[1], A. F. Cabrera[1], V. Bilovol[3], M. Rentería[1] and M. Weissmann[4]

[1]Departamento de Física – IFLP (CONICET La Plata),
Fac. de Cs. Exactas, UNLP, CC 67, 1900 La Plata, Argentina
[2]Universidad Nacional del Noroeste de la Provincia de Buenos Aires (UNNOBA)
Monteagudo 2772, Pergamino, CP 2700, Buenos Aires, Argentina
[3]Laboratorio de Sólidos Amorfos, INTECIN, Fac. de Ingeniería,
Universidad Nacional de Buenos Aires, Paseo Colón 850, C1063ACV,
Buenos Aires, Argentina
[4]Depto. de Física, Comisión Nacional de Energía Atómica,
Av. del Libertador 8250, 1429 Buenos Aires, Argentina

ABSTRACT

In this work we review our theoretical investigation of TM-doped rutile TiO_2 system (TM: Mn, Fe, Co, Ni, and Cu) for different concentrations and distributions of these impurities. The effect of oxygen vacancies on the magnetic moments and the magnetic alignment for pure and doped TiO_2 is studied as a function of vacancy concentration and distribution.

We also review the experimental work done in our laboratory using X-ray absorption, magnetometric techniques and Mössbauer spectroscopy.

The combined theoretical and experimental studies enabled us to characterize the local structure around the TM impurities in rutile TiO_2 unambiguously and support the conclusion that oxygen vacancies are fundamental for the ferromagnetic response of these systems, explaining the diversity of magnetic moments observed experimentally for samples grown in different conditions.

1. INTRODUCTION

Titanium dioxide in its two phases, rutile and anatase, is one of the most investigated and most widely used metal oxides for several technological applications. This oxide is used in heterogeneous catalysis, as a photocatalyst, in solar cells for the production of hydrogen and energy, as a gas sensor, as a white pigment in paints and cosmetics products, as a corrosion-protective coating, as an optical coating, in ceramics applications and in electronic devices. It also plays an important role in Earth Sciences and in bio-implants, among others. A list of applications can be found, for example, in Ref. 1. TiO_2 is also of significant interest in the emerging area of spintronics. The goal of this research field is to extend the properties and applications of established electronic devices by making use of the spin of electrons in addition to their charge to convey information. Such spin-polarized electronic devices could be much smaller, consume less energy, and be more efficient that actual electronic devices which are based on electron charge only. In addition to its potential utility, the study of spin-polarized transport is revealing new fundamental physics [2, 3, 4].

Magnetic devices, such as reader heads for magnetic data storage which use the giant magnetoresistance effect, are based largely on ferromagnetic metals. The development of magnetic semiconductors would open new possibilities. Magnetism and semiconducting properties are known to coexist in some ferromagnetic semiconductors, such as europium chalcogenides [5] and ferrimagnetic or ferromagnetic semiconducting spinels [6, 7], that have a periodic array of magnetic elements. Often such magnetic semiconductors are difficult to grow and have low Curie temperatures (T_C), typically well below 100 K. An alternative is offered by the so-called *diluted magnetic semiconductors* (DMS) in which non-magnetic semiconductors are doped with magnetic atoms. DMS's based on II-VI compound semiconductors such as Mn-doped CdTe and ZnSe [8] were the first to be studied. However, this family of DMS has a T_C in the order of 1 K and is therefore not useful for practical applications. DMS based on III-V semiconductor compounds, such as Mn-doped GaAs [9] have T_C as high as 110 K but this is still far from room temperature. Room-temperature ferromagnetism has been observed in other Mn-doped compounds, such as $CdGeP_2$, [10] $ZnGeP_2$, [11] and ZnO [12].

In 2001, Matsumoto *et al.* [13] reported that Co-doped anatase TiO_2 films grown by a combinatorial pulsed-laser-deposition (PLD) molecular beam epitaxy (MBE) technique can keep ferromagnetic order up to 400 K. The measured saturated magnetic moment per Co atom was 0.32 μ_B/Co for a Co concentration up to 8 at. %. More recently, Co-doped anatase TiO_2 films grown by the oxygen-plasma assisted (OPA-MBE) method were reported by Chambers *et al.* [14] to have a magnetic moment of 1.26 μ_B/Co. It is interesting to note that Matsumoto *et al.* observed magnetic behavior in Co-doped TiO_2 anatase, but not in the rutile phase, suggesting that the magnetization of the Co atoms could depend on the local structure around it. However, in 2002, Park *et al.* were able to grow ferromagnetic Co-doped rutile TiO_2 rutile films by sputtering. T_C was estimated to be above 400 K [15]. The measured magnetic moment was 0.94 μ_B/Co and the Co concentration was 12 at. %. Since the rutile structure is thermodynamically more stable than the anatase one, it could have a higher potential for technical applications if thin films could be grown easily.

This variety of results motivated intensive experimental and theoretical [3, 4, 16-34] studies of the structural, magnetic, and electronic properties of doped TiO_2 but the microscopic mechanism of long-range magnetic order is still an open problem. Carrier-induced interaction between the magnetic atoms was suggested as the reason for ferromagnetism in III-V based DMS [35]. However, subsequent reports raised concerns about this idea, due to the possibility of ferromagnetic metal clustering under different growth conditions [12, 36]. Furthermore, it has been suggested that the strong interaction between Co and oxygen vacancies in Co-doped TiO_2 plays a key role for the explanation of its high T_C [16,17, 37-41]. Experimental results also suggest that oxygen defects may be responsible for the observed ferromagnetism in Fe-doped TiO_2 samples [42]. In some recent papers [43, 44] a possible hint on the origin of oxide ferromagnetism is given. In contrast to mass magnetization (assuming that the whole sample mass is ferromagnetic) it is suggested that the ferromagnetic part of the sample volume is small, and concentrated possibly in the near surface regions. In fact, the defect concentration and not the doping by magnetic ions seems to be the limiting factor for the appearance of magnetism. These defects produce an impurity band inside the oxide band gap and a Stoner type criterion may be used to predict magnetic behaviour. The dopants contribute to this picture but are not the determinant factor.

The location and distribution of the impurities in the host lattices is also still unclear. While Soo *et al.* [45] found through extended X-ray absorption fine structure spectroscopy (EXAFS) that the local environment surrounding Co impurities remains anatase-like, Kim *et al.* [46] claim that Co tends to cluster and form metallic Co. At this point it is important to clarify what is meant by DMS. In the words of S. A. Chambers [4]: "*A true DMS is one in which the magnetic dopant spins retain remnant alignment under the influence of spin polarized free carriers, which are introduced by either the magnetic dopant or by some secondary electronic dopant. It is critical to distinguish true ferromagnetic semiconductors from those which merely show magnetic hysteresis. The latter can result from other physical causes, such as phase separation leading to magnetic secondary phase formation*".

As a conclusion, the field is marked by enough activity and controversy, and the use of these systems in spintronic devices has today two types of limitations: the first one is the preparation of a homogeneous solid solution of the dopant in the oxide host. The second one is the understanding of the mechanism associated with the ferromagnetic ordering.

In this work we will summarize our theoretical and experimental results obtained in transition metal (TM) doped rutile TiO_2. In order to characterize these samples we used X-ray absorption experiments, ^{57}Fe Mössbauer spectroscopy and magnetometry. The combination of all these results with electronic structure *ab initio* calculations enabled us to give an exhaustive characterization of the local structure around TM impurities in rutile TiO_2. This point is very important because conjectures were usually based on macroscopic experimental results and relatively little attention was given to the local structure around the dopants. We believe that the local environment of the dopant may be essential to understand the mechanisms giving rise to magnetic order in these compounds. Another objective of this work is to determine the role played by the concentration and distribution of the impurities and oxygen vacancies on the magnetic moments and in the magnetic order. Finally, our results are compared with theoretical and experimental results reported in the literature. The validity of different models proposed to understand the origin of the magnetism and the high Curie temperature of these systems is also discussed.

This article is organized as follows. In the next section we present a review of experimental results reported by other authors as well as our own results concerning structural, magnetic and Mössbauer hyperfine parameters focusing in those that could be compare with calculated ones. Section 3 contains a brief description of our theoretical approach used and their results and the comparison with experimental ones. Finally, in section 4 we summarize our conclusions.

2. EXPERIMENTAL SECTION

2.1 Experimental details

In this section we present our own results and also those of other authors. Concerning our samples, we obtained by mechanical alloying Fe-doped rutile TiO_2 with iron concentration between 2.5 and 15 at. % starting from TiO_2+α-Fe_2O_3 (hematite group) and TiO_2+FeO (*wurtzite* group). Details on sample preparation are described in references 47 and 48. The samples were characterized by X-ray diffraction (XRD), extended X-ray absorption fine structure (EXAFS) spectroscopy and Mössbauer and magnetization versus applied magnetic field techniques. The ^{57}Fe Mössbauer spectra (MS) were taken at room temperature (RT) under constant acceleration of the source. The isomer shifts quoted are referred to α-Fe. The magnetic measurements were carried out with a PPMS (Physical Properties Measurements System) equipment at RT. EXAFS measurements were carried out at RT in transmission mode at the Fe *K*-edge, using a Si(111) monochromator at the XAFS2 beamline of LNLS (Campinas, Brazil).

2.2 Co- and Fe-doped rutileTiO_2

Cobalt and iron have possibly been the most frequently used dopants to explore the magnetic behavior of rutile TiO_2. The samples have been prepared by a variety of methods and different dopant concentrations in different forms such as thin films and powders. In the case of films, the rutile phase is obtained mainly when depositing on saphire [21, 24, 42, 49-52] or on rutile TiO_2 single-crystal substrates [53-55]. In the case of powders, it is obtained mainly by mechanical alloying [47, 56-60]. Transition metal (TM) doped rutile was also obtained by implantation on TiO_2 rutile single-crystals [61-63], by vaporization-condensation [64] or by solid state reaction [65] methods.

While some works report that the TM ion substitutes Ti^{4+} in rutile structure [42, 49, 50, 54, 55, 59, 64, 66] others indicate the presence of precipitates of metallic TM, TM oxides and/or TM-Ti oxides [51, 56, 62, 63]. Generally, the observation of precipitates occurs for high TM concentrations (higher than 5 at. % for Fe and 10 at. % for Co). In some cases, such as in reference 42, tiny cobalt associations were detected but not necessarily metal clusters.

In cases where a TM ion substitutes for a Ti one in the rutile structure, it was reported that the incorporation is followed by the formation of oxygen vacancies [24, 32, 53, 57] and by an expansion of the rutile cell [58]. In the case of Co, the preferred oxidation state is 2+ but in the case of iron it is both 3+ and 2+.

As we mentioned before, we obtain Fe doped TiO_2 rutile with iron concentrations between 2.5 and 15 at. % by mechanical alloying. We found by EXAFS measurements that Fe ions substitute Ti ones in the rutile TiO_2 structure. Iron ions are in 3+ and 2+ oxidation states depending on the impurity concentration and sample fabrication conditions. We observed that Fe-O average distances (beyond 1.98 Å) are larger than Ti-O average distances in the rutile phase (1.95 Å) and we found a strong correlation between the number of oxygen nearest neighbors, the Fe^{2+} fraction and the Fe-O average distance: the number of oxygen nearest neighbors decreases and the average distance increases when the Fe^{2+} fraction increases (see Figure 1).

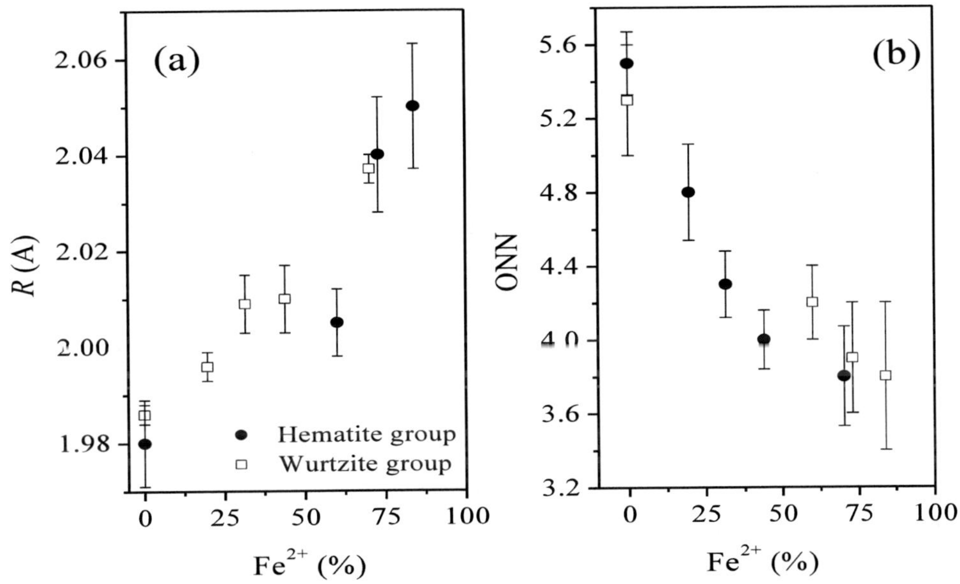

Figure 1. a) Average Fe-O distance and b) number of oxygen nearest neighbors, both obtained by EXAFS analysis versus Fe^{2+} fraction obtained by Mössbauer spectroscopy for Fe-doped TiO_2 rutile samples.

Table I. Parameters obtained from Mössbauer spectra fitting reported in the literature for Fe-doped TiO_2 rutile samples. δ: Isomer shift and ΔQ: quadrupole splitting

Ref.	Fabrication conditions	Fe 3+			Fe 2+		
		δ (mm/s)	ΔQ (mm/s)	Area (%)	δ (mm/s)	ΔQ (mm/s)	Area (%)
[59]	Milling in Ar	0.4	0.56	21	0.98	2.22	79
	Milling in air	0.38	0.65	72	0.99	2.25	28
	Milling in O_2	0.43	0.61	100	-	-	-
[56]	Milling in vacuum	0.37	0.55	16	0.99 0.96	2.49 1.84	44 40
[64]	Vaporization-condensation method	0.5 0.5	0.46 0.87	33 67	-	-	
[65]	PLD deposited thin film 10^{-1} Torr	0.36	0.63	100			

The spectroscopy based on the Mössbauer effect [67] is a very interesting technique in order to compare experimental data with those obtained by electronic structure *ab initio* calculations. In this work we will compare the experimental *quadrupole splittings* (ΔQ) and the isomer shifts (δ) hyperfine parameters with the calculated ones. The first one is a "fingerprint" of the local symmetry around the probe nucleus and the second one provides information on the charge density of *s*-electrons at the nucleus and also on the local chemical bond.

Several works have used Mössbauer spectroscopy to characterize Fe-doped rutile TiO_2 [47, 52, 56-60, 64, 65]. In all of them, only paramagnetic interactions have been detected at room temperature (RT) for pure Fe-doped rutile TiO_2. The reported parameters are shown in Table I.

The hyperfine interactions were associated with ions in 2+ and 3+ oxidation states [56, 59] for samples milled in air and argon atmospheres, whereas in samples obtained in O_2 atmosphere only an interaction associated with Fe^{3+} ions is detected [59]

The relative area of Fe^{2+} interaction increases with decreasing oxygen partial pressures. Furthermore, studies in samples prepared by vaporization-condensation method, show only a paramagnetic interaction with isomer shift and quadrupole splitting corresponding to Fe^{3+} ions in two different sites at 80 K [64]. In the case of the film, only an interaction associated to Fe^{3+} is detected. It can be seen that in all systems the values of the hyperfine parameters associated with Fe^{3+} or Fe^{2+} are similar.

Table II. Mössbauer fitting parameters of our Fe-doped rutile TiO_2 samples at room temperature

$TiO_2 + Fe_2O_3$						
	Fe^{3+}			Fe^{2+}		
at. %	ΔQ_1 (mm/s)	δ_1 (mm/s)	Area (%)	ΔQ_2 (mm/s)	δ_2 (mm/s)	Area (%)
2.5	0.61_3	0.39_2	30	2.13_4	1.00_3	70
5.0	0.66_4	0.39_4	80	2.21_2	0.95_4	20
7.0	0.69_2	0.38_3	100	-	-	-
12.5	0.72_4	0.39_5	56	2.11_3	0.98_3	44
15.0	0.69_3	0.39_3	68	2.21_3	0.96_2	32
$TiO_2 + FeO$						
	Fe^{3+}			Fe^{2+}		
2.5	0.56_3	0.41_2	60	-	-	-
	1.09_3	0.39_3	41	-	-	-
5.0	0.68_2	0.39_1	40	2.06_4	0.98_2	60
7.5	0.76_4	0.41_1	27	2.07_2	1.00_3	73
10	0.87_4	0.38_4	16	1.99_2	1.01_2	84

Table II shows the Mössbauer parameters obtained from the fitting of our Fe-doped rutile samples (hematite and wurtzite groups) spectra. In agreement with the cited references we get only paramagnetic contributions for both groups of samples. Neither contributions of bcc-Fe nor magnetic iron oxides were detected. Hence, for both groups of samples, the spectra were fitted with two doublets, one of them with isomer shift belonging to Fe^{2+} ($\delta \cong 1$ mm/s) and the other one to Fe^{3+} ($\delta \cong 0.4$ mm/s). In the sample doped with 2.5 at. % of Fe (Wurtzite group), two doublets corresponding to Fe^{3+} were required to reproduce the spectra.

Figure 2 shows the hyperfine parameters as a function of iron concentration. It can be seen that ΔQ_1 (Fe^{3+}) increases with Fe concentration in both groups. The isomer shifts do not show significant variations between both groups. Values of ΔQ_2 (Fe^{2+}) are between 1.99 and 2.21 mm/s in good agreement with those reported in the other works (see Table I).

Concerning the magnetic properties, pure rutile TiO_2 is paramagnetic at room temperature but when it is doped with transition metals its magnetic behavior can change. The literature is extensive in this topic and the reported results are diverse. The magnetic behaviors of these samples could be paramagnetic, ferromagnetic, superparamagnetic or antiferromagnetic. Nowadays it is not clear under which preparation conditions each behavior is displayed. In order to give a panorama of this investigation a brief summary on the magnetic response in the rutile TiO_2 doped with Fe and Co transition metals will be presented.

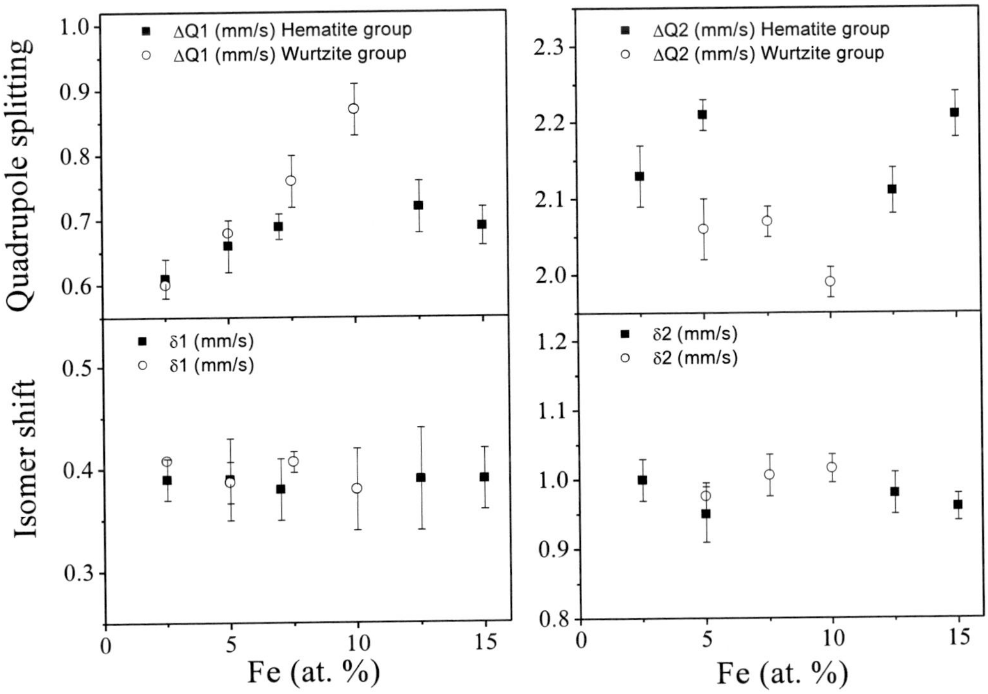

Figure 2. Quadrupole splitting and isomer shift of the Fe doped TiO_2 rutile samples obtained by mechanical alloying (hematite and wurtzite).

Some authors, like Balcells *et al.* [64] and Nomura *et al.* [52] have reported paramagnetism in Fe-doped samples prepared by different techniques. In both cases the doped oxide remains paramagnetic independently of the preparation method and forms of the sample. The oxidation state of the Fe ions was 3+ for the nano-particles with 13 at. % of Fe [64] and for the films with 6 at. % of Fe [52]. On the contrary, there are no reports on paramagnetic Co-doped rutile oxide.

Table III. Results reported in the literature for TM-doped rutile TiO_2 and our results for the saturation magnetization M_s in units of μ_B per TM ion. g is the spectroscopic splitting factor taken as 2 and J is total angular moment. TT: thermal treatment

TM	Preparation conditions	Concentration (at. %)	Ms (μ_B/TM)	TM magnetic moment (gJ)	Ref.
Fe	Vaporization condensation method	13		5	[64]
Fe (films)	PLD on Silicon substrates	8	0.09		[66]
		12.6	0.051		
Fe (films)	Spin – on technique on sapphire substrates	5	0.09		[42]
		5+ TT at 500C in vacuum	0.14		
		5+ TT at 600C in vacuum	0.48		
Fe (powders)	Solid state reaction	3	0.0625		[68]
Fe (powders)	Milling in vacuum	2.7	0.14		[56]
Fe (powders)	Milling	5	0.015 (Ar) 0.013(air) 0.009 (O_2)		[59]
Fe (films)	PLD	2-8	2.3-2.4		[50]
Co(films)	Magnetron sputtering	2.6	0.54		[69]
		17	0.04		
Co (films)	Spin-on technique	3+ TT at 600 C in vacuum	0.68		[39]
Co (films)	PLD	12	0.31		[70]
Co (films)	RT deposition on TiO2 (100)	5	0.9		[53]
Co (films)	Sputtering	6	0.94		[71]
Co (films)	Laser molecular beam epitaxy	3	1		[21]
		5	1.1		
		10	1		
			1.5		
Co/Fe (powders)	milling	9.5/0.5	1.02		[74]

TM	Preparation conditions	Concentration (at. %)	Ms (μ_B/TM)	TM magnetic moment (gJ)	Ref.
CoFe (powders)	Solid state reaction	5/5	0.24		[68]
Co (films)	Laser molecular beam epitaxy	5	1		[72]
Co (single crystals)	High temperature ion implantation	3	0.6		[73]
Fe (wurtzite group)	Milling in air	2.5	0.12	4.92	This work, [74]
		5	0.13	4.8	
		7.5	0.083	4.44	
		10	0.079	4.34	
Fe (hematite group)	Milling in air	2.5	0.33	3.4	
		5	0.14	3.86	
		7	0.07	3.28	
		12.5	0.05	3.08	
		15	0.04	3.8	

In the literature, several works have reported ferromagnetism for Co-, Fe- or Co/Fe-doped samples independently of the preparation method or even by using the same method but with different experimental conditions. In Table III and Figure 3 a) and b) the saturation magnetization (M_S) reported values are presented. It can be observed that M_S diminishes for both Fe and Co when the concentration of the TM increases.

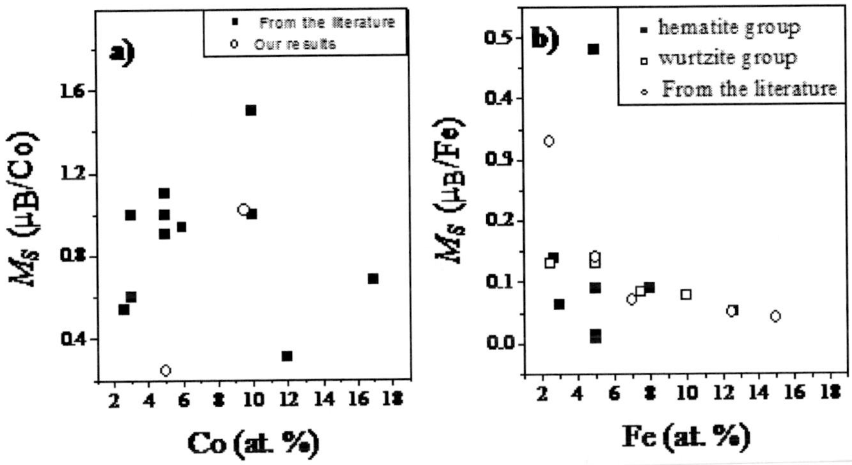

Figure 3. Magnetization saturation at room temperature versus the TM concentration.

In general the magnetic moments reported for Fe-doped oxides are lower than those of Co ones independently of the condition and preparation method. For the Co dopant the M_S are spread between 0.02 and 1.5 μ_B/Co, while for Fe dopant the dispersion is smaller and the values are between 0.01 and 0.5 μ_B/Fe. In our samples we found a mixture of paramagnetic and ferromagnetic like behavior. We separate both contributions by fitting *M* vs *H* cycles with an appropriate function [75]. The obtained saturation magnetization values at room temperature (see Figure 3b) are in good agreement with the reported ones. Furthermore, we observed that as the relative percentage of Fe^{2+} iron increases the saturation magnetization decreases.

In Table III we also report the TM magnetic moment estimated as *gJ* (*J* values were obtained by fitting 5 K hysteresis loops using a Brillouin function for the paramagnetic contribution). The general trend indicates that when the dopant concentration increases the magnetic response diminishes independently of the preparation technique and the experimental conditions. In all the results reported, the oxygen vacancies may play a crucial role for the appearance of ferromagnetism. The effect of doping seems to be to increase the number of vacancies. In some cases the strong interaction between TM and vacancies is invoked to explain the magnetic moment values.

3. THEORETICAL SECTION

3.1. The System under Study: Rutile TiO_2 Structure

Titanium dioxide crystallizes in three major different structures: anatase (tetragonal, *$I4_1/amd$*), rutile (tetragonal, *$P4_2/mnm$*), and brookite (rhombohedrical, *Pbca*) [1, 76], being rutile the most stable configuration at room temperature. Other structures exist as well. For example, cotunnite TiO_2 has been grown at high pressures and is one of the hardest polycrystalline materials known, see Ref. 1 and references therein.

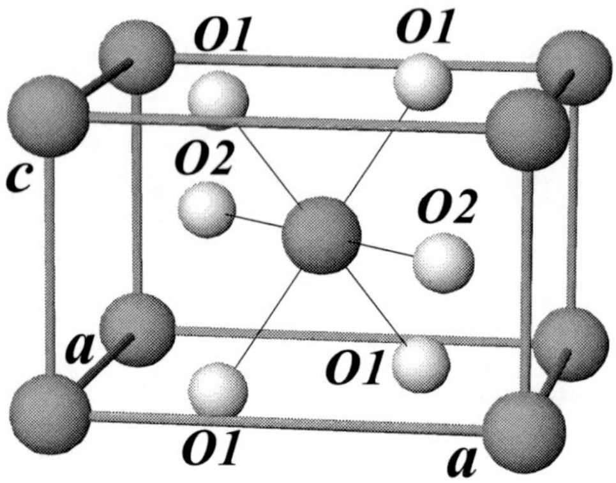

Figure 4. Unit cell of rutile TiO_2 (Ti gray balls, O: white balls).

The unit cell of rutile TiO_2 contains two Ti atoms at positions (0, 0, 0) and (1/2, 1/2, 1/2) and four oxygen atoms at positions (u, u, 0; ½ + u, ½ + u, ½) with u=0.30493$_7$ and a = b = 4.5845 Å, c = 2.9533 Å [77]. The structure has only one crystallographic Ti site. These Ti atoms are surrounded by a slightly distorted octahedron of oxygen atoms, with a rectangular basal plane (O1) at a distance of 1.94 Å from Ti and two vertex atoms (O2) at 1.98 Å (see Figure 4). For Oxygen there is only one crystallographic site, and each oxygen atom is surrounded by three Ti.

3.2. Method of calculation

We performed *ab initio* electronic structure calculations to determine the self-consistent potential and the charge density inside the rutile TiO_2 cell. Our aim here is to obtain from first-principles calculations electronic and magnetic structures and relaxation effects. In the case of the Fe impurity we also calculated the hyperfine parameters at the impurity that replaces Ti in rutile TiO_2 taking properly into account the structural and electronic effects introduced by the impurity and/or oxygen vacancies in the host lattice. We simulated these non-periodic systems considering a periodically repeated large unit cell where some TM-atoms replace Ti in the rutile TiO_2 host. The super-cell (SC) considered here consists of eight unit-cells of rutile TiO_2. If only one Ti atom is replaced the composition is $Ti_{0.9375}TM_{0.0625}O_2$, which represents about $x = 6$ at.% of doping. The resulting 48-atoms SC (called 48A-SC in the following) has dimensions $a' = 2a = b' = 2b = 9.1690$ Å, $c' = 2c = 11.8132$ Å (Figure 5 a). Calculations were also performed in the SC consisting of two unit cells of TiO_2 stacked along the c-axis with one Ti atom replaced by the magnetic impurity (Figure 5 b). The resulting 12A-SC has dimensions $a' = a$, $b' = b$, $c' = 2c$. The impurity concentration in this SC is 25% ($x = 0.25$). For checking purposes, SC´s of dimensions $a' = b' = 2a$, $c' = 3c$ and $a' = b' = 3a$, $c' = 2c$ were also considered.

To solve the scalar-relativistic Kohn-Sham equations, calculations based on the Density Functional Theory (DFT, Ref. 78) have been performed with the *ab initio* full-potential augmented plane waves plus local orbitals (APW + lo) method [79] as embodied in the WIEN2K code [80]. Exchange and correlation effects were treated within density-functional theory using the local spin density approximation (LSDA) [81]. However, the application of LSDA to solids can also lead to results which are in bad agreement with experiment, e.g., the band gap of semiconductors and insulators is underestimated or even absent. For this reason, and to compare results, calculations using the modified Becke-Johnson functional (MBJ) were performed [82]. This exchange and correlation potential yields band gaps with an accuracy comparable to approaches which are orders of magnitude more expensive in the computational aspect. The LDA + U [83] formalism with the same value of U for the $3d$ states of Ti and of the TM impurities was also employed, but we have to mention here that this approximation predicts a large splitting of the impurity levels, giving rise to large magnetic moments if the value of U is chosen to predict the correct band gap of TiO_2. Moreover, LDA + U must be complemented with a fixed spin moment calculation, as it does not always find the lower energy minimum directly. Additionally, contrary to the conventional model where the charge remains localized at the defect, scanning tunnelling microscopy and resonant photoelectron diffraction show that the charge is delocalized over the surrounding titanium atoms, as is predicted by LSDA [84, 85]. For this reason we present here only the LSDA and MBJ results and leave the comparison between LSDA and LDA + U for a future work.

In the APW+lo method no shape approximation on either the potential or the electronic charge-density is made.

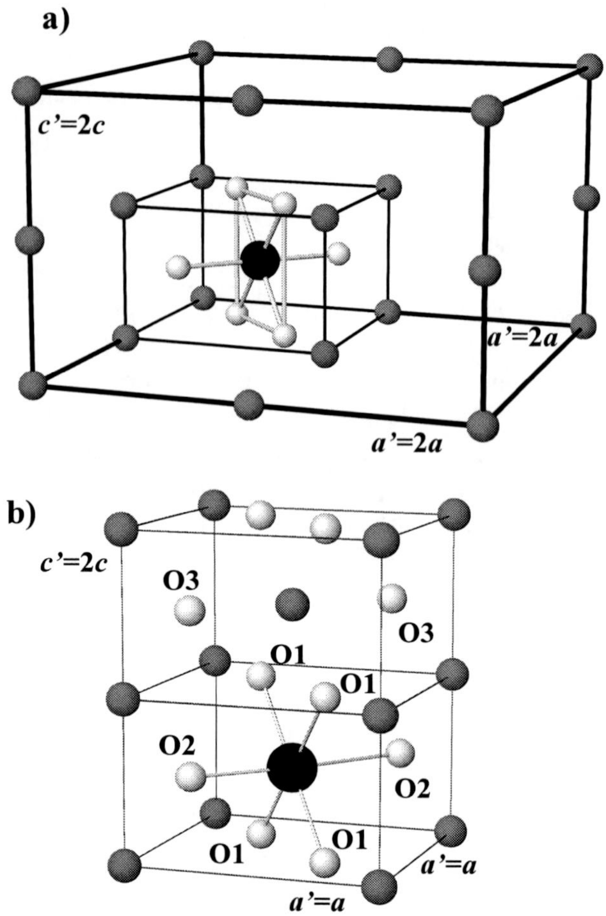

Figure 5. a) 48-atoms super-cell ($Ti_{15}TMO_{32}$) used in the calculations (Ti, gray balls, O, white balls; impurity black balls). Not all O and Ti atoms are shown for clearness. b) 12-atoms super-cell (Ti_3TMO_8). The different types of Ti-neighbors (O1, O2, O3) are indicated.

For methodological purposes the unit-cell is divided into non-overlapping spheres with radius R_i and an interstitial region. The parameter RK_{MAX}, which controls the size of the basis-set, was set to 6.0 (R is the smallest muffin tin radius and K_{MAX} the largest wave number of the basis set). We also introduced local orbitals to include Ti-3*s* and 3*p*, O-2*s* and TM-3*p* levels. Integration in the reciprocal space was performed using the tetrahedron method taking from 50 to 100 *k*-points in the first Brillouin zone. Once self-consistency of the potential was achieved, quantum-mechanically-derived forces were obtained and the ions were displaced according to a Newton-damped scheme, and then the new positions for the atoms were obtained (for details, see Ref. 86). The procedure was repeated until the forces on the ions were below a tolerance value taken as 0.01 eV/Å. In all the calculations we decided to keep the lattice parameters fixed at the experimental ones (although volume changes associated with the concentration of impurities and/or vacancies may be present) so as to make all the results comparable.

To check the accuracy of the present results we performed several additional calculations. For selected systems, we increased the basis set from RK_{MAX} = 5.0 to RK_{MAX} = 7.0, we increase the number of *k*-points from 50 to 200 and studied the effect of the muffin-tin radii on the properties of the system. Magnetic moments, electric field gradients (EFG), isomer shifts and inter-atomic distances can be obtained with adequate precision using $RK_{MAX} = 6.0$ and 50 k-points (the inter-atomic distances, EFG components, magnetic moments and isomer shifts are converged in less than 0.01 Å, $0.1x10^{21}$ V/m^2, 0.02 μ_B and 0.05 mm/s, respectively). Structural relaxation, leading to forces on the atoms smaller than 0.01 eV/Å, is the most important factor for convergence of the electronic properties, and they do not change by increasing the number of plane waves or of k-points. This is not the same for the total energy that requires a larger number of plane waves and of k-points for convergence. However, the energy differences are well converged for the employed parameters.

3.3. Theoretical Results

3.3.1. Systems with one substitutional TM ($Ti_{15}TMO_{32}$)

As a first step in our study, we replace a Ti atom by a transition metal (TM) impurity. We begin the discussion with the cases of Mn, Fe, Co, Ni, and Cu. Spin polarized and non-spin polarized calculations show that the magnetic solutions have lower energies than the non-magnetic ones, hence we will center our work on the magnetic phases.

We find that the presence of substitutional TM impurities in the rutile TiO_2 host produce a structural distortion, changing the TM-nearest oxygen neighbors (ONN) bond lengths. As can be seen in Table IV, the bond lengths TM-O1 and TM-O2 are reduced by about 0.05 Å. These structural distortions are nearly isotropic and their magnitudes are small and essentially independent of the impurity considered. This is quite different from the highly anisotropic expansion found in Fe-doped anatase TiO_2 using the same method of calculation [87]. Comparing these structural relaxations with those reported in Ref. 16 for the case of the 12A-SC, we see that the bond-lengths TM-O1 are nearly the same, but the TM-O2 bond-lengths are larger in the case of the 12A-SC.

This difference is originated by the small impurity-impurity separation in the smallest SC. Calculations performed in SCs larger than the 48A-SC show that no significant variations should be expected if larger SCs were considered. The results for the magnetic moments are summarized in Figure 6 and Table IV. They are almost independent of the local structural distortions and of the impurity concentration. Spin-polarization occurs mainly at the impurity sites, but the ONN can be polarized up to 0.06μ_B/ONN.

In the case of Cu, the structural distortions are anisotropic, the Cu-O1 bond lengths are shorter while the Cu-O2 distances are enlarged by about 0.04 Å (see Table IV). Concerning the magnetic moment we find that, in the case of the 12A-SC, the lowest energy corresponds to a non-polarized state, but a magnetic solution with a magnetic moment of 0.95 μ_B/SC differs from the lowest-energy state in less than $2x10^{-4}$ eV. In the 48A-SC, however, the lowest energy corresponds to a magnetic state with μ_B = 0.95 μ_B/SC. This concentration dependence and the existence of two nearly degenerate solutions show that the variety of results obtained experimentally for different samples also appears in the *ab initio* calculations, even though these are performed for bulk materials and at 0K. A similar dependence of the magnetic moment on the concentration of Cu was also found in calculations for ZnO [88].

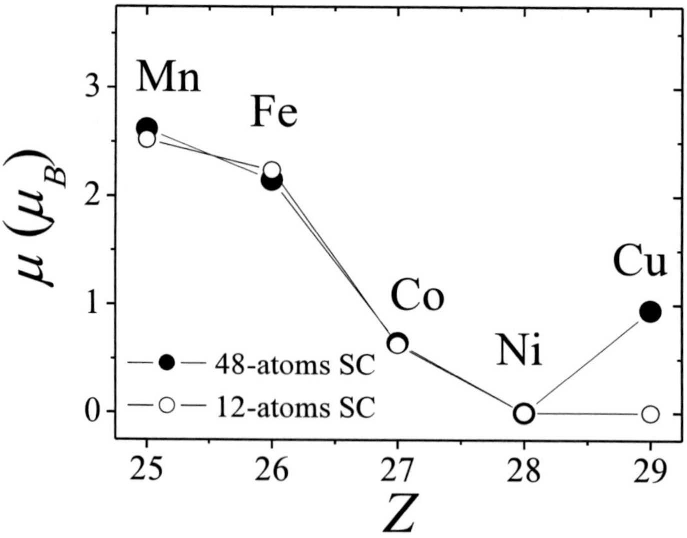

Figure 6. Calculated magnetic moments at the impurity sites as a function of the atomic number Z of the TM impurity for two impurity dilution: 25% (12A-SC) and 6.25% (48A-SC).

Table IV. Calculated relaxed bond lengths *d* between the TM-impurity and the two types of neighbor oxygen atoms in rutile TiO_2 (O1 and O2) for impurity concentrations of 25% (Ti_3TMO_8) and 6.25% ($Ti_{15}TMO_{32}$), magnetic moments in the muffin-tin sphere of the impurities (μ^{imp}) and in the super-cell (μ^{SC})

System	d(TM-O$_1$) [Å]	d(TM-O$_2$) [Å]	μ^{imp} [μ_B]	μ^{SC} [μ_B]
TiO_2	1.94	1.98	0.00	0.00
12-atoms SC				
Ti_3MnO_8	1.91	1.95	2.52	3.04
Ti_3FeO_8	1.89	1.96	2.24	2.58
Ti_3CoO_8	1.90	1.93	0.63	1.03
Ti_3NiO_8	1.91	1.95	0.00	0.00
Ti_3CuO_8	1.93	2.02	0.00	0.00
48-atoms SC				
$Ti_{15}MnO_{32}$	1.90	1.92	2.51	3.00
$Ti_{15}FeO_{32}$	1.89	1.91	2.15	2.49
$Ti_{15}CoO_{32}$	1.90	1.90	0.65	1.00
$Ti_{15}NiO_{32}$	1.90	1.91	0.00	0.00
$Ti_{15}CuO_{32}$	1.92	2.02	0.37	0.95

This result indicates that magnetic ions are not essential to obtain ferromagnetic response and also that is not due to impurity clustering. Another difference between Cu and Mn, Fe, Co, and Ni is that in the case of Cu the magnetic moment is much more delocalized.

In Figure 7 we present the charge and spin densities for Co and Cu in the rutile structure along the (110) and (001) planes, where the difference in spin localization is evident. The hybridization is also very different for the two TM and this raises the question if the mechanism for long-range ferromagnetic order could be the same for Co, Mn, Fe, Ni, and Cu.

To further study the magnetization and exchange coupling, we show in Figures 8 and 9 the density of states (DOS) for pure and doped rutile TiO_2 predicted by the APW+lo calculations. The overall band structure of pure rutile TiO_2 is consistent with previous works (see, for example, Ref. 89) and the semiconducting nature of this material was found: a gap between the valence and conduction bands appears, but the usual problem of the LDA gap underestimation is observed, the minimum gap being 2.0 eV. Experimentally, the gap of TiO_2 was found to be 3.0 eV [90], see Figure 8. LDA + U predicts a band gap of 2.5 eV for $U = 5.4$ eV, meanwhile MBJ predicts a gap in very good agreement with the experimental results, 2.8 eV. From the analysis of the partial DOS, we found that the valence band corresponds essentially to O-2p with a small mixture with Ti-3d states.

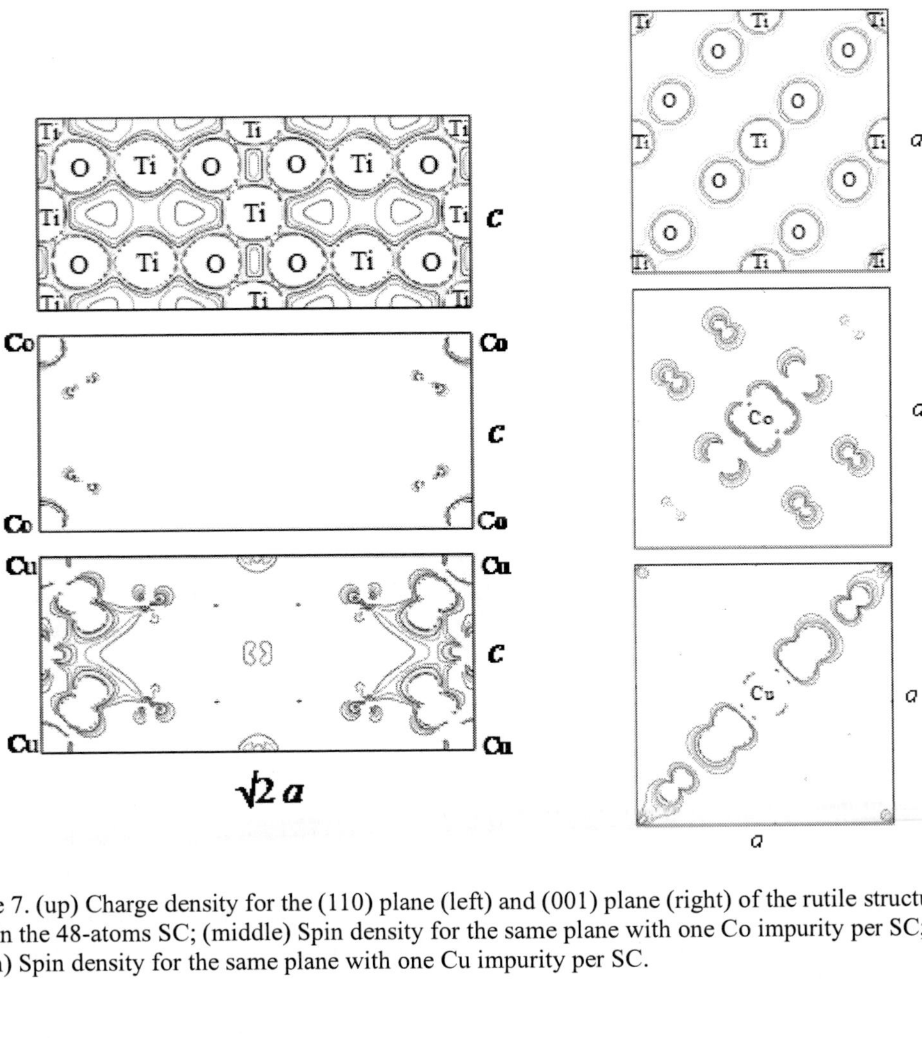

Figure 7. (up) Charge density for the (110) plane (left) and (001) plane (right) of the rutile structure of TiO_2 in the 48-atoms SC; (middle) Spin density for the same plane with one Co impurity per SC; (down) Spin density for the same plane with one Cu impurity per SC.

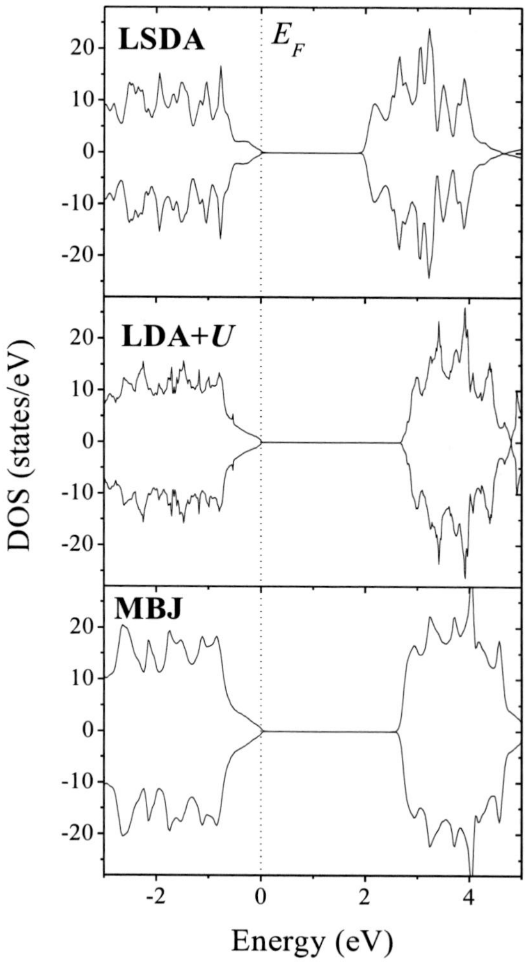

Figure 8. Total density of states (DOS) of rutile TiO_2 calculated using the LSDA, the LDA+U and the MBJ approximations for the exchange and correlation potential. Energies are referred to the Fermi level (E_F), denoted by a vertical dotted line.

The conduction band is almost entirely constituted of Ti-3d levels, a clear indication of the ionic nature of this oxide. In the case of doped TiO_2 (see Figure 9) the impurity states are mainly located in the energy-gap of TiO_2, and contain d features from the impurities along with contributions from the neighboring O1- and O2-p states. The impurity-states are localized along the z direction but also extend in the xy plane, showing that the oxygen atoms could play an important role in mediating magnetic ordering among impurity atoms. For this concentration the systems doped with Mn and Ni show a reduced band gap, while those doped with Fe, Co, and Cu are metallic. The partial DOS projected in the different muffin-tin spheres indicate that the contributions from the neighbouring O1- and O2-p states increase with the number of d electrons. MBJ predicts a semiconductor gap for all cases, as shown in Figure 10. Similar results were obtained in the case of LDA+U (DOS not shown here).

Comparing the calculations with the experimental results we find that the magnetic moments for the different impurities are in clear discrepancy with the experiments. The magnetic moment per Fe atom is larger than those of Co, in contradiction with the results of Figure 3 and Tables I-III.

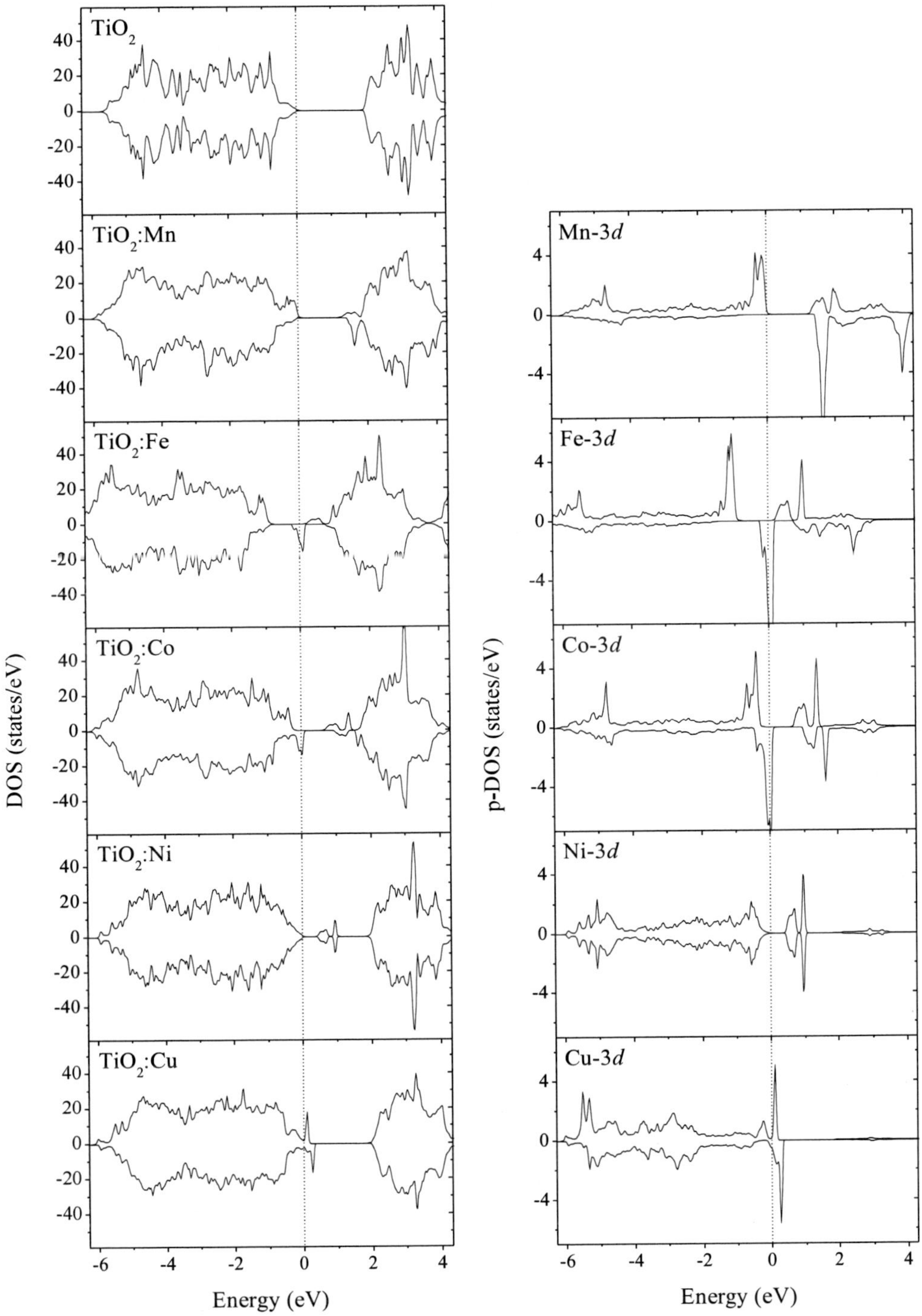

Figure 9. Calculated total DOS of the systems $Ti_{15}TMO_{32}$ (TM: Mn, Fe, Co, Ni, and Cu) and 3*d*-projection at the impurity muffin-tin spheres. In all cases, the DOS correspond to the 48-atoms SCs (impurity dilution 6.25%) and LSDA calculations. Energies are referred to the Fermi level.

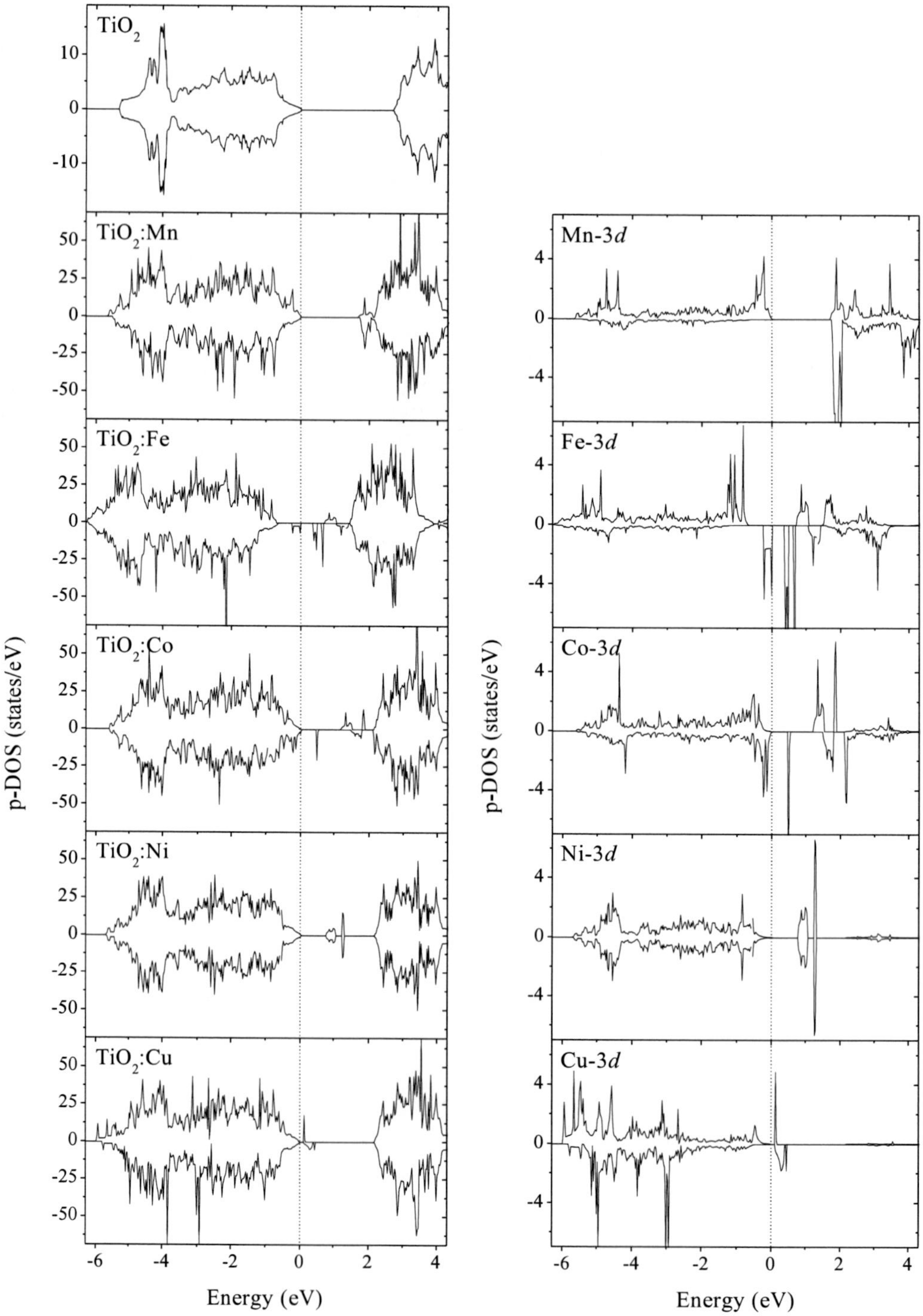

Figure 10. Calculated total DOS of the systems $Ti_{15}TMO_{32}$, (TM: Mn, Fe, Co, Ni, and Cu) and 3*d*-projection at the impurity muffin-tin spheres. In all cases, the DOS correspond to the 48-atoms SCs (impurity dilution 6.25%) and MBJ calculations. Energies are referred to the Fermi level.

The result for Ni (a null magnetic moment) is not in agreement with the experimental result obtained by Hong *et al.* (0.25 μ_B/Ni, Ref. 66). In the case of Cu impurities, a magnetic moment in the order of 1.5 μ_B/Cu was obtained from the magnetization curves [40]. The average Fe-ONN relaxed bond-lengths (1.90 Å) are in poor agreement with the EXAFS results (Fe-ONN bond-lengths larger than 1.98 Å). Further, our *ab initio* calculations predict an isomer shift δ of 0.11 mm/s (see Table V), which is far from the experimental results (Tables I and II). The ΔQ value does not agree with the results obtained from the fitting (Tables V, I and II). Thus, we conclude that a simple substitutional doping cannot account for the magnetic, Mössbauer and EXAFS experimental results. Considering that EXAFS predicts a TM coordination smaller than 6, in the following we will explore a second scenario that includes the formation of oxygen vacancies along with substitutional TM-doping.

Table V. Calculated hyperfine parameters ΔQ and δ at a substitutional Fe impurity. In order to compare with the experimental results, the absolute value of ΔQ must be considered

System	$\|\Delta Q\|$ [mm/s]	δ [mm/s]	$\|\Delta Q^{average}\|$ [mm/s]	$\delta^{average}$ [mm/s]
$Ti_{15}FeO_{32}$	0.85	+0.10	0.85	+0.10
$Ti_{15}FeO_{31}$ vacancy at O1	1.72	+0.47		
$Ti_{15}FeO_{31}$ vacancy at O2	0.38	+0.40	0.88	+0.44
$Ti_{15}FeO_{31}$ vacancy at O3	0.56	+0.45		

3.3.2. Study of the System $Ti_{16}O_{31}$ (One Oxygen Vacancy in TiO_2)

Different experimental and theoretical works suggest that oxygen vacancies play an important role in the ferromagnetism of doped TiO_2 [16-19, 22, 28, 32, 34, 37, 38, 91]. The experimental results obtained by Suryanarayanan et al. in Fe- and Co-doped TiO_2 films [39, 42] and Duhalde et al. in Cu-doped TiO_2 films [40] point out the importance of the oxygen content in the sample growth atmosphere. In addition, ferromagnetism was discovered in thin films of pure TiO_2 and other nonmagnetic oxides [92-105]. One of the mechanisms proposed to explain the observed ferromagnetism is based in the presence of intrinsic oxygen vacancies, which act as donors, leading to n-type doping of the material. This hypothesis is supported by Zhou et al. [99], who showed that ferromagnetism can be induced in pure TiO_2 single crystals by oxygen ion irradiation. By combining X-ray diffraction, Raman scattering, and electron-spin resonance spectroscopy, a defect complex, i.e. Ti3+ ions on the substitutional sites accompanied by oxygen vacancies, were identified in irradiated TiO2. The Ti3+ ions "provide" local 3d moments which are decisively related with the observed ferromagnetism [99].

Let us first study the effect of vacancies in undoped TiO_2. Under the usual growth conditions, an oxygen-deficient phase of rutile $TiO_{2-\delta}$ is found with δ typically in the order of 0.01 or less. For larger values of δ, the system changes to other crystallographic structures, the Magneli phases, which are more stable [106]. For our study, we use the 48A-SC previously described removing an oxygen atom and taking into account the structural distortions produced by the vacancy. For this SC the vacancy concentration is x=0.03, a good compromise between the experimental vacancy concentration, computational cost and accuracy of the calculations (results for the 12-atoms SC were presented in Ref. [16]).

Because TiO_2 is an ionic crystal an oxygen vacancy produces a large structural relaxation, the Ti nearest neighbors to the vacancy repel each other and leave a big hole in the place of the vacancy. This distortion is much larger than those found for the TM impurities: some Ti atoms are displaced as much as 0.2 Å away from their original positions. The distances Ti-vacancy site are enlarged from 1.94 Å and 1.98 Å to 2.13 Å and 2.20 Å. This result is in very good agreement with those obtained by Cho *et al.* [107] for a larger SC (3*a*x3*b*x5*c*) and by von Oertzen and Gerson [108]. Also in agreement with the results of Cho *et al.*, we find that the electron population of the three Ti atoms surrounding the oxygen vacancy increases from the stoichiometric case.

In a previous work, using the 12A-SC [16] we found that there were two possible degenerate solutions, one nonmagnetic and the other having a magnetic moment of approximately 1.7 μ_B/SC, due to polarization of *d*-levels of Ti atoms. Now using the 48A-SC we find the same result and a magnetic moment of 1.8μ_B/SC. It is worth mentioning that the magnetic case appears only when the structural relaxation is taken into account. These results are in agreement with those found by von Oertzen and Gerson [108] and by Lindan *et al.* [109].

To conclude the study of the system $TiO_{2-\delta}$ we calculate the formation energy (E^{vac}) of an oxygen vacancy using:

$$E^{vac} = E^{tot}(\text{vac}) - E^{tot}(\text{st}) + \mu(\text{O}) ,$$

where E^{tot}(vac) and E^{tot}(st) are the total energies of the supercell containing one vacancy and of the stoichiometric SC, respectively, and μ(O) is the chemical potential of the oxygen atom. Assuming that μ(O) is half the total energy of an oxygen molecule, the vacancy formation energy is 5.90 eV. The calculated vacancy formation energy reported by Chen *et al.* [18] and Cho *et al.* [107] are 2.71 eV and 4.44 eV, respectively. This discrepancy is fairly typical of calculations based on DFT and the models employed to obtain the chemical potential of the oxygen atom.

3.3.3. Systems with One Substitutional TM and Oxygen Vacancies ($Ti_{15}TMO_{32-\delta}$)

We study the oxygen deficient systems $Ti_{15}TMO_{31}$ considering three possibilities for removing an oxygen atom from the SC: (i) from the Ti-contained octahedron (O3) in Figure 5, (ii) from the impurity-contained octahedron removing a basal O atom (O1), and (iii) from the impurity-contained octahedron removing a vertex O atom (O2). In all cases, we find that the structural relaxations induced by the oxygen vacancies are larger than those produced by the impurity (see Tables IV and VI). Comparing the total energies of the three cases (and even larger Fe-oxygen vacancies separations) we find that the oxygen vacancies near the impurities are more stable than those near Ti for all the cases studied (Figure 11). Similar results were obtained in the case of Fe in rutile SnO_2 [105] and Fe in anatase TiO_2 [87].

Figure 11 shows that the energy required to form an oxygen vacancy decreases from 5.90 eV in the undoped system to much smaller values in the doped ones. This effect is more important if the vacancies are close to the impurities and if the number of *d* electrons increases. Also, the energy required to substitute a Ti atom by an impurity is strongly reduced when there is an oxygen vacancy in the SC. This reduction is larger for Cu and for nearest-neighbor impurity-vacancy pairs (see Figure 11). The difference in the values of the substitution and vacancy formation energies with respect to those previously published [16] originates in the different impurity concentrations and in the fact that the chemical potentials of the impurity atoms were obtained as a first approximation from their corresponding metals. Therefore we predict that doped systems will have more vacancies than undoped ones, and that oxygen vacancies will tend to be close to the impurities. We also calculated the system with two oxygen vacancies in the SC and found that the energy required to form two vacancies near an impurity is larger than that necessary to create one vacancy in the first shell of neighbours of two different impurities. For this reason, we discard the existence of vacancy clusters around the impurity atoms. However, in case each impurity had already one vacancy close to it a new vacancy would prefer to go there and not to a Ti octahedron.

Table VI. Distance (in Å) from the TM-impurity to its oxygen nearest neighbors in the relaxed structures of $Ti_{15}TMO_{31}$. The number in brackets indicates the multiplicity

System	Vacancy at O1	Vacancy at O2	Vacancy at O3
$Ti_{15}MnO_{31}$	1.93 1.96 1.98 [2] 2.02	1.94 [4] 2.09	1.99 2.00 2.03 [4]
$Ti_{15}FeO_{31}$	1.94 1.95 1.98 [2] 2.02	1.94 [4] 2.11	1.99 2.00 2.03 [4]
$Ti_{15}CoO_{31}$	1.93 1.94 1.97 [2] 2.00	1.90 1.91 [4]	1.99 2.00 2.04 [4]
$Ti_{15}NiO_{31}$	1.93 1.94 1.99 [2] 2.01	1.94 [4] 2.12	1.99 2.00 2.03 [4]
$Ti_{15}CuO_{31}$	1.91 1.96 1.99 [2] 2.08	1.97 [4] 2.04	2.01 [2] 2.03 [2] 2.05 [2]

Concerning the magnetic properties of the systems $Ti_{15}TMO_{31}$ we found that the magnetic solutions have lower energy than the nonmagnetic ones, the magnetic moment being due mostly to polarization of the impurity atoms. Table VII shows that the presence of a vacancy close to the impurity increases the magnetic moment of the unit cell when there already was a magnetic moment, sometimes induces a magnetic moment for Ni and favours the spin-polarized solution in the case of Cu. These results support the idea that ferromagnetism strongly depends on the oxygen deficiency, and the different magnetic moments observed in samples grown by different methods and/or different conditions can thus be understood. It is interesting to note that covalency is larger in bonds between Cu and its ONN than for the others TM, and for this reason the magnetic moment is not so localized in this case. In view of this, it seems now questionable to describe the new type of magnetism with a Heisenberg model Hamiltonian, and also to estimate Curie temperatures with it.

Figure 12 shows the DOS of pure and doped reduced rutile TiO_2 with one oxygen vacancy in O2. The main effect of the vacancy is to move the Fermi level towards the conduction band, so that many of the impurity states inside the gap are now occupied. There is some experimental evidence that the ferromagnetism in these systems is carrier mediated [110], which is supported by the calculated half-metallic DOS. However, the half-metallic density of states obtained by LSDA calculations does not persist if a beyond-LSDA method like LDA + U or MBJ are used. In effect, calculations performed with the LDA + U and the MBJ approximations predict semiconductor behavior (see Figure 13), in agreement with other experimental results [21].

To explain ferromagnetism in substituted zinc oxide, Venkatesan and Coey [31, 111] proposed that high Curie temperatures are found whenever the impurity band overlaps unoccupied 3d states of the host oxide.

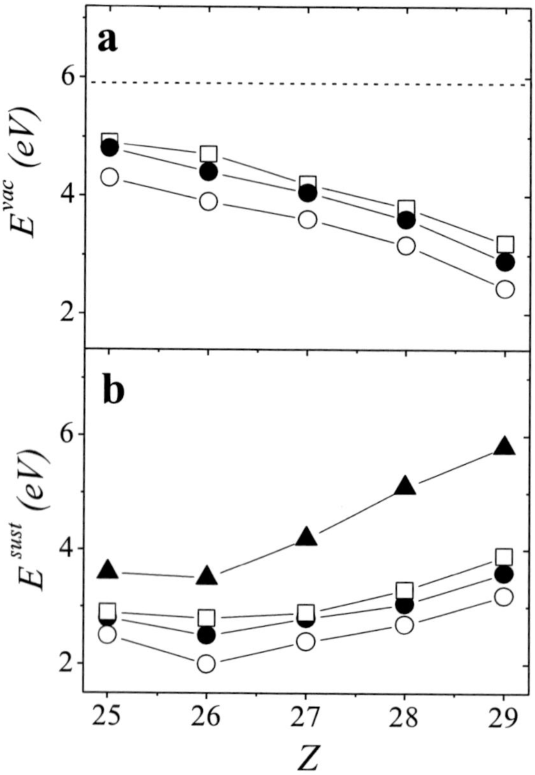

Figure 11. Vacancy formation energies (a) and substitution energies in the presence of vacancies (b) for the different impurities, represented by their atomic number Z. Open circles: vacancy at O1; solid circles: vacancy at O2; open squares vacancy at O3. Dashed line in (a) is the energy necessary to generate an oxygen vacancy in pure rutile TiO_2. Solid triangles in (b) are the substitution energies in stoichiometric TiO_2. Solid lines are just to guide the eyes.

According to our calculations this is not the case for ideal Ti11-xTMxO2, in which the impurity states are in the band gap. However, the presence of oxygen vacancies moves the Fermi level to the conduction band, giving some support to this model. Thus, more detailed computational analyses of the electronic structure with methods that include correlation effects are needed to determine the origin of the ferromagnetism and to explain the high Curie temperatures.

We now compare the predictions of this impurities + oxygen vacancies model with the experimental results. Regarding the hyperfine parameters at the Fe sites in $Ti_{15}FeO_{31}$, a vacancy located at O1 or O2 produces an increment of the isomer shift from 0.11 mm/s (without vacancies) up to 0.44 mm/s and the average of the quadrupole splitting is 0.88 mm/s, in agreement with the Mössbauer results (see Tables I, II and V). Furthermore, for both configurations the average Fe-O_{NN} bond-length is 1.98 Å (Table VI), in good agreement with the distances obtained from the EXAFS analysis (Fig. 1 a) for Fe).

Table VII. Magnetic moments (in units of μ_B) in the muffin tin spheres of the impurities (μ^{imp}) and in the SC (μ^{SC}) for the different impurities and vacancy positions studied

System	Vacancy at O1	Vacancy at O2	Vacancy at O3
	μ^{imp}/μ^{SC}	μ^{imp}/μ^{SC}	μ^{imp}/μ^{SC}
$Ti_{15}MnO_{31}$	3.60 / 5.00	3.02 / 3.10	3.55 / 5.00
$Ti_{15}FeO_{31}$	3.20 / 4.00	2.10 / 2.20	3.45 / 4.2
$Ti_{15}CoO_{31}$	2.25 / 3.00	0.75 / 1.00	2.33 / 3.00
$Ti_{15}NiO_{31}$	1.37 / 2.00	0.00 / 0.00	1.42 / 2.00
$Ti_{15}CuO_{31}$	0.55 / 1.00	0.55 / 1.00	0.59 / 1.00

Very recently, electron paramagnetic resonance (EPR) and electron-nuclear double resonance (ENDOR) were used to characterize Cu^{2+} ions substituting for Ti^{4+} ions in TiO_2 crystals having the rutile structure [112]. This study provides evidence that the Cu^{2+} ions have an adjacent doubly ionized oxygen vacancy and suggests that other divalent cation impurities in TiO_2 such as Co^{2+} and Ni^{+} may also have an adjacent oxygen vacancy for charge compensation.

In conclusion, the experimental results support the idea of TM-impurities substitutionally located at cationic sites but with an oxygen coordination smaller than six.

3.3.4. Systems with Two Substitutional TM ($Ti_{14}TM_2O_{32}$)

In the previous paragraphs ferromagnetic alignment was assumed since only one impurity per unit cell was considered. To be able to determine if the impurities align ferromagnetically (FM) or antiferromagnetically (AF), we performed calculations with two impurities, initially with no oxygen vacancies, in the 48A-SC for different impurity distributions in order to find the preferred configurations and the dependence of the magnetic moment on the impurity distribution. Figure 14 shows the three impurity-impurity configurations studied. There are other possible configurations in the 48A-SC, but with a larger impurity-impurity separation and they were not considered.

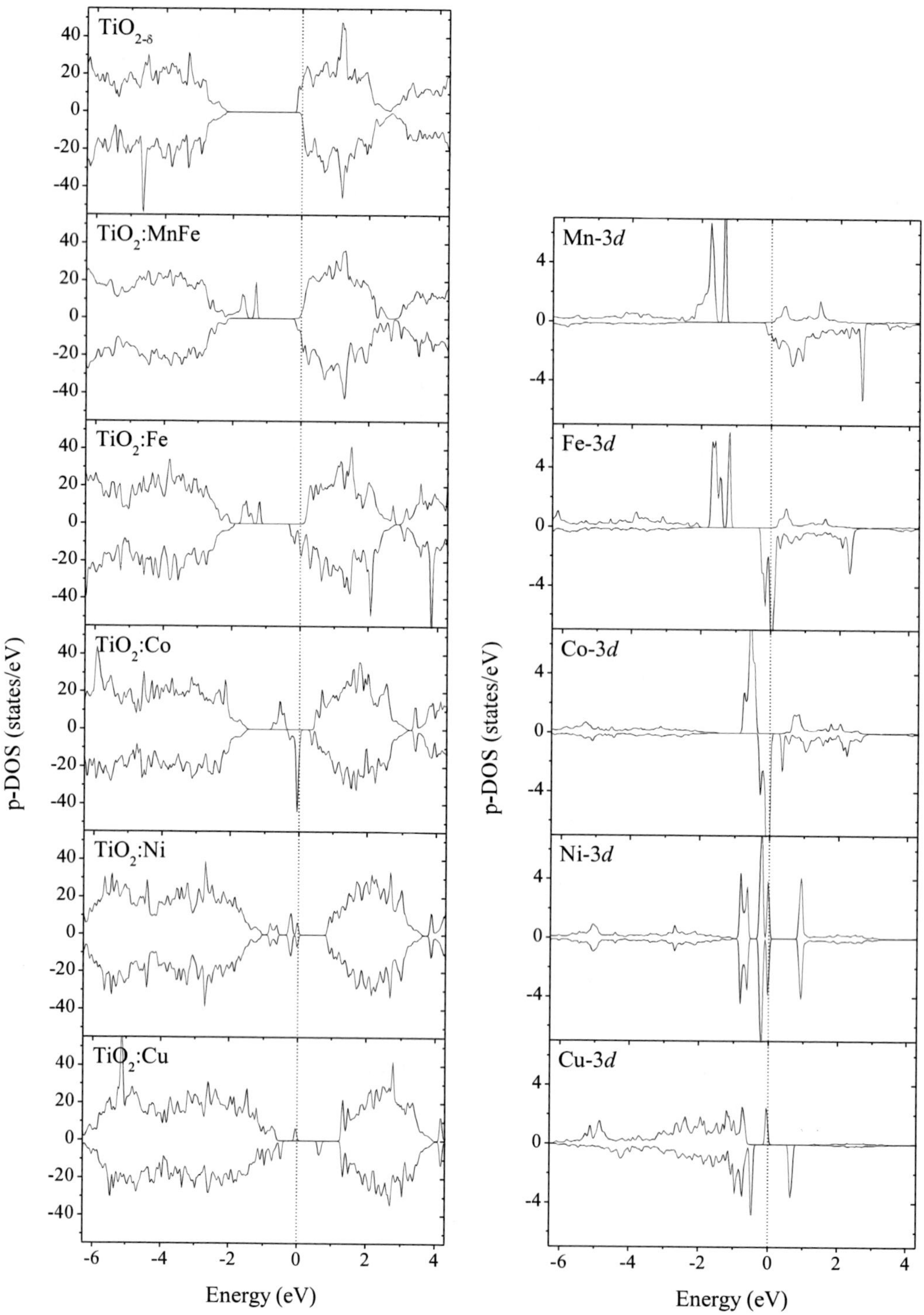

Figure 12. Calculated total DOS of the oxygen-deficient ($Ti_{15}TMO_{31}$) systems and 3*d*-projection at the impurity muffin-tin spheres. Energies are referred to the Fermi level. The results correspond to LSDA calculations.

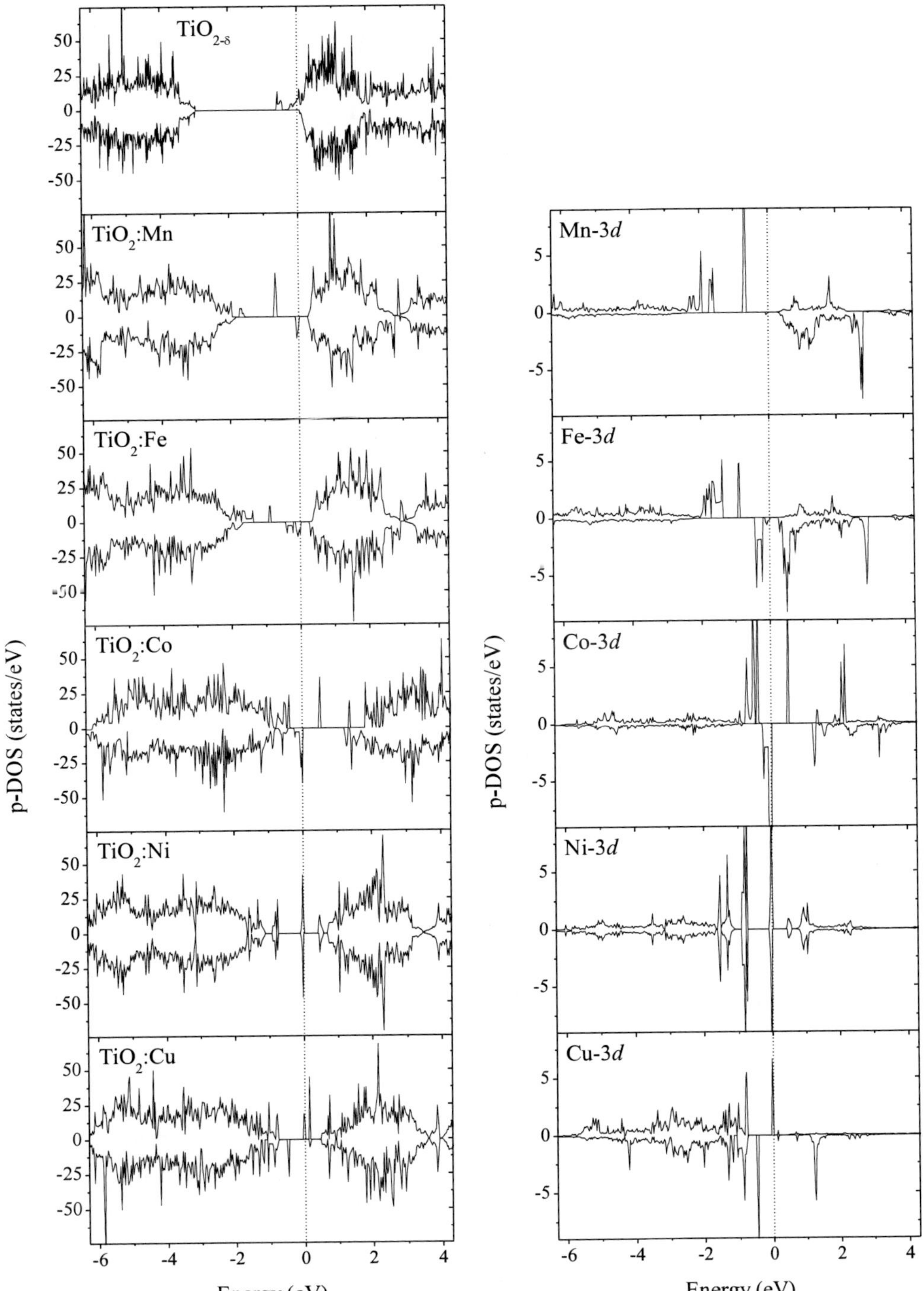

Figure 13. Calculated total DOS of the oxygen-deficient ($Ti_{15}TMO_{31}$) systems and 3*d*-projection at the impurity muffin-tin spheres. Energies are referred to the Fermi level. The results correspond to MBJ calculations.

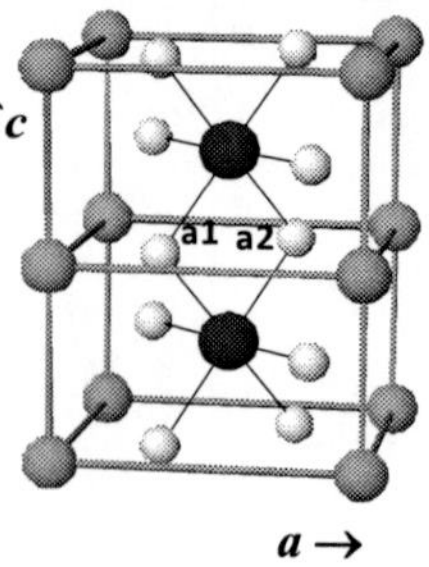

Configuration a

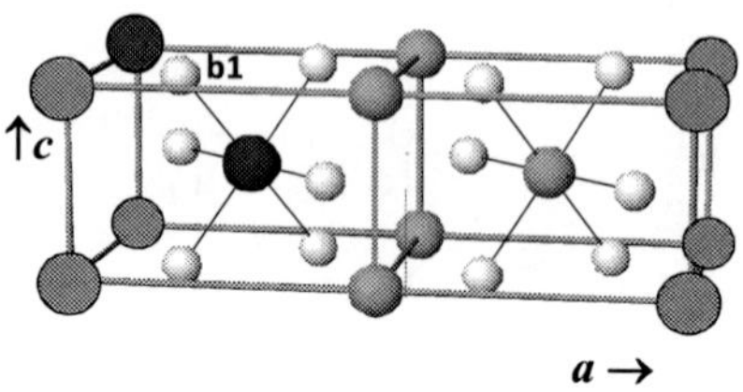

Configuration b

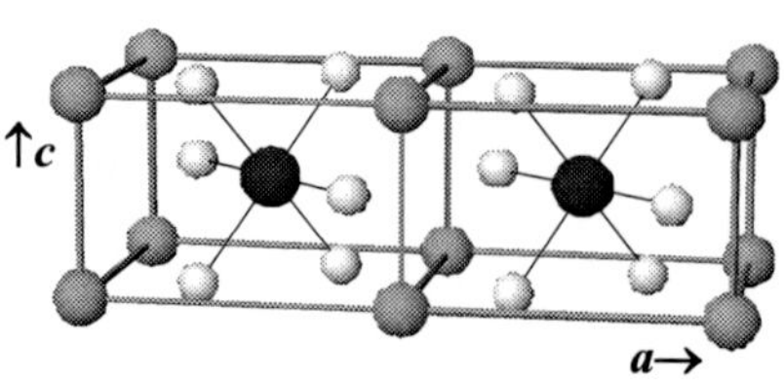

Configuration c

Figure 14. 48A-SCs with two impurity atoms (black balls) in three different impurity distributions within the rutile TiO_2 host. For simplicity, only a portion of the SC is showed.

Table VIII. Calculated total energies, referred to the nonmagnetic case, and local magnetic moments for Mn, Fe and Co in rutile TiO_2 for the different configurations of two TM studied (see Figure 14)

Impurity	Configuration	E_{ferro} (eV)	E_{aferro} (eV)	μ_{ferro} (μ_B)	μ_{aferro} (μ_B)	Magnetic ordering
Mn	*a*	-1.9747	-2.0024	+2.49	+2.20/-2.20	AF
	b	-1.8224	-1.8768	+2.46	+2.40/-2.44	AF
	c	-2.5568	-2.5636	+2.45	+2.44/-2.44	AF
Fe	*a*	-0.5631	-0.7983	+1.57	+2.20/-2.20	AF
	b	-0.7296	-0.7725	+1.58	+1.88/-1.88	AF
	c	-0.5658	-0.5304	+1.81	+1.68/-1. 68	F
Co	*a*	-0.1265	-0.0030	+0.63	+0.16/-0.16	F
	b	-0.1088	-0.0050	+0.62	+0.52/-0.52	F
	c	-0.1455	-0.0952	+0.65	+0.57/-0.57	F

Our results show that if the impurity atoms are located at the shortest possible distance from each other the system has its lowest energy. This corresponds to impurities located along the *c*-axis of the rutile structure, where they are at 2.90 Å. Table VIII shows the energies for the FM and AF cases and the local magnetic moment of each configuration. In the case of Co impurities the most stable structures are those ferromagnetically aligned; the magnetic moments in the Co-spheres are independent of the impurity distribution and very similar to those obtained in Sec. 3.3.1. In the case of Mn impurities the local magnetic moments are independent of the configuration and of the magnetic ordering considered, and they are four times larger than those for Co. The most stable phases in this case are antiferromagnetically aligned, as one would expect for Mn.

In the case of Fe impurities, we find that for two of the three configurations the most stable phase is antiferromagnetic. In consequence, for this impurity, the possibility of obtaining a stable ferromagnetic semiconductor with a large magnetic moment depends on the distribution of the impurities in the TiO_2 lattice and a lower magnetization than that of Co doped samples is expected. Similar results were obtained by Chen *et al.* [18], but the magnetic ordering is opposite to that obtained by us for the configurations (b) and (c). This may be due to the fact that a different SC was used in Ref. 18, and also that the relaxation procedure was different. A similar study was performed by Mallia and Harrisom, but using the B3LYP hybrid exchange functional [20]. Their calculations, for Fe dopants with a concentration of 0.125 and impurities far apart from each other, gave large local magnetic moments that do not couple to form a stable ferromagnetic state at room temperature. This result differs from ours and the discrepancy can be associated to the different method of calculation.

For two Ni impurities no magnetic moment was found for any distribution. For Cu, the stable phases are antiferromagnetically aligned, being the differences between ferromagnetic and antiferromagnetic alignments in the order of 1.10^{-3} eV. It is important to mention that there is an increase of the magnetic moment from 1.0 μ_B/SC up to 2.8 μ_B/SC when two Cu impurities are considered instead one (see Table IV and VIII). This result is different to those obtained for Mn, Fe and Co impurities as in these cases, when two impurities are introduced, the magnetic moment per SC doubles. This difference is possibly due to the delocalization of the magnetic moment in the case of Cu-doped TiO_2.

3.3.5. Systems with Two Substitutional TM and Oxygen Vacancies ($Ti_{14}TM_2O_{32-\delta}$).

Since the presence of oxygen vacancies increases or induces a magnetic moment at the impurity site, we will now explore the effect of oxygen vacancies on the magnetic alignment. To do this, we study the three configurations showed in Figure 14, but considering one or two oxygen vacancies. In Figure 14 we see that for configuration (a) the two TM ions are linked by two TM-O-TM paths ($d_{TM\text{-}TM}$ = 2.95 Å in the un-relaxed structure). For configuration (b) the two TM are linked by only one TM-O-TM path with ($d_{TM\text{-}TM}$ = 3.57 Å in the un-relaxed structure). In the case of configuration (c) the TM atoms are linked by three separate TM-O-Ti-O-TM chains ($d_{TM\text{-}TM}$ = 4.59 Å). The question here is to determine the location of the oxygen vacancies. Experimentally, precise estimation of the location and concentration of oxygen vacancies is very difficult and to understand the localization and role of oxygen vacancies computational modelling is necessary. However, the magnetic behaviour is far more complicated since the concentration and distribution of impurities and vacancies all affect the magnetic interactions so that the properties of the material will probably be an average of different impurity-defect configurations. A more realistic model calculation would require SCs with many impurities and vacancies in different distributions. To deal with this problem, we start by considering only one oxygen vacancy, and we calculate the total energy of the different impurity-oxygen vacancy configurations (taking into account the structural distortions induced by both impurities and by the oxygen vacancy). A similar study was later performed for two oxygen vacancies. Comparing the results we conclude that:

- One oxygen vacancy in configuration (a) prefers to be in one of the two TM-O-TM paths (sites a1 or a2 in Figure 14). Two oxygen vacancies prefer to occupy both TM-O-TM paths (sites a1 and a2 in Figure 14).
- One oxygen vacancy in configuration (b) prefers to be in the TM-O-TM path (site b1 in Figure 14). When there are two oxygen vacancies one prefers to be in the TM-O-TM path and the other as a first neighbour of one of the TM impurity.
- One oxygen vacancy in configuration (c) prefers to be near one of the impurities as in the unit cell with only one TM impurity. When a second oxygen vacancy is present, the stable configuration is that in which each vacancy is located as a first neighbour of each of the impurities.

In the following, we will discuss the results corresponding to the lower energy configurations. For Mn, we found that the most stable phases are antiferromagnetically aligned, i.e., the presence of the oxygen vacancies does not affect the magnetic alignment and only small reductions of the differences E^{anti}-E^{ferro} were found. For Co, similarly to the cases without oxygen vacancies, the most stable structures are ferromagnetically aligned, but the difference in energy between the ferromagnetic and antiferromagnetic solutions increase with respect to the case without oxygen vacancies. In other words, the oxygen vacancies reinforce the ferromagnetic alignment of Co atoms.

For Fe, without oxygen vacancies the stable structures are antiferromagnetic, except at a separation distance of 4.59 Å. When one vacancy is located in site a1 or a2, we found that the system is also antiferromagnetic. For this case, the difference ΔE=E^{anti}-E^{ferro} is larger than for the case without oxygen vacancies. But, at the same time, this configuration is energetically the least favourable with respect to the other configurations. When two vacancies are located at sites a1 and a2 the lowest energy corresponds to a ferromagnetic state (see Table IX). Another ferromagnetic state was obtained for one impurity at site b1. In configuration (c), when one or two oxygen vacancies are considered, the alignments are ferromagnetic with ΔE similar to those obtained in the configuration without oxygen vacancies. For this case calculations performed using LDA+U predicted a qualitatively similar behaviour.

Finally, in the case of Cu, the energies of the ferromagnetic and antiferromagnetic states are the same for the three configurations and the different oxygen vacancies location studied (the energy difference is smaller than the convergence error). This result can explain the ferromagnetic signal observed experimentally. A similar result was obtained in the case of Ni impurities.

The conclusion of these calculations is that oxygen vacancies play an important role in the magnetic properties, enhancing the magnetic moments and favouring the ferromagnetic coupling between TM dopants. Additionally, the *ab initio* study suggests that the overall magnetic properties of TM-doped rutile TiO_2 are probably an average over many possible defect configurations and that ferromagnetism in these systems is a complex phenomenon, which involves the interplay between oxygen vacancies and doping impurities.

Table IX. Difference between the energies of the ferromagnetic and antiferromagnetic solutions ($\Delta E = E_{aferro} - E_{ferro}$) for the different impurities-oxygen vacancies configurations studied in $Ti_{14}Fe_2O_{32-\delta}$

Configuration	Vacancy site	ΔE (eV)	Magnetic ordering
a	*No vacancies*	-0.12	AF
	Site a1	-0.28	AF
	Sites a1 and a2	+0.21	F
b	*No vacancies*	-0.03	AF
	Site b1	+0.01	F
c	*No vacancies*	+0.02	F
	One FeO_5 octahedron	+0.02	F
	Two FeO_5 octahedrons	+0.03	F

3.3.6. Interstitial Impurities

There is a last point to discuss. Some Mössbauer experiments for Fe-doped TiO_2 require a second interaction to reproduce the experimental spectra, in addition to the interaction characterized by an isomer shift in the order of 0.40 mm/s that is associated to Fe atoms at substitutional sites with one oxygen vacancy as a first neighbour. This second interaction is characterized by a high isomer shift in the order of 1.0 mm/s and appears only in samples prepared by high-energy methods like ball-milling and ion implantation. Since none of the previous calculations predict a higher isomer shift, we consider the possibility that this interaction may correspond to interstitial Fe.

To study this problem we initially calculate the effect of a Ti interstitial located at the empty octahedral site, i.e., the centre of the (010) face (or equivalently the (100) face) in Fig. 4 (the atomic position is 0;1/2;1/2 or 1/2;0;1/2). Cho *et al.* [107] have performed calculations changing the position of the interstitial Ti atoms along the z axis and confirmed that octahedral sites are the most stable positions for it. This interstitial Ti atom has six ONN, 2 at 1.66 Å and the other four at 2.22 Å. After the structural relaxation, the closest Ti atoms relax outward from the interstitial site by 0.26 Å while the nearest four O atoms are pulled in toward the cation interstitial by 0.21 Å. These interstitial Ti – ONN bond-lengths are very similar to those corresponding to that of Ti at the regular sites of the structure (1.94 and 1.98 Å), showing that the interstitial atoms try to reconstruct their natural coordination symmetry. The calculated formation energy of an interstitial defect is 7.80 eV for the interstitial Ti, in good agreement with the result of Cho *et al.* (7.09 eV) [107]. Spin polarized calculations were performed, but no magnetic moment was found.

Next, we consider a Fe interstitial and verify that the octahedral site at the centre of the (010) face is also the most stable position. After relaxation, the bond-lengths Fe-ONN are 1.87 Å and 1.96 Å, very similar to those obtained for a substitutional Fe (see Table IV). The formation energy of an interstitial Fe is 6.4 eV, 3.0 eV larger than for the substitutional position. Concerning hyperfine properties, we found an isomer shift in the order of 0.6 mm/s. The spin-polarized calculations predict a null magnetic moment.

After that, we studied the role of oxygen vacancies and found that the preferential site for the vacancy is close to the interstitial Fe ion. Similar to the substitutional case, the presence of the oxygen vacancies reduces the formation energy, but this energy is still larger than that for the substitutional Fe. A magnetic moment appears, that depends on the site of the vacancy (at 1.87 or 1.96 Å in the unrelaxed structure), and is mainly originated by the polarization of Ti atoms. Concerning the isomer shift, when vacancies are present it grows up to 0.65 mm/s, still far from the experimentally observed value of 0.90-1.0 mm/s. More studies, considering different configurations of substitutional and interstitial Fe and oxygen vacancies are needed in order to assign this second interaction to interstitial Fe sites.

CONCLUSIONS

We have presented here a theoretical and experimental investigation of TM-doped rutile TiO_2. In order to characterize these samples we used X-ray absorption experiments, ^{57}Fe Mössbauer spectroscopy, *ab initio* calculations and magnetometry. Results from TM K-edge near-edge and extended X-ray absorption fine structure confirm that TM^{3+} replaces Ti^{4+} in the TiO_2 rutile structure increasing the metal-anion bond length. Our calculations, assuming different structural scenarios, allow us to assign the two interactions observed in Mössbauer experiments to Fe^{3+} replacing Ti^{4+} in rutile with oxygen vacancies close to the dopant and to interstitial impurities, also with an oxygen vacancy in its close vicinity. Calculations including vacancies reproduce well the TM-O_{NN} distances obtained from EXAFS results.

Ab initio studies of the systems $Ti_{1-x}TM_xO_{2-\delta}$ for different substitutional impurity concentrations and distributions in the rutile structure reveals that magnetic moments appear for Mn, Fe, and Co, but not for Ni. They are originated on the impurity *d* states, which are hybridized with *p*-oxygen states, and a related feature appears in the host band energy gap. The values of these magnetic moments are almost independent of the impurity concentration when no oxygen vacancies are considered. The magnetic ordering results antiferromagnetic for Mn and also for Fe in some geometrical distributions but ferromagnetic for Co.

When oxygen vacancies are introduced along with the TM impurities our results show that the energetically preferred vacancy location is that of a nearest neighbour to the impurity. Additionally, we found that the strong interaction between oxygen vacancies and impurities increases the local magnetic moment in the cases of Mn, Fe, and Co and induces a magnetic behaviour in the cases of Ni and Cu. An important result of these calculations is that doping lowers the formation energy of vacancies, so that doped systems will have more vacancies than the undoped ones. These results support the hypothesis that oxygen vacancies play an important role in the magnetic properties of doped rutile TiO_2, and can explain the diversity of magnetic moments observed experimentally for samples grown in different conditions. Moreover, we showed that the oxygen vacancies enhance the ferromagnetic coupling between impurities. Our results show that only large scale simulations can reveal the complex behaviour of reduced TM-doped TiO_2. Such information is difficult to obtain purely by experimental means where the sample character cannot be strictly controlled. This point suggests that, for the technological application of these doped oxides, an "engineering of defects" could be needed. The overall magnetic properties of TM-doped TiO_2 are probably a statistical average of different defect configurations corresponding to a real disordered system. For realistic modelling SC larger that the present ones would be needed, so as to account for the different types and concentration of defects and reach a better understanding of the experimental results.

ACKNOWLEDGMENTS

This work was partially supported by Agencia Nacional de Promoción Científica y Tecnológica (ANPCyT) under Grant No. PICT98 03-03727, Consejo Nacional de Investigaciones Científicas y Técnicas (CONICET) under Grants PEI6174, PIP6032, PIP0002, PIP6005, Fundación Antorchas, Argentina, and the Third World Academy of Sciences (TWAS), Italy, under Grant No. RGA 97-057. This research made use of the HP-Parallel-Computing Bose Cluster and the HUGE cluster, Aarhus University, Aarhus, Denmark. The magnetic measurements were performed using the RN3M facilities, Argentina.

REFERENCES

[1] U.Diebold, *Surf. Sci. Reports* 48, 53 (2003).
[2] S. Das Sarma, *Am. Sci.* 89 516 (2001).
[3] R. Janisch, P. Gopal and N. A. Spaldin, *J. Phys.: Condens. Matter* 17, R657 (2005).
[4] S. A. Chambers, *Surf. Sci. Reports* 61, 345 (2006).
[5] A. Mauger and C. Godart, *Phys. Rep.* 141, 51 (1986).
[6] L. Treitinger, H. Göbel and H. Pink, *Mater. Res. Bull.* 11 1375 (1976).
[7] E. Steigmeier and G. Harbeke, Phys. *Kondens. Mater.* 12, 1 (1970).
[8] J. K. Furdyna and J. Kossut, DMSs, *Semiconductors and Semimetals* (Academic Press, New York, 1988), Vol. 25.
[9] H. Ohno, A. Shen, F. Matsukura, A. Oiwa, A. Endo, S. Katsumoto, and Y. Iye, *Appl. Phys. Lett.* 69, 363 (1996).
[10] G. A. Medvedkin, T. Ishibashi, T. Nishi, K. Hayata, Y. Hasegawa, and K. Sato, *Jpn. J. Appl. Phys.*, Part 2 39, L949 (2000).
[11] S. Cho, S. Choi, G. B. Cha, S. C. Hong, Y. Kim, Y. J. Zhao, A. J. Freeman, J. B. Ketterson, B. J. Kim, Y. C. Kim, and B. C. Choi, *Phys. Rev. Lett.* 88, 257203 (2002).
[12] P. Sharma, A. Gupta, K. V. Rao, F. J. Owens, R. Sharma, R. Ahuja, J. M. Osorio Guillen, B. Johansson, and G. A. Gehring, *Nat. Mater.* 24, 673 (2003).
[13] Y. Matsumoto, M. Murakami, T. Shono, T. Hasegawa, T. Fukumura, M. Kawasaki, P. Ahmet, T. Chikyow, S. Koshihara, and H. Koinuma, *Science* 291, 854 (2001).
[14] S. A. Chambers, S. Thevuthasan, R. F. C. Farrows, R. F. Marks, J. U. Thiele, L. Folks, M. G. Samant, A. J. Kellock, N. Ruzycki, D. L. Ederer, and U. Diebold, *Appl. Phys.* Lett. 79, 3467 (2001); S. A. Chambers, *Mater. Today* 5, 34 (2002).
[15] W. K. Park, R. J. Ortega-Hertogs, J. S. Moodera, A. Punnoose, and M. S. Seehra, Appl. *Phys. Lett.* 91, 8093 (2002).
[16] L. A. Errico, M. Rentería and M. Weissmann, *Phys. Rev.* B 72, 184425 (2005).

[17] V. I. Anisimov, M. A. Korotin, I. A. Nekrasov, A. S. Mylnikova, A. V. Lukoyanov, J. L. Wang, and Z. Zeng, *J. Phys. Cond. Matter* 18, 1695 (2006).
[18] J. Chen, P. Rulis, L. Ouyang, S. Satpathy, and W. Y. Ching, *Phys. Rev.* B 74, 235207 (2006)
[19] B. Vodungbo , Y. Zheng , M. Marangolo , D. Demaille and J. Varalda, J. *Phys. Cond. Mat.* 19,116205 (2007).
[20] G. Mallia and N. M. Harrison, *Phys. Rev.* B 75, 165201 (2007).
[21] T Fukumura, H Toyosaki, K Ueno, M Nakano, and M Kawasaki, *New J. Phys.* 10, 055018 (2008).
[22] L. Sangaletti, M. C. Mozzati, G. Drera, P. Galinetto, C. B. Azzoni, A. Speghini, and M. Bettinelli, *Phys. Rev. B* 78, 075210 (2008).
[23] A. I. Rykov, K. Nomura, J. Sakuma, C. Barrero, Y. Yoda, and T. Mitsui, *Phys. Rev. B* 77, 014302 (2008).
[24] S. Ogale, D. Kundaliya, S. Mehraeen, L. Fu, S. Zhang, A. Lussier, J. Dvorak, N. Browning, Y. Idzerda and T. Venkatesan, *Chem. Mat.* 20, 1344 (2008).
[25] L. Sangaletti, F. Federici Canova, G. Drera, G. Salvinelli, M. C. Mozzati, P. Galinetto, A. Speghini, and M. Bettinelli, *Phys. Rev.* B 80, 033201 (2009).
[26] S. Li and P. Jena, *Phys. Rev.* B 79, 201204 (R) (2009).
[27] W. Yan, Z. Sun, T. Yao, Z. Pan, Z. Li, Q. Liu, and S. Wei, *J. Appl. Phys.* 106, 123918 (2009).
[28] B. Ali, A. K. Rumaiz, A. Ozbay, E. Nowak, and S. I. Shah, *Solid State Comm.* 149, 2210 (2009).
[29] H. Nakano, N. Hasuike, K. Kisoda, K. Nishio, T. Isshiki, and H. Harima, *J. Phys. Cond. Matter* 21, 064214 (2009).
[30] B. Z. Jiang, T. L. Phan, D. S. Yang, K. W. Lee, and S. C. Yu, *Solid. State Comm.* 150, 1932 (2010).
[31] J. M. D. Coey , P. Stamenov , R. D. Gunning, M. Venkatesan, and K. Paul, *New J. of Physics* 12, 053025 (2010).
[32] X. B. Qin, P. Zhang, L. H. Liang, B. Z. Zhao, R. S. Yu, B. Y. Wang, and W. M. Wu, *J. Phys. Conf. Series* 262, 012051 (2011).
[33] V. Pore, M. Dimri, H. Khanduri, R. Stern, J. Lu, L. Hultman, K. Kukli, M. Ritala, and M. Leskelä, *Thin Solid Films* 519, 3318 (2011) and referentes there in.
[34] J. Varalda, C. A. Dartora, A. J. A. de Oliveira, W. A. Ortiz, B. Vodungbo, M. Marangolo, F. Vidal, Y. Zheng, G. G. Cabrera, and D. H. Mosca, *Phys. Rev.* B 83, 045205 (2011).
[35] H. Munekata, H. Ohno, S. von Molnar, A. Segmüller, L. L. Chang, and L. Esaki, *Phys. Rev. Lett.* 63, 1849 (1989).
[36] S. R. Shinde, S. B. Ogale, S. Das Sarma, J. R. Simpson, H. D. Drew, S. E. Lofland, C. Lanci, J. P. Buban, N. D. Browning, V. N. Kulkarni, J. Higgins, R. P. Sharma, R. L. Greene, and T. Venkatesan, *Phys. Rev.* B 67, 115211 (2003).
[37] M. S. Park, S. K. Kwon, and B. I. Min, *Phys. Rev.* B 65, 161201 (R) (2002).
[38] H. Weng, X. Yang, J. Dong, H. Mizuseki, M. Kawasaki, and Y. Kawasoe, *Phys. Rev.* B 69, 125219 (2004).
[39] R. Suryanarayanan, V. M. Naik, P. Kharel, P. Talagala, and R. Naik, *Solid State Commun.* 133, 439 (2005).

[40] S. Duhalde, M. F. Vignolo, F. Golmar, C. Chiliotte, C. E. Rodríguez Torres, L. A. Errico, A. F. Cabrera, M. Rentería, F. H. Sánchez, and M. Weissmann, *Phys. Rev.* B 72, 161313 (R) (2005).
[41] M. Weissmann and L. A. Errico, *Physica* B 398 , 179 (2007).
[42] R. Suryanarayanan, V. M. Naik, P. Kharel, P. Talagala, and R. Naik, *J. Phys.: Condens. Matter* 17, 755 (2005).
[43] M. Khalid, A. Setzer, M. Ziese, P. Esquinazi, D. Spemann, A. Pöppl, and E. Goering, *Phys. Rev.* B 81, 214414 (2010).
[44] J. M. D. Coey, P. Stamenov, R. D. Gunning, M. Venkatesan, and K. Paul, *New Journal of Phys.* 12, 053025 (2010).
[45] Y. L. Soo, G. Kioseoglou, S. Kim, Y. H. Kao, P. Sujatha, J. Parise, R. J. Gambino, and P. I. Gouma, *Appl. Phys. Lett.* 81, 655 (2002).
[46] J. Y. Kim, J. H. Park, B. G. Park, H. J. Noh, S. J. Oh, J. S. Yang, D. H. Kim, S. D. Bu, T. W. Noh, H. J. Lin, H. H. Hsieh, and C. T. Chen, *Phys. Rev. Lett.* 90, 017401 (2003).
[47] C. E. R. Torres, A. F. Cabrera, M. B. F. Van Raap, and F. H. Sánchez, *Physica* B 354, 67 (2004).
[48] A. M. Mudarra Navarro, V. Bilovol, C. E. Rodríguez Torres, and A. F. Cabrera, Physica B, in press (2011).
[49] Z. Wang, J. Tang, L. Tung, W. Zhou, and L. Spinu, *J. Appl. Phys* 93, 7870 (2003).
[50] Z. Wang, W. Wang, J. Tang, L. D. Tung, L. Spinu, and W. Zhou, *Appl. Phys. Lett.* 83, 518 (2003).
[51] S. R. Shinde, S. B. Ogale, J. S. Higgins, H. Zheng, A. J. Millis, V. N. Kulkarni, R. Ramesh, R. L. Greene, and T. Venkatesan, *Phys. Rev. Lett.* 92, 166601 (2004).
[52] K. Nomura, H. Eba, K. Sakurai, A. Rykov, and T. Hasegawa, *Thin Solid Films* 515, 8649 (2007).
[53] Q. Liu, Y. Jiang, W. Yan, Z. Sun, Z. Pan, T. Yao, Z. Wu, and S. Wei, *J. Phys.: Conference Series* 190, 012107 (2009).
[54] Y. J. Kim, S. Thevuthasan, T. Droubay, A. S. Lea, C. M. Wang, V. Shutthanandan, S. A. Chambers, R. P. Sears, B. Taylor, and B. Sinkovic, *Appl. Phys Lett* 84, 3531 (2004).
[55] W. Yan, Z. Sun, Z. Pan, Q. Liu, T. Yao, Z. Wu, Ch. Song, F. Zeng, Y. Xie, T. Hu, and S. We, *Appl. Phys. Lett.* 94, 042508 (2009).
[56] P. Xiaoyan, J. Dongmei, L. Yan, and M. Xueming, *J. Mag. Mag. Mat* 305, 388 (2006).
[57] A. F. Cabrera, C. E. Rodríguez Torres, L. A. Errico, and F. H. Sánchez, *Physica* B 384, 345 (2006).
[58] A. M. Mudarra Navarro, V. Bilovol, A. F. Cabrera, C. E. Rodríguez Torres, and F. H. Sánchez, *Physica* B 404, 2838 (2009).
[59] L. Fan, J. Dongmei, L. Yan and M. Xueming, *Physica* B 403, 2193 (2008)-
[60] V.Bilovol, A.M.Mudarra Navarro, W. T. Herrera, D. R. Sánchez, E. M. Baggio-Saitovich, C. E. Rodríguez Torres, F. H. Sánchez, and A. F. Cabrera, *Hyperfine Interact.* 195, 155 (2010).
[61] N. Akdogan, A. Nefedov, H. Zabel, K. Westerholt, H. Becker, Ch. Somsen, Ş. Gök, A. Bashir, R. Khaibullin and L. Tagirov, J. Phys. D: *Appl. Phys.* 42, 115005 (2009).
[62] R. I. Khaibullin, Sh. Z. Ibragimov, L. R. Tagirov, V. N. Popok, and I. B. Khaibullin, *Nuc. Instr. and Meth. in Phys. Res.* B 257, 369 (2007).
[63] J. V. Pinto, M. M. Cruz, R. C. da Silva, E. Alves, and M. Godinho, *J. Mag. Mag. Mat.* 294, e73 (2005).

[64] L. Balcells, C. Frontera, F. Sandiumenge, A. Roig, and B. Martínez, J. Kouamb, and C. Monty, *Appl. Phys. Lett.* 89, 122501 (2006).

[65] K. Nomura, K. Inaba, S. Iio, T. Hitosugi, T. Hasegawa, Y. Hirose, and Z. Homonnay, *Hyperfine Interact.* 168, 1065 (2006).

[66] N. H. Hong, J. Sakai, and W. Prellier, *J. Mag. Mag. Mat.* 281, 347 (2004).

[67] See, e.g., G. Schatz and A. Weidinger, *Nuclear Condensed Matter Physics—Nuclear Methods and Applications* (Wiley, Chichester, 1996); E. N. Kaufmann and R. J. Vianden, Rev. Mod. Phys. 51 161 (1979); H. Frauenfelder and R. M. Steffen, in *Alpha-, Beta-, and Gamma-Ray Spectroscopy*, edited by K. Siegbahn (North Holland, Amsterdam, 1966), Vol. 2.

[68] D. H. Kim, S. I. Woo, S. H. Moon, H. D. Kim, B.Y. Kim, J. H. Cho, Y. G. Joh, and E. C. Kim, *Solid State Commun.* 136, 554 (2005).

[69] W.B. Mi, E.Y. Jiang, and H. L. Bai, *J. Mag. Mag. Mat* 321, 2472 (2009).

[70] N. Hoa Hong, J. Sakai, W. Prellier, and A. Ruyter, *J. Phys.D: Appl.Phys* 38 , 816 (2005).

[71] W. Kyu Park, R. J. Ortega-Hertogs, J. S. Moodera, A. Punnoose, and M. S. Seehra, *J. Appl. Phys.* 91, 8093 (2002).

[72] Y. Matsumoto, R. Takahashi, M. Murakami, T. Koida, X. Fan, T. Hasegawa, T. Fukumura, M. Kawasaki, S. Koshihara, and H. Koinuma, *Jpn. J. Appl. Phys.* 40, L1204 (2001).

[73] V. Shutthanandan, S. Thevuthasan, S. M. Heald, T. Droubay, M. H. Engelhard, T. C. Kaspar, D. E. McCready, L. Saraf, S. A. Chambers, B. S. Mun, N. Hamdan, P. Nachimuthu, B. Taylor, R. P. Sears, and B. Sinkovic, *App. Phys. Lett* 84, 4466 (2004).

[74] A. M. Mudarra Navarro, Ph.D. Thesis, Universidad Nacional de La Plata (2010); V. Bilovol, Ph.D. Thesis, Universidad Nacional de La Plata (2010).

[75] S. Duhalde, C. E. Rodríguez Torres, M. F. Vignolo, F. Golmar, C. Chillote, A. F. Cabrera, and F. H Sánchez, *J. of Physics*: Conference Series 59, 479 (2007).

[76] G. V. Samsonov, *The oxides Handbook*, IFI/Plenum Press, New York (1982).

[77] J. Hill and C. J. Howard, J. Appl. Crystallogr. 20, 467 (1987).

[78] P. Hohenberg and W. Kohn, *Phys. Rev.* 136, B864 (1964); W. Kohn and L. J. Sham, ibid. 140, A1133 (1965).

[79] E. Sjöstedt, L. Nordström, and D. J. Singh, *Solid State Commun.* 114, 15 (2000); G. K. H. Madsen, P. Blaha, K. Schwarz, E. Sjöstedt, and L. Nordström, Phys. Rev. B 64, 195134 (2001); see also S. Cottenier, *Density Functional Theory and the Family of (L)APW-Methods: A Step-by-Step Introduction* (KU Leuven, Belgium, 2002) (http://www.wien2k.at/reg_user/textbooks).

[80] P. Blaha, K. Schwarz, G. Madsen, D. Kvasnicka, and J. Luitz, *WIEN2k, An Augmented Plane Wave Plus Local Orbitals Program for Calculating Crystal Properties* (Karlheinz Schwarz, Technical Universität Wien, Austria, 1999).

[81] J. P. Perdew and Y. Wang, *Phys. Rev.* B 45, 13244.

[82] F. Tran and P. Blaha, *Phys. Rev. Lett.* 102, 226401 (2009).

[83] V. I. Anisimov, I. V. Solovyev, M. A. Korotin, M. T. Czyzyk, and G. A. Sawatzky, *Phys. Rev.* B 48, 16929 (1993).

[84] P. Krüger, S. Bourgeois, B. Domenichini, H. Magnan, D. Chandesris, P. Le Fèvre, A. M. Flank, J. Jupille, L. Floreano, A. Cossaro, A. Verdini, and A. Morgante, *Phys. Rev. Lett.* 100, 055501 (2008).

[85] T. Minato, Y. Sainoo, Y. Kim, H. S. Kato, K. I. Aika, M. Kawai, J. Zhao, H. Petek, T. Huang, W. He, B. Wang, Z. Wang, Y. Zhao, J. Yang, and J. G. Hou, *J. Chem. Phys.* 130, 124502 (09).

[86] L. A. Errico, G. Fabricius, M. Rentería, P. de la Presa, and M. Forker, Phys. Rev. Lett. 89, 055503 (2002); L. A. Errico, G. Fabricius, and M. Rentería, *Phys. Rev.* B 67, 144104 (2003).

[87] C. E. Rodríguez-Torres, A F Cabrera, L A Errico, C Adán, F G Requejo, M. Weissmann and S J Stewart, *J. Phys.: Condens. Matter* 20, 135210 (2008).

[88] L. M. Huang, A. L. Rosa, and R. Ahuja, *Phys. Rev.* B 74, 075206 (2006).

[89] P. I. Sorantin and K. Schwarz, *Inorg. Chem.* 31, 567 (1992).

[90] J. Pascual, J. Camassel, and H. Mathieu, *Phys. Rev.* B 18, 5606 (1978).

[91] S. A. Chambers, S. M. Heald, and T. Droubay, *Phys. Rev.* B 67, 100401(R) (2003).

[92] Q. Zhao, P. Wu, B. L. Li, Z. M. Lu, and E. Y. Jiang, *J. Appl. Phys.* 104, 073911 (2008).

[93] C. Sudakar, P. Kharel, R. Suryanarayanan, J. S. Thakur, V.M. Naik, R. Naik, and G. Lawes, *J. Mag. Mag. Mat.* 320 L31 (2008).

[94] N. H. Hong, J. Sakai, N. Poirot, and V. Brizé, Phys. Rev. B 73, 132404 (2006); N. H. Hong, J. Sakai, F. Gervais *J. Mag. Mag. Mat.* 316, 214 (2007).

[95] A. Sundaresan, R. Bhargavi, N. Rangarajan, U. Siddesh, and C. N. R. Rao, *Phys. Rev.* B 74, 161306 (r) (2006).

[96] P. Galinetto, M. C. Mozzati, C. Vercesi, L. Malavasi, Y. A. Diaz Fernandez, F. Rossella , G. Drera, and L. Sangaletti, *J. Phys. Conf. Series* 200, 072030 (2010).

[97] G. Drera, M. C. Mozzati, P. Galinetto, Y. Diaz-Fernandez, L. Malavasi, F. Bondino, M. Malvestuto, and L. Sangaletti, *Appl. Phys. Lett.* 97, 012506 (2010).

[98] A.Rumaiz, B. Ali, A. Ceylan, M. Boggs, T. Beebe, and S. Ismat Shah, *Solid State Commun.* 144, 334 (2007).

[99] S. Zhou, E. Čižmár, K. Potzger, M. Krause, G. Talut, M. Helm, J. Fassbender, S. A. Zvyagin, J. Wosnitza, and H. Schmidt, *Phys. Rev.* B 79, 113201 (2009).

[100] D. Kim, J. Hong, Y. R. Park, and K. J. Kim, *J. Phys. Cond. Matter* 21, 195405 (2009).

[101] K. Yang, Y. Dai, B. Huang, and Y. P. Fena, *Phys. Rev.* B 81, 033202 (2010).

[102] Ch. Das Pemmaraju and S. Sanvito, Phys. Rev. Lett. 94, 217205 (2005).

[103] H. Weng and J. Dong, *Phys. Rev.* B 73, 132410 (2006).

[104] N. H. Hong, N. Poirot, and J. Sakai, *Phys. Rev.* B 77, 033205 (2008).

[105] C. E. Rodríguez Torres, L. Errico, F. Golmar, A. M. Mudarra Navarro, A. F. Cabrera, S. Duhalde, F. H. Sánchez, and M. Weissmann, *J. Mag. Mag. Mat.* 316, 219 (2007).

[106] Y. Le Page and M. Marezio *J. Solid State Chem.* 53, 13 (1984).

[107] E. Cho and S. Han, H. S. Ahn, K. Lee, S. K. Kim, and Ch. S. Hwang, *Phys. Rev.* B. 73, 193202 (2006).

[108] G. U. von Oertzen and A. R. Gersonb, *J. Phys. Chem. Solids* 68, 324 (2007).

[109] P. J. D. Lindan, N. M. Harrison, M. J. Gillan, and J. A. White, *Phys. Rev.* B 55, 15919 (1997).

[110] V. R. Singh, Y. Sakamoto, T. Kataoka, M. Kobayashi, Y. Yamazaki, A. Fujimori, F. H. Chang, D. J. Huang, H. J. Lin, C. T. Chen, H. Toyosaki, T. Fukumura, and M. Kawasaki, *J. Phys.: Condens. Matter* 23, 176001 (2011).

[111] J. M. D. Coey, K. Wongsaprom, J. Alaria, and M. Venkatesan, *J. Phys. D: Appl. Phys.* 41, 134012 (2008).

[112] A. T. Brant, Shan Yang , N. C. Giles, M. Zafar Iqbal, A. Manivannan, and L. E. Halliburton, *J. Appl. Phys.* 109, 073711 (2011).

In: Rutile: Properties, Synthesis and Applications
Editor: It-Meng (Jim) Low

ISBN 978-1-61942-233-9

Chapter 9

DEVELOPMENT AND CHARACTERIZATION OF RuO_2-BASED ELECTRODES FOR ENVIRONMENTAL APPLICATION

***Josimar Ribeiro*[1,2] *and Adalgisa Rodrigues De Andrade*[1]**

[1]Departamento de Química da Faculdade de Filosofia
Ciências e Letras de Ribeirão Preto
Universidade de São Paulo, Av. Bandeirantes, 3900,
14040-901 Ribeirão Preto, SP, Brazil
[2]Departamento de Química
Centro de Ciências Exatas/UFES
Av. Fernando Ferrari, 514 – Goiabeiras – Vitória/ES,
29075-910, Brazil

ABSTRACT

Remediation of water polluted by toxic organic compounds has been the subject of many investigations. Many researchers are currently seeking alternatives for the degradation of these toxic organic compounds. On the basis of the data obtained by our research group, we will focus on recent results regarding the preparation of ruthenium-based electrodes for application in the electrochemical degradation of toxic organic compounds. Thermally prepared ruthenium, titanium, and tantalum metallic oxides are natural candidates when properties such as high corrosion resistance and physical-chemical stability under high positive potentials are required. DSA®-type electrodes are frequently prepared by either thermal decomposition (TD) of the precursor metal salts or thermal decomposition of polymeric precursors, DPP. Physico-chemical characterization is an essential part of all the investigations involving electrode materials. The various aspects of characterization include the chemical composition and compositional homogeneity of the materials, which can be evaluated by scanning electron microscopy (SEM), energy dispersive spectroscopy (EDS) and transmission electron microscopy (TEM). Impurities that may affect the properties should also be assessed. Additionally, analysis of the structure by X-ray diffraction (XRD), atomic coordination, and ultrastructure X-ray absorption spectroscopy(XAS) should be accomplished in order to reveal the crystallinity of the crystal system and unit cell. Electrochemical properties (e.g., durability and electrocatalytic activity) of electrocatalytic coatings are associated with the microstructure of the active layer. The aim of this work is to report on a different method for the production of DSA®-type electrodes that contain different amounts of RuO_2, TiO_2, and Ta_2O_5, and to describe their morphology and physical-chemical characterization. The performance of these electrodes as anode for organic compounds oxidation is presented.

1. General Aspects

The history of dimensionally stable anodes, DSA®, merges with the importance of the use of RuO_2 as electrode material and may be recollected from a series of articles produced by Trasatti [1, 2] and Hayfield [3-5]. Briefly, the first patent for DSA® appeared in 1965 [6], and the good response in chloro-alkali industry prompted the widespread fundamental investigation and use of this material for other important classes of reactions. In this context, an increasing number of studies on the oxidation of organic substances by DSA® have been reported [7-10]. The purpose of the electrochemical remediation of organic pollutants is to search for conditions for direct or indirect electrochemical oxidation where oxidation and/or complete combustion of the target compound are achieved. The direct electron transfer between the electrode and the organic compounds present in the solution normally produces mild oxidation. Moreover, indirect oxidation leads to the formation strong oxidizing species on the surface of the electrode. Examples of such species are hydroxyl radical, • OH, ($E^o_{OH/H2O}$ = 2.80 V vs. NHE) responsible for the oxidation of the organic material mainly via hydrogen abstraction and hydroxylation, and active chlorine compounds, such as hypochlorite [11, 12].

Several research groups have devoted efforts to the preparation and characterization of ruthenium-based oxides [1, 2]. Normally, the catalytic component in the traditional DSA®-based material is RuO_2, which is responsible for conferring metallic character to the mixture and is mixed with valve metals (TiO_2, Ta_2O_5) or other materials, in order to modulate or stabilize the oxide coatings [2,15]. In an attempt to find materials that present both stability and catalytic activity in the domain of the oxygen evolution reaction (OER), researchers have proposed several changes to material compositions, while RuO_2 should be kept as the main catalytic material. Thus, a wide variety of coatings can be found in the literature:

- Mixtures containing ternary oxides: $Ti/Ru_{0.3}Sn_xTi_{(1-x)}O_2$ [14]; $Ti/RuO_2 + TiO_2 + CeO_3$ [16]; $Ti/RuO_2TiO_2PtO_x$ [17].
- Binary mixtures: $Ti/RuO_2 - Nb_2O_5$ [18] and $Ti/RuO_2/Co_3O_4$ [19].

Generally, in order to choose a good electrode material, one must search for high mechanical strength and physico-chemical stability under drastic operational conditions (high current density and high positive potential). In this context, the mixture consisting of RuO_2, as the main catalytic material, and oxides, such as TiO_2, SnO_2, and Ta_2O_5, displays attractive properties [2, 14, 15, 20]. Trasatti [1] has stated that $Ti/RuO_2 - TiO_2$ surely is not the most prominent material in terms of intrinsic catalytic properties, but this weakness can be overcome by the higher operational lifetime obtained with this mixture. When it comes to industrial application, the operational time of an anode during electrolysis is one of the most important properties, and it depends mainly on the operating conditions. Few materials can operate in strongly oxidizing (high applied potential) and acidic environments.

A difficult task for electrocatalysis and materials science is to find the theoretical basis for the development of new electrode materials. A relationship can be established only if factors responsible for the electrocatalytic properties are identified. It is known that geometric and electronic factors generally govern the catalytic reaction that is taking place on oxides Characterization and investigation of the surface properties of thin films such as double layer capacitance, particle size, crystallographic structure, and lifetime aid understanding of the observed electrochemical behavior.

The morphological characteristics of oxide films are associated with changes in the surface area of the electrode material, which in turn is a consequence of various factors such as thickness, density, and segregation of one component, among others. Since the electronic properties are ruled by the atomic interaction occurring between elements present in the oxides mixture, formation of solid solutions and electron affinity, for instance, must be considered. Therefore, for comparison of different electrode materials it is important to understand the contribution of individual morphological and electronic effects.

2. Ruthenium Oxide

It is frequently reported that oxide films based on RuO_2 have large double layer capacitance values due to their large surface area and to the occurrence of reversible solid-state redox transition in the potential region of double layer charging [1, 13, 21, 22]. However, the use of Ti/RuO_2 films under severe operating conditions leads to its total deactivation within a short time period. Some authors have proposed that this is often caused by formation of an insulating TiO_x film between the substrate and the conductive oxide layer [13, 22-24], while others have claimed that deactivation stems mainly from progressive ruthenium loss from the film by erosion/corrosion [15, 25-27].

When the Ru content in the RuO_2/TiO_2 mixture is less than 20 atomic %, there is an increase in film resistivity [28-30]. Therefore, the deactivation process can be due to both processes; i.e., ruthenium loss and insulating film growth on the support base.

As mentioned earlier, RuO_2 electrodes have high storage charge. In addition, they present transport capacity in both internal and external active sites [31]. Hence, they display a large capacitive current coupled with a faradaic current. This property allows one to use these materials as electrochemical capacitors (supercapacitors) [32], which can be applied as energy source in electric vehicles, digital mobile telecommunication devices, and lasers; power supply for cameras; and power source as 'backup' for computer memories and medical instrumentations [32-34].

The mechanism of energy storage in electrochemical capacitors involves two processes: (i) charge separation in the interface between the solid electrode and the electrolyte, and (ii) the faradaic reaction occurring near the surface of the solid electrode. Process (i) is called the capacitance of the electric double layer, whereas process (ii) is known as the pseudocapacitance process associated with the reaction between the Ru active sites and H^+ ions in the acidic electrolyte, as represented by equation (1.1) [1, 13, 23, 32, 34].

$$RuO_x(OH)_y + \delta H^+ + \delta e^- \leftrightarrow RuO_{x-\delta}(OH)_{y+\delta} \quad (1)$$

Crystalline RuO_2 and ruthenium oxide hydrate, $RuO_2.xH_2O$, have high power density / energy because of their excellent specific capacitance (380-720 F/g), not to mention that a cycle can be obtained (*e.g.* 240 cycles between V *vs.* -0,2-1,0 Ag/AgCl; $\nu = 20$ mV s^{-1}, in 0.5 mol dm^{-3} H_2SO_4) [34]. In addition, ruthenium oxide exhibits high conductivity and good electrochemical reversibility compared to other electrode materials [35, 36]. Evidence of its high pseudocapacitance values compared to other alternative metal oxides such as NiO, MnO_2, and CoO_x can be easily found in the literature [37].

Besides the above mentioned applications, the high stability and catalytic activity of RuO_2-based DSA® make them the best choice for application in reactions such as organic synthesis [38], water electrolysis [39], and oxygen reduction [40]. Coating electrodes with metal oxides may be an attractive strategy when it comes to avoiding the phenomenon of Pt poisoning in fuel cells. This is because it is believed that the metal oxide ions have multiple oxidation states, thereby being able to promote formation of large quantities of -OH and -O species. The latter act as oxidizing agents during oxidation of small organic fuels in fuel cell applications. RuO_2 supported on Pt nanoparticles has afforded significantly increased methanol oxidation. Furthermore, it has been shown to play an important role by donating oxygen species, thereby promoting complete CO oxidation to CO_2 [41].

3. Methods for Preparation of RuO_2 Thin Films for Application in Organic Oxidation

In order to obtain thin oxide films with high catalytic activity and long electrolysis lifetime, some special features concerning the preparation route should be taken into account . One can orient the properties and characteristics of the electrodes to be synthesized by introducing changes to the material engineering process. Indeed, there are many literature papers describing modifications to preparation methods [14, 42-44] and calcination procedures [45]. In this context, electrodes can be prepared by the traditional method; i.e., thermal decomposition of precursor metal chloride salts, which will be denominated standard decomposition (SD) hereafter [1-6, 13, 15, 16, 46]; the spray-pyrolysis method [47]; the sol-gel method [43, 44, 48-50], and the thermal decomposition of polymeric precursors (DPP) [14, 18, 20, 23, 51-52]. In the OER domain, the SD-electrodes may exhibit relatively low lifetime, mainly due to erosion / corrosion of the electrode active layer, which leads to loss of its electrochemical properties and results in overvoltage elevation during the experiment [13, 23]. To overcome this problem, researchers have introduced new preparation procedures, such as the DPP, which considerably increase electrode lifetime and improve their electrochemical properties [14, 18, 20, 23, 51]. The sol-gel technique has also been claimed to produce electrodes with better performance [43, 44, 48-50].

Since the 1960s, the SD method has been adopted for the preparation of many thin films [1-6, 13, 15, 16, 46]. This method is based on the deposition of thin layers of a precursor salt solution (chlorides of the desired metals), followed by successive calcinations at elevated temperatures, usually in the 350 – 550 °C range, which culminates in the respective oxides. The use of chloride salts solutions has become the most prominent method for DSA preparation. Its claimed main advantage is the ease of preparation, as well as the prompt availability and low cost of the employed salts. However, it can imply in coatings poisoned by the remaining chlorides and in loss of volatile material during the calcination process [53, 54].

The main feature of the thermal treatment is the formation of solids with the well-known "mud-cracked" morphology, responsible for the large surface area of these materials [1-6] (Figure 1).

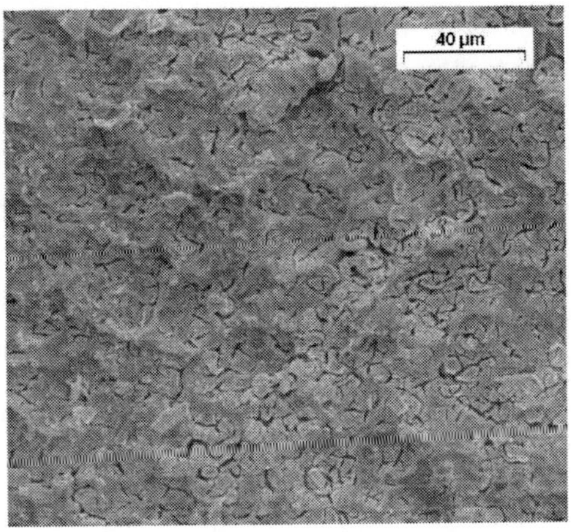

Figure 1. SEM image of the Ti/(RuO_2) 30% + (Ta_2O_5) 70% thin film with nominal thickness of 2 μm.

Figure 2 depicts a representative SEM–EDS image for the X-ray scanned in the line over the cross-section of the Ti/RuO_2 – Ta_2O_5 thin film embedded in an acrylic resin. It can be seen that the thin film consists of three different regions. Region 1 corresponds to the acrylic resin used to recover the thin film; region 2 consists of a thick RuO_2 and Ta_2O_5 oxide layer whose average thickness is around 8 μm; and region 3 is attributed to the metallic titanium base.

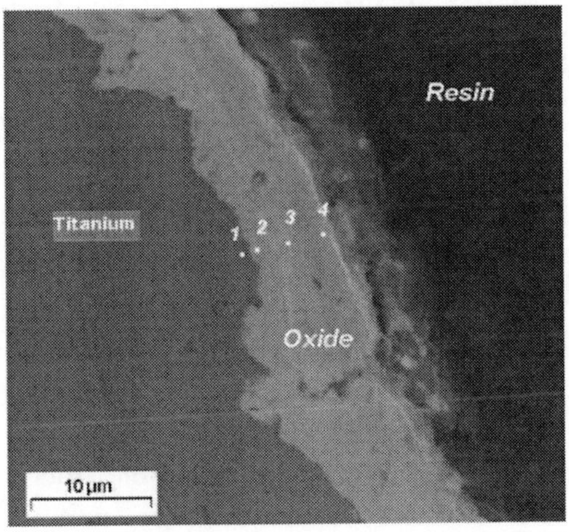

Figure 2. SEM image of the Ti/(RuO_2) 50% + (Ta_2O_5) 50% thin film with nominal thickness of 10 μm. For analysis of the oxide transversal section, thin films were embedded in an acrylic resin. The ensemble was then submitted to polishing, and the surface was covered with gold for the SEM and EDS analyses.

Table 1 summarizes the results obtained at different points of the thin film presented in Figure 2. There is segregation of tantalum oxide to the film surface, as previously observed for TiO_2 in the case of the $RuO_2 + TiO_2$ composition [34].

Another approach used for producing oxide coatings is based on the procedure described by Pechini [14, 18, 52]. In this method, a polymeric chain is formed before the calcination process, and the metal atoms (M) are then "trapped" in this matrix. The main advantage of this route is that evaporation and loss of volatile species, such as Sn, is avoided, not to mention that excellent control of film composition can be achieved [53-55]. A polyester thin film can be obtained by dip-coating the electrode or brushing the substrate into a resin solution consisting of ethylene glycol, citric acid, and the desired metal chlorides. Details on the preparation procedure can be found elsewhere [14, 18, 23]. When the polymeric film is subjected to high temperatures, namely above 400 °C, the oxide coating is obtained. At this stage, the organic material is eliminated as CO_2 and water. Figure 3 illustrates the scheme for the oxide film preparation route, which furnishes uniform films with homogeneous surfaces and allows for better control of the oxide composition. Furthermore, when this method is employed for production of ceramic powders, it generates particles with a very small size; i.e., powders with large surface area [52]. These characteristics make this method a promising way of obtaining electrodes composed of oxide films that can be used for electrocatalysis or other purposes [14, 18, 26, 56, 57].

Table 1. Experimental EDS composition of the Ti/(RuO_2) 50% + (Ta_2O_5) 50% oxide layers with nominal thickness of 10 μm. The thin film was embedded in an acrylic resin*

Metal	Point 1 (%)	Point 2 (%)	Point 3 (%)	Point 4 (%)
Ru	14.5	49.1	36.2	20.2
Ti	77.9	10.4	4.3	2.9
Ta	7.6	40.6	59.5	76.8

* The ensemble was then submitted to polishing, and the surface was covered with gold for the SEM and EDS analyses.

4. Application of RuO_2-Based Material in Organic Oxidation Reaction (OOR)

From the beginning of the 1990s, there has been increasing interest in the use of ruthenium-based electrodes for the electrochemical degradation of toxic effluents. Additionally, many electrochemical processes have been proposed for the oxidation of organic compounds because it is easy to operate such electrodes [54-59]. Phenol has been utilized as model compound for OOR by many research groups [55, 60]. Other classes of compounds that have also received special attention are herbicides [9] and dyes [10], among other frequently investigated pollutants. Considering environmental issues, the main goal in electrocatalyst material design aiming at the oxidation of organic compounds is to find non-toxic and low-cost anode materials that yield the best substrate conversion to CO_2, or that at least lead to an environmentally acceptable total organic carbon (TOC) content.

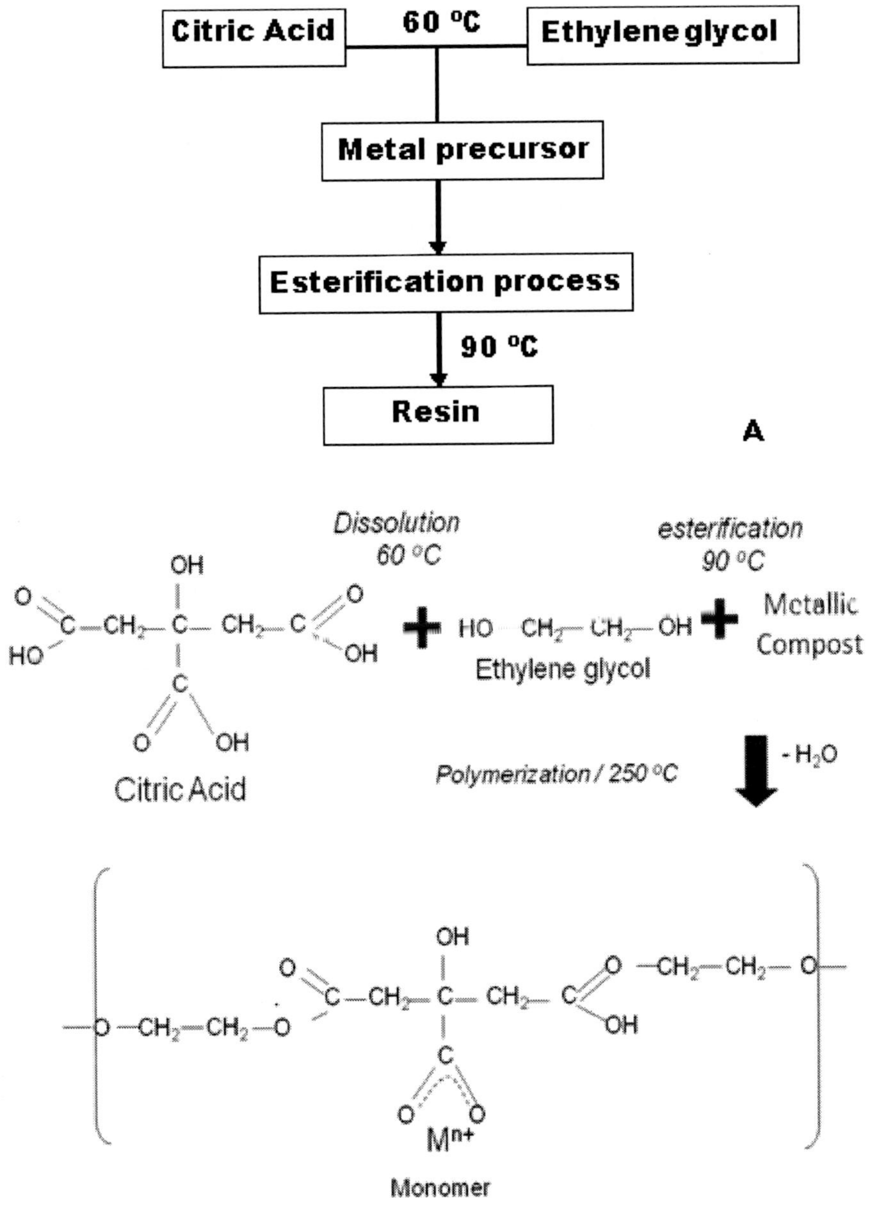

Figure 3. (A) Resin preparation from the metal precursor and (B) schematic chemical reactions during calcination of the resin by the DPP process.

Also, it is important to investigate the toxicity of the generated intermediates, which must be as low as possible. In this regard, the challenge has been to find electrode materials where the OOR is maximum. The competition between OER, an undesirable side reaction, and the OOR can be understood by taking the mechanism generally proposed for oxide electrodes into account [61]:

$$MO_x + H_2O \rightarrow MO_x(^{\bullet}OH) + H^+ + e \quad (2)$$

$$MO_x(^{\bullet}OH) \rightarrow MO_{x+1} + H^+ + e \quad (3)$$

$$R + MO_{x+1} \rightarrow RO + MO_x \quad (4)$$
$$MO_{x+1} \rightarrow \frac{1}{2} O_2 + MO_x + H^+ + e \quad (5)$$

where R stands for the organic compounds and MO_x corresponds to the RuO_2 compound class.

Based on what has been stated above, there is continuous competition between OOR (reaction 4) and OER (reaction 5). This is explained by the fact that RuO_2-based materials are very active for OER, showing low overpotential for this reaction; i.e., around 1.2 V versus RHE in acid medium. Therefore, mild oxidation is the preferable pathway instead of electrochemical incineration (CO_2 formation). However, the lower activity toward OOR presented by RuO_2 electrodes as compared to more active materials such as Boron Doped Diamond (BDD) [10] can be overcome by either choosing the right electrolyte conditions or coupling the electrochemical reaction with a photochemical process [7, 10]. Achievement of these proper conditions and the fact that oxide electrodes cost 10 % the price of BDD make the use of DSA-based anodes a good option, especially when scaling-up of the electrolytic process to large applications is desirable. In this context, in the next section the procedures that are usually employed for improvement of the OOR in ruthenium-based electrodes are discussed.

5. Changing the Composition of the Supporting Electrolyte: the Use of Chloride Medium

In order to improve the ruthenium oxide performance against OOR, one may recollect the original application of DSA materials in the chlorine industry and apply their excellent performance in this medium for the oxidation of undesirable organic compounds. The mechanism for the chlorine evolution reaction (ClER) on metal oxide electrodes is described in equations 6 and 7 [2]:

$$2Cl^- \leftrightarrows Cl_{2(el)} + 2e^- \quad (6)$$
$$Cl_{2(el)} \leftrightarrows Cl_{2(sol)} \quad (7)$$

Under mild chlorine concentration and proper pH conditions [9,10,62,63] adsorbed chlorine, $Cl_{2(el)}$, will desorb from the electrode and form $Cl_{2(sol)}$ The latter further reacts with water, to form reactive species such as hypochlorous acid (HClO) and hypochlorite (ClO^-):

$$Cl_{2(sol)} + H_2O \rightarrow HClO + Cl^- + H^+ \quad (8)$$
$$HClO \leftrightarrows H^+ + ClO^- \quad (9)$$

Table 2 lists some literature results obtained in chlorine medium. Figure 4 shows the improvement that was achieved by simply changing the supporting electrolyte, clearly demonstrating that total oxidation is frequently attained in chloride medium. Since this type of salt is often employed in the production of many compounds, treatment of industrial effluents can be easily performed with maximum degradation rate.

When electrolysis is conducted in chloride medium, one must be aware of the formation of organochloride compounds (RCl). Some papers investigating the influence of chlorine concentration on RCl production can be found in the literature [9,10]. The good news is that the generated RCl is rapidly degraded as the electrolysis progresses, thereby yielding effluents with lower toxicity.

Table 2. Percentage of COD for the indirect electrooxidation of selected RuO_2-based electrodes with active chlorine

Electrode composition	Organic compound	Experimental condition	Current density/ mA cm^{-2}	TOC decay / %	Ref.
Ti/RuO_2	Basic blue	C_R 0.8 g L^{-1}, C_{Cl^-} 1.2 g L^{-1}	20	94	[10]
$Ti/Ru_{0.3}Ti_{0.7}O_2$ $Ti/Ru_{0.3}Sn_{0.7}O_2$ $Ti/Ru_{0.3}Pb_{0.7}O_2$	glyphosate	C_R 1.0 g L^{-1}, C_{Cl^-} 2.2 g L^{-1}	50	100	[9]
$Ti/RuSnTiO_2$	Anionic surfactants	C_R 1.0 g L^{-1}, C_{Cl^-} 2.2 g L^{-1}	75	100	[63]
$Ti/Ru_{0.3}Ti_{0.7}O_2$	Humic acid	C_R 0.1 g L^{-1}, C_{Cl^-} 50 mmolL^{-1}	40	46	[64]
$Ti/Ru_{0.3}Ti_{0.7}O_2$	Tannery wastewater	COD 2.5 g L^{-1}, C_{Cl^-} 5.5 g L^{-1}	40	100	[65]
$Ti/Ru_{0.3}Ti_{0.7}O_2$; $Ti/Ru_{0.3}Sn_{0.7}O_2$; $Ti/Ru_{0.3}Pb_{0.7}O_2$	Reactive blue 4 reactive orange 16	C_R 0.1 g L^{-1}, C_{Cl^-}10 mmol L^{-1}	50	75 67 55	[66]

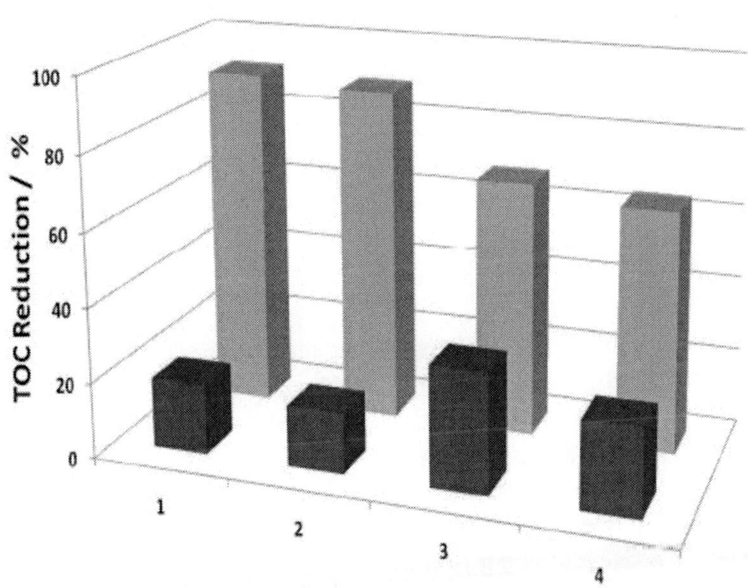

Figure 4. Total organic carbon profile of the organic compounds in Na_2SO_4 medium (black) and 0.01 Cl^- (gray) for the $Ti/Ru_{0.3}Ti_{0.7}O_2$ anode. (1) glyphosate, (2) glyphosate commercial sample, (3) reactive blue, and (4) reactive orange 16, i_{ap} = 50 mA cm^{-2}, time for 1 and 2 = 4 hours, for 3 and 4 = 120 s.

6. Electrochemical Photo-Assisted Abatement

The long lifetime of the Ti/RuO_2-TiO_2 electrode allied with the TiO_2 photoactivity under UV-light enables combination of electrochemical and photochemical degradation. Recently, Martinez-Huille and Brillas [10] have introduced this treatment in an extensive review of dyes wastewater cleanup. Several examples have indicated that 100 % color removal can be achieved within a short time period. We have investigated three different electrode compositions consisting of a mixture of two semiconductors, namely $Ti/Ru_{0.1}Ti_{0.9}O_2$, $Ti/Ru_{0.1}Ti_{0.7}Sn_{0.1}O_2$, and $Ti/Ru_{0.1}Ti_{0.3}Sn_{0.6}O_2$ [67]. The electrolysis data evidenced that the double semiconductor system (TiO_2, SnO_2) improves the lignin degradation process. Lignin was also employed as probe compound for evaluation of Ti/RuO_2-TiO_2 electrodes operating via UV-assisted electrochemical process, and 25 % TOC removal was obtained [68].

Conclusion and Perspectives for the Future

Among several catalytic oxides, RuO_2 displays special characteristics. One of its features is the large number of Ru oxidation states in the potential region located between the end of the hydrogen evolution reaction, HER, and the beginning of the oxygen evolution reaction, OER. The Ru(VI)/Ru(VII) (ruthenate/perruthenate) redox couple is an excellent OOR mediator due to its good reversibility and stability. Despite their favourable properties, ruthenium-based anodes are poorly efficient for OOR due to competition with OER. However, this situation can be drastically changed if NaCl or another homogeneous oxidant is electrogenerated on the electrode surface, thereby enhancing organic compound degradation toward less oxidizable and more biodegradable compounds. Photoassisted electrochemical treatment might also contribute to boosting application of this technology to OOR. Indeed, various improvements in terms of electrode composition and preparation procedure have been introduced, aiming to increase the electrode lifetime and activity toward organic oxidation. These modifications have contributed to the recent advances in electrochemical technology, making it a viable technique for reduction of wastewater toxicity.

References

[1] Trasatti, S. Electrocatalysis: Understanding the success of DSA®. *Electrochim. Acta* 2000, *45*, 2377-2385.

[2] Trasatti, S. Physical electrochemistry of ceramic oxides. *Electrochim. Acta* 1991, *36*, 225-241.

[3] Hayfield, P. C. S. Development of the noble metal/oxide coated titanium electrode. Part I: The beginning of the story. *Platin. Met. Rev.* 1998, *42*, 27-33.

[4] Hayfield, P. C. S. Development of the noble metal/oxide coated titanium electrode Part II: The move to platinum/iridium to ruthenium oxide electrocatalyst. *Platin. Met. Rev.* 1998, *42*, 46-55.

[5] Hayfield, P. C. S. Development of the noble metal/oxide coated titanium electrode Part III: Coated titanium anodes in widely ranging oxygen evolving situations. *Platin. Met. Rev.* 1998, *42*, 116-122.

[6] Beer, H. B. Electrode for use in electrolytic process, British Patent, 1,195,871, 1967.

[7] Pelegrino, R. L.; Di Iglia, R. A.; Sanches, C. G.; Avaca, L. A.; Bertazzoli, R. Comparative study of commercial oxide electrodes performance in electrochemical degradation of organics in aqueous solutions. *J. Braz. Chem. Soc.* 2002, *13*, 60-65.

[8] Lanza, M. R. V.; Bertazzoli, R. Selection of a commercial anode oxide coating for electro-oxidation of cyanide. *J. Braz. Chem. Soc.* 2002, *13*, 345-351.

[9] Aquino Neto, S.; De Andrade, A. R. Electrochemical degradation of glyphosate formulations at DSA anodes in chloride medium: an AOX formation study. *J. App. Electrochem.* 2009, *39*, 1863-1870.

[10] Martinez-Huitle, C. A. ; Brillas, E. Decontamination of wastewaters containing synthetic organic dyes by electrochemical methods: a general review. *App. Catal. B: Environ.* 2009, *87*, 105-145.

[11] Rajeshwar, K; Ibanez, J. G.; *Fundamentals and Application in Pollution Abatement*, Academic Press: San Diego, 1997.

[12] Brillas, E.; Cabot, P. L.; Casado, J. Chemical degradation methods for wastes and pollutants environmental and industrial applications. In *M. Tarr* Ed.; Marcel Dekker: New York, 2003, p. 235-304.

[13] Trasatti, S, *Electrodes of Conductive Metallic Oxides..*; Lodi, G., Eds.; *Part A and Part B;* Elsevier: Amsterdam, 1980 and 1981.

[14] Forti, J. C.; Olivi, P.; De Andrade, A. R. Characterization of DSA-type coatings with nominal composition $Ti/Ru_{0.3}Ti_{(0.7-x)}Sn_xO_2$ prepared via a polymeric precursor. *Electrochim. Acta.* 2001, *47*, 913-920.

[15] Ribeiro, J.; De Andrade, A. R. Characterization of RuO_2-Ta_2O_5 coated titanium electrode: microstructure, morphology, and electrochemical investigation. *J. Electrochem. Soc.* 2004, *151*, D106-D112.

[16] De Faria, L. A.; Boodts, J. F. C.; Trasatti, S. Effect of composition on the point of zero charge of RuO_2+TiO_2+CeO_2 mixed oxides. *Col. Surf. A-Physicochem. Eng. Aspects* 1998, *132*, 53-59.

[17] Lassali, T. A. F.; De Castro, S. C., Boodts, J. F. C. Structural, morphological and surface properties as a function of composition of Ru + Ti + Pt mixed-oxide electrodes. *Electrochim. Acta* 1998, *43*, 2515-2525.

[18] Terezo, A. J.; Pereira, E. C. Preparation and characterization of Ti/RuO_2-Nb_2O_5 electrodes obtained by polymeric precursor method. *Electrochim. Acta* 1999, *44*, 4507-4513.

[19] Da Silva, L. M.; De Faria, L. A.; Boodts, J. F. C. Electrochemical impedance spectroscopic (EIS) investigation of the deactivation mechanism, surface and electrocatalytic properties of $Ti/RuO_2(X) + Co_3O_4(1 - X)$ electrodes. *J. Electroanal. Chem.* 2002, *532*, 141-150.

[20] Ribeiro, J.; Moats, M. S.; De Andrade, A. R. Morphological and electrochemical investigation of RuO_2-Ta_2O_5 oxide films prepared by the Pechini-Adams method. *J. App. Electrochem.* 2008, *38*, 767-775.

[21] Tavares, A. C.; Trasatti, S. Ni + RuO_2 co-deposited electrodes for hydrogen evolution. *Electrochim. Acta* 2000, *45*, 4195-4202.

[22] Hine, F.; Yasuda, M.; Noda, T.; Yoshida, T.; Okuda, J. Electrochemical behavior of the oxide-coated metal anodes. *J. Electrochem. Soc.* 1979, *126*, 1439-1445.

[23] Ribeiro, J.; De Andrade, A. R. Investigation of the electrical properties, charging process, and passivation of the $RuO_2 - Ta_2O_5$ oxide films. *J. Electroanal. Chem.*, 2006, *592*, 153-162.

[24] Vercesi, G. P.; Rolewicz, J.; Comninellis, C.; Hinden, J. Characterization of DSA-type oxygen evolving electrodes. Choice of base metal. *Thermochim. Acta*, 1991, *176*, 31-47.

[25] Tilak, B. V.; Birss, V. I.; Wang, J.; Chen, C. P.; Rangarajan, S. K. Deactivation of thermally formed Ru/Ti oxide electrodes - An AC impedance characterization study. *J. Electrochem. Soc.* 2001, *148*, D112-D120.

[26] Klementeva, V. S.; Uzbekov, A. A. Dissolution of Ruthenium and Titanium at different ratios of their oxides in the coating of Ruthenium-Titanium oxide anodes. *Sov. Electrochem.* 1985, *21*, 736-738.

[27] Iwakura, C.; Inai, M.; Manabe, M.; Tamura, H. The cause of the activity loss of Titanium-supported Ruthenium dioxide electrodes during the anodic evolution of oxygen. *Denki Kagaku*, 1980, *48*, 91-96.

[28] Vallet, C. E. Computer-simulations of the semiconductor-metal transition in Ru/Ti and Ir/Ti mixed oxides. *J. Electrochem. Soc.* 1991, *138*, 1234-1238.

[29] Duvigneaud, P. H.; Coussement, A. Effect of Chlorine on solid-solution formation in Ruthenium Titanium-dioxide coatings. *J. Solid State Chem.* 1984, *52*, 22-31.

[30] Trasatti, S.; Buzzanca, G. Ruthenium dioxide: A new interesting electrode material. Solid state structure and electrochemical behavior. *J. Electroanal. Chem.* 1971, *29*, 1-5.

[31] Ardizzone, S.; Fregonara, G.; Trasatti, S. Inner and outer active surface of RuO_2 electrodes. *Electrochim. Acta* 1990, *35*, 263-267.

[32] Conway, B. E. Transition from supercapacitor to battery behavior in electrochemical energy-storage. *J. Electrochem. Soc.* 1991, *138*, 1539-1548.

[33] Yong-gang, W.; Xiao-gang, Z. Preparation and electrochemical capacitance of RuO_2/TiO_2 nanotubes composites. *Electrochim. Acta*, 2004, *49*, 1957-1962.

[34] Hu, C.-C.; Huang, Y.-H. Cyclic voltammetric deposition of hydrous ruthenium oxide for electrochemical capacitors. *J. Electrochem. Soc.* 1999, *146*, 2465-2471.

[35] Tilak, B. V.; Chen, C. P.; Birss, V. I.; Wang, J. Capacitive and kinetic characteristics of Ru-Ti oxide electrodes: influence of variation in the Ru content. *Canadian J. Chem.-Revue Canadienne De Chimie,* 1997, *75*, 1773-1782.

[36] Birss, V. I.; Myers, R.; Angersteinkozlowska, H.; Conway, B. E. Electron-microscopy study of formation of thick oxide-films on Ir and Ru electrodes. *J. Electrochem. Soc.* 1984, *131*, 1502-1510.

[37] Kim, I. H.; Kim, K. B. Ruthenium oxide thin film electrodes for supercapacitors. *Electrochem. Solid State Lett.*, 2001, *4*, A62-A64.

[38] Zanta, C. L. P. S.; Andrade, A. R.; Boodts, J. F. C. Solvent and support electrolyte effects on the catalytic activity of Ti/RuO_2 and Ti/IrO_2 electrodes: oxidation of isosafrole as a probe model. *Electrochim. Acta*, 1999, *44*, 3333-3340.

[39] Shieh, D. T.; Hwang, B. J. Oxygen evolution on Pt-Modified Ruthenium oxide electrodes investigated by CV techniques. *J. Electroanal. Chem.* 1995, *391*, 77-91.

[40] Anastasijevic, N. A.; Dimitrijevic, Z. M.; Adzic, R. R. Oxygen reduction on a ruthenium electrode in alkaline electrolytes. *J. Electroanal. Chem.* 1986, *199*, 351-364.

[41] Villullas, M. H.; Mattos-Costa, F. I.; Bulhões, L. O. S. Electrochemical oxidation of methanol on Pt nanoparticles dispersed on RuO_2. *J. Phys. Chem. B* 2004, *108*, 12898-12903.

[42] De Oliveira-Sousa, A.; Da Silva, M. A. S.; Machado, S. A. S.; Avaca, L. A.; De Lima-Neto, P. Influence of the preparation method on the morphological and electrochemical properties of Ti/IrO_2-coated electrodes. *Electrochim. Acta* 2000, *45*, 4467-4473.

[43] Terezo, A. J.; Pereira, E. C. Preparation and Characterisation of Ti/RuO_2 anodes obtained by sol-gel and conventional routes. *Mater. Letters* 2002, *53*, 339-345.

[44] Mattos-Costa, F. I.; De Lima-Neto, P.; Machado, S. S.; Avaca, L. A. Characterisation of surfaces modified by sol-gel derived $Ru_xIr_{1-x}O_2$ coatings for oxygen evolution in acid medium. *Electrochim. Acta* 1998, *44*, 1515-1523.

[45] Pelegrino, R. R. L.; Vicentin, L. C.; De Andrade, A. R.; Bertazzoli, R. Thirty minutes laser calcination method for the preparation of DSA® type oxide electrodes. *Electrochim. Commun.* 2002, *4*, 139-142.

[46] Makgae, M. E.; Theron, C. C. ; Przybylowicz, W. J.; Crouch, A. M. Preparation and surface characterization of Ti/SnO_2-RuO_2-IrO_2 thin films as electrode material for the oxidation of phenol. *Mater. Chem. Phys.* 2005, *92*, 559-564.

[47] Correa-Lozano, A.; Comninellis, C.; De Battisti, A. Physical properties of SnO_2 – Sb_2O_5 films prepared by the spray pyrolysis technique. *J. Electrochem. Soc.* 1996, *143*, 203-209.

[48] Diaz-Flores, L. L.; Ramirez-Bon, R.; Mendoza-Galvan, A.; Prokhorov, E.; Gonzalez-Hernandez, J. Impedance spectroscopy studies on SnO_2 films prepared by the sol-gel process. *J. Phys. Chem. Solids* 2003, *64*, 1037-1042.

[49] Oliveira, M. M.; Schnitzler, D. C.; Zarbin, A. J. G. $(Ti,Sn)O_2$ mixed oxides nanoparticles obtained by the sol-gel route. *Chem. Mater.* 2003, *15*, 1903-1909.

[50] Schottner, G. Hybrid sol-gel derived polymers: Applications of multifunctional materials. *Chem. Mater.* 2001, *13*, 3422-3435.

[51] Rodrigues, E.; Olivi, P. Preparation and characterization of Sb-doped SnO_2 films with controlled stoichiometry from polymeric precursors. *J. Phys. Chem. Solids* 2003, *64*, 1105-1112.

[52] Pechini, M. P. Method of preparing lead and alkaline earth Titanates and Niobates and coated method using the same to form a capacitor, in US Patent 1967, 3,330, 697, 1- 6.

[53] Lassali, T. A. F.; Bulhões, L. O. S.; Abeid, L. M. C.; Boodts, J. F. C. Surface characterization of thermally prepared, Ti-supported, Ir-based electrocatalysts containing Ti and Sn. *J. Electrochem. Soc.* 1997, *144*, 3348-3354.

[54] Comninellis, C.; Vercesi, G. P. Problems in DSA coating deposition by thermal-decomposition. *J. Appl. Electrochem.* 1991, *21*, 136-142.

[55] Comninellis, C.; Pulgarin, C. Anodic-oxidation of Phenol for waste-water treatment. *J. Appl. Electrochem.* 1991, *21*, 703-708.

[56] Ribeiro, J.; Purgato, F. L. S.; Kokoh, K. B.; Léger, J.-M. De Andrade, A. R. Application of Ti/RuO_2-Ta_2O_5 electrodes in the electro-oxidation of ethanol and derivants: reactivity versus electrocatalytic efficiency. *Electrochim. Acta* 2008, *53*, 7845-7851.

[57] Malpass, G. R. P.; Motheo, A. J. Screening process for activity determination of conductive oxide electrodes for organic oxidation. J. Braz. Chem. Soc. 2008, *19*, 672-678.

[58] Savall, A. Electrochemical treatment of industrial organic effluents. *Chimia* 1997, *49*, 23-27.

[59] De Andrade, A. R.; Donate, P. M.; Alves, P. P. D.; Fidellis, C. H. V.; Boodts, J. F. C. Ethanol electro-oxidation in ruthenium-oxide-coated titanium electrodes. *J. Electrochem. Soc.* 1998, *145*, 3839-3843.

[60] Cestarolli, D. T.; De Andrade, A. R. Electrochemical oxidation of phenol at $Ti/Ru_{0.3}Pb_{(0.7-x)}Ti_xO_y$ electrodes in aqueous media. *J. Electrochem. Soc.* 2007, *154,* E25-E30.

[61] Simond, O.; Schaller, V.; Comninellis, C. Theoretical model for the anodic oxidation of organics on metal electrodes. *Electrochim. Acta* 1997 42, 2009-2012.

[62] Cornell, A.; Hakansson, B.; Lindberg, G. Ruthenium-based dimensionally stable anode in chlorate electrolysis, effects of electrolyte composition and the anode potential. *J. Electrochem. Soc.* 2003, *150*, D6-D12.

[63] Panizza, M.; Delucchi, M.; Cerisola, G. Electrochemical degradation of anionic surfactants. *J. Appl. Electrochem.* 2005, *35*, 357-361.

[64] Barbosa, C. P.; Malpass, G. R. P.; Miwa, D. W.; Gomes, L.; Bertazzoli, R.; Motheo, A. J. Electrochemical removal of Cu-II in the presence of Humic acid. *J. Braz. Chem. Soc.* 2010, *21*, 651-658.

[65] Szpyrkowicz, L.; Kaul, S. N.; Neti, R. N. Tannery wastewater treatment by electro-oxidation coupled with biological process. *J. Appl. Electrochem.* 2005, *35*, 381-390.

[66] Da Silva R. G.; Aquino-Neto, S.; De Andrade, A. R. Electrochemical degradation of reactive dyes at different DSA® compositions. *J. Braz. Chem. Soc.*, 2011, *22*, 126-133.

[67] Pelegrini, R.; Reyes, J.; Duran, N.; Barbosa, P. P.; De Andrade, A. R. Photoelectrochemical degradation of lignin *J. Appl. Electrochem.* 2000, 30, 953-958.

[68] Pinhedo, L.; Pelegrini, R.; Bertazolli, R.; Motheo, A. J. Photoelectrochemical degradation of humic acid on a $(TiO_2)_{0.7}(RuO_2)_{0.3}$ dimensionally stable anode *Appl. Catalysis B: Environ.* 2005, *57*, 75-81.

In: Rutile: Properties, Synthesis and Applications
Editor: It-Meng (Jim) Low

ISBN 978-1-61942-233-9

Chapter 10

MINERAL CHEMISTRY AND POTENTIAL APPLICATIONS OF NATURAL-MULTI-DOPED HYDROTHERMAL RUTILE FROM PORPHYRY COPPER DEPOSITS

Osvaldo M. Rabbia* and Laura B. Hernández
Instituto de Geología Económica Aplicada – GEA
Universidad de Concepción
Casilla 160-C, Barrio Universitario, Concepción
Chile

ABSTRACT

Rutile is by far the most frequent form of titanium dioxide or titania (TiO_2) found in nature, and is widely distributed among different rock types of the Earth's crust. Hydrothermal rutile associated to sub-volcanic geological environments, such as porphyry copper deposits, is formed under oxidizing and water-rich-dominated conditions at relatively high-temperatures (~400-700°C). This kind of rutile is usually enriched in a great variety of trace elements or dopants, including W^{6+}, Mo^{6+}, Nb^{5+}, Ta^{5+}, Sb^{5+}, Si^{4+}, Zr^{4+}, Sn^{4+}, Th^{4+}, U^{4+}, Fe^{3+}, V^{3+}, Cr^{3+}, Sc^{3+}, Mn^{2+}, Cu^{2+}, H^{+} among others. Most of these cations are located in the Ti^{4+} octahedral site of the rutile structure entering via direct or coupled substitutions. Although crystallographically controlled, the distribution of these impurities in the rutile lattice is very irregular; displaying enriched sectors with metal content up to several weight percent (zoned crystals). These impurities generate defect energy levels in the band gap, altering the band structure, and enhance, in many cases, the light absorbance and broaden the energy spectral response from ultraviolet to visible light. Calculated density values (~4.31 g/cm^3) as well as the expected ferromagnetic response of these natural rutiles are slightly higher than found for the impurity-free equivalents.

The performance of the natural-multi-doped rutile in photocatalytic oxidation cells (POC), microbial fuel cells (MFC) and other applications has not been entirely evaluated,

* E-mail address: rabbia@udec.cl

even when it could be a promising alternative considering the broad spectrum of different dopants present in this mineral phase combined with its abundance in nature and low recovery costs. Natural multi-doped rutile can potentially be recovered, using existing low-cost technologies, as a by-product from large-scale mining operations, such as those associated with giant world-class porphyry Cu-(Mo-Au) deposits.

This chapter aims to provide an overview of the origin, mineral chemistry and processes controlling the trace element composition of natural multi-doped hydrothermal rutile. The economic importance of porphyry-type ore deposits as a source for natural rutile and its recovery as a by-product are also briefly discussed. Finally, an outline of the hydrothermal rutile applications as semiconductor highlights the necessity for further mineralogical studies on natural catalysts, a new developing research field in Earth Sciences.

Keywords: TiO_2, rutile, mineral chemistry, porphyry copper deposits, semiconductor, photocatalysis.

INTRODUCTION

The transition metal titanium is the ninth most abundant element in the Earth's crust, making up 0.43 wt% of its mass (Rudnick and Gao 2003). Its tetravalent form (Ti^{4+}) dominates under all geologically relevant conditions (Waychunas 1987, Watson and Harrison 2005, Watson et al. 2006). There are three main naturally occurring titanium dioxide (TiO_2) polymorphs: rutile, anatase and brookite, where rutile is the most frequent. Although it is a less common metastable polymorph, the TiO_2 (B) phase also occurs in nature (Marchand et al. 1980). Rutile is widely distributed as an accessory mineral in different rock types of the Earth's crust (Meinhold 2010). It is traditionally considered as a highly stable mineral in most geological environments, even at the elevated pressures and temperatures of the Earth's mantle. Its durability, refractory character and low solubility in aqueous fluids are related to the strong covalent character of the Ti-O bond that dominates the structure of this mineral. These characteristics are shared by most of the covalent materials (Laing 1993).

Even though rutile is broadly distributed in different geological settings, in this chapter we mostly focus on natural hydrothermal rutile from economic ore mineralizations, in particular from porphyry Cu-(Mo-Au) deposits (Figure 1). This choice is based on the availability of this geological resource and the low recovery cost of rutile from these deposits that along with its particular mineral chemistry enhances its potential applicability as a cost-effective alternative catalyst.

Porphyry copper (± Mo ± Au) deposits are large natural hydrothermal systems (up to about 100 Km^3), where multi-doped rutiles are generated. From a geochemical point of view, these systems represent huge sulfur anomalies in the Earth's crust, where sulfur is 10 times higher than copper (Hunt 1991). From an economical point of view, however, they are considered to be a major metal source. At present, porphyry copper systems supply nearly three-quarters of the world's copper, half the molybdenum, about one-fifth of the gold, and most of the rhenium along with minor amounts of other metals (Sillitoe 2010).

An important feature of porphyry copper deposits is their large size, containing up to about 200 MTn of copper (e.g.: Río Blanco-Los Bronces, central Chile; Sillitoe 2010).

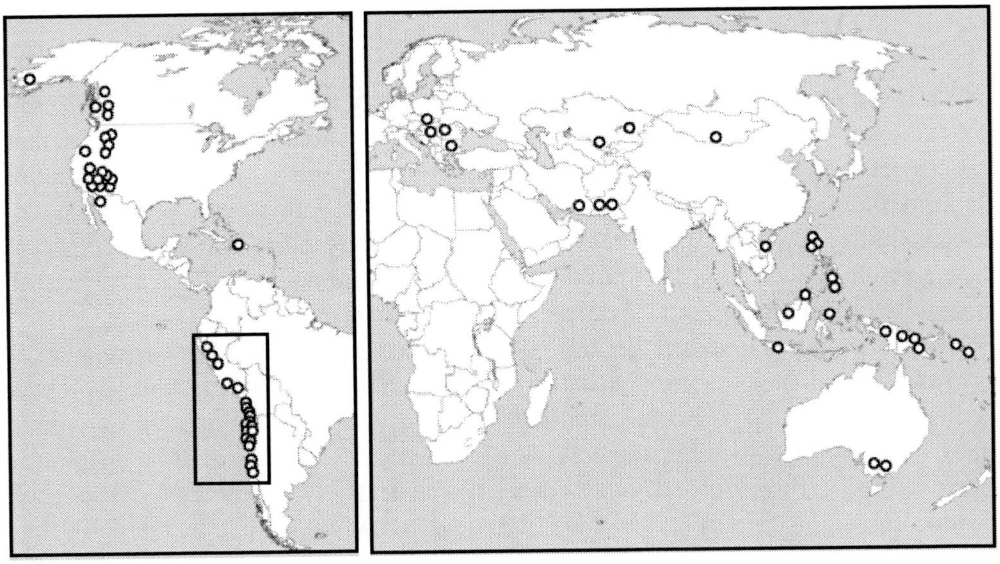

Figure 1. Worldwide location of porphyry Cu-Mo-Au systems considered as potential source of natural multi-doped hydrothermal rutile. Internal box indicates location of Chile-Peru Andean mining district, which along with the SW-USA and SE-Asia, are the world most important mining areas.

In these huge ore deposits, rutile is commonly not recovered and is usually disposed along with other gangue materials in mine tailings. The estimated amount of hydrothermal multi-doped-rutile present in these systems is about half of their copper content (Rabbia 2002). Rutiles associated with porphyry copper deposits are mainly formed at temperatures between about 400 and 700°C, at pressures of a few hundred MPa (near surface environment), with oxygen fugacity above nickel-nickel oxide (NNO) buffer and under water-rich-dominated conditions (Rabbia et al. 2009). Despite the common presence of metal-complexing agents (e.g.: Cl^-, SO_4^{2-}, HS^-, S_3, HO^- and CO_3^{2-} among others) in porphyry Cu-(Mo-Au) systems, rutile is formed *in situ* due to breakdown of pre-existing Ti-bearing phases during hydrothermal alteration processes. This suggests that titanium behaves rather conservatively or as an immobile element, at least at the deposit scale, in this upper crustal environment. This is a key behaviour of titanium that helps to explain the origin, distribution and chemical composition of natural hydrothermal rutiles.

The diversity of fluid and precursor mineral compositions involved in the rutile-forming reactions and its residual nature are responsible for its broad range of impurities (trace metals). Indeed, it is this wealth and variety of impurities that makes hydrothermal rutiles potentially useful as alternative natural catalyst. Recent studies have demonstrated that natural impure rutile may serve as a cost-effective alternative catalyst to the synthetic equivalents in photocatalytic oxidation (POC) applications as well as to the extremely high-cost noble metals in microbial fuel cells (MFC) uses (Wong and Wong 2007, Lu et al. 2010).

The combination of all aforementioned factors has opened the possibility of using natural-multi-doped hydrothermal rutile in large-scale applications. However, the mineral chemical characterization from a catalytic perspective is still deficient, particularly for those rutiles from porphyry copper deposits. The ultimate goal of this chapter is to draw attention to the potentiality of using natural mineral phases as inexpensive, abundant and efficient catalyst semiconductors.

ORIGIN OF NATURAL HYDROTHERMAL RUTILE

Natural rutile may form through different ways. It may crystallize either directly from a hydrous silicate melt (supra-*solidus* process) as a primary magmatic phase, like in pegmatites (e.g.: Černý et al. 1981) and evolved peraluminous granitic magmas (Carruzzo et al. 2006), or result from fluid-assisted metamorphism or metasomatism (sub-*solidus* processes). In the latter case, rutile may be formed either *in situ* by mineral reactions, as in granulite-type of rocks (e.g.: Bohlen and Liotta 1986), or by precipitation from aqueous fluids as a vein-filling phase, implying Ti migration (e.g.: Zack 1999, Luvizotto and Zack 2009).

The process that generates rutile in porphyry copper deposits corresponds to an extreme case of fluid-dominated metasomatism (hydrothermalism), where the role of fluids in destabilizing pre-existing Ti-bearing minerals is key to the *in situ* formation of rutile. The chemical stability of Ti-bearing minerals is particularly affected by high-temperature and SO_2-rich oxidizing fluids (e.g.: Udubasa 1982, Force 1991, Haggerty 1991). This explains why rutile formation in porphyry systems is coeval with the main mineralogical changes induced by hot fluid circulation (hydrothermal alteration; Figure 2) and why it results from the breakdown of pre-existing Ti-bearing minerals (Williams and Cesbron 1977, Czamanske et al. 1981, Beane and Titley 1981, Force 1991). In porphyry Cu-(Mo-Au) systems, the main rutile parental minerals include phases such as sphene ($CaTiSiO_5$; with ~35-40 wt% of TiO_2), ilmenite ($FeTiO_3$; with ~48-53 wt% of TiO_2), titano-magnetite (Fe_3O_4-$TiFe_2O_4$; with ~0-36 wt% of TiO_2), biotite ($K(Mg,Fe)_3\,[AlSi_3O_{10}]\,(OH,F)_2$; with ~4-6 wt% of TiO_2) and amphibole ($Ca_2[Mg_4\,(Al,Fe,Ti)]\,Si_7AlO_{22}(OH)_2$; with ~0.5-2.0 wt% of TiO_2). Most of the whole-rock TiO_2 content is, however, frequently controlled by only a few phases. Based on the relative abundance of these mineral phases and their titanium contents, sphene and biotite are commonly the most important source of hydrothermal rutile in porphyry systems (Rabbia 2002).

Field-based evidence suggests that titanium in hydrothermal rutiles is almost entirely inherited from the host rock. This is consistent with the general insolubility shown by this element at near-surface conditions (e.g., Ayers and Watson 1993), implying that TiO_2 content in the host rock is one of the factors that limit the amount of total residual rutile generated during hydrothermal alteration processes. Most rocks from porphyry Cu-(Mo-Au) deposits contain relatively low amounts of Ti (~0.3 to 0.6 wt% TiO_2), except for those systems where ferromagnesian silicate-rich (mafic) wallrocks (with up to ~1.0 to 1.3 wt% TiO_2) are present (e.g.: the El Teniente copper deposit, in Central Chile; Skewes et al. 2002). Mafic rocks also have Ti-bearing phases (e.g.: pyroxenes) different from those typically dominating in most porphyry systems; which may also imply different chemical stabilities for the precursor minerals from which rutile may form.

Hydrothermal fluid composition is another factor controlling rutile formation. During the lifespan of an ore-forming system, different fluids with contrasting pH, chlorine content, oxygen fugacity and temperature may interact with the ore host rocks. These fluids have different reactivity selectively affecting mineral stability. Under high temperature (above ~500 °C), variable salinity and close to neutral pH conditions (known as the potassic alteration stage in porphyry copper mineralizations), SO_2-bearing aqueous fluids promote the breakdown of sphene and ilmenite to form rutile as well as other Ti-free hydrothermal mineral phases. In contrast to sphene and ilmenite, primary magmatic biotite re-equilibrates under these hydrothermal conditions by adjusting its composition via exsolution-like processes. As a result, minor quantities of tiny (few microns across) needle-like ("sagenitic") rutile is formed inside a Ti-poor (2-4 wt% of TiO_2) hydrothermal biotite (Figure 3).

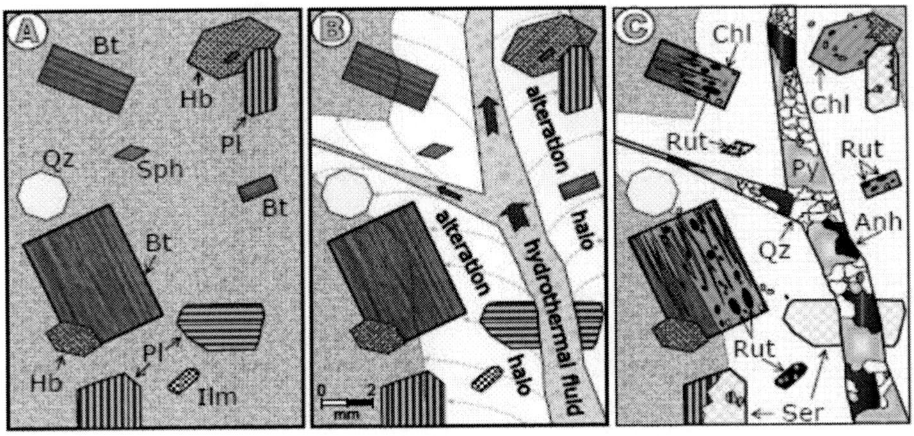

Figure 2. Schematic representation of natural hydrothermal rutile formation. A. Unaltered (fresh) porphyritic rock with phenocrysts (large crystals) included in a fine grained matrix. Ti-bearing minerals are biotite (Bt), ilmenite (Ilm), sphene (Sph) and hornblende (Hb). Although sphene and ilmenite are not usually cogenetic, both are included as example of precursor minerals from which rutile forms by hydrothermal breakdown. Plagioclase (Pl) and quartz (Qz) are the main Ti-free phases. B. Flux of hydrothermal fluids through fresh rock generates an alteration halo, due to intergranular fluid percolation around fluid filled fractures. C. Most minerals interacting with hydrothermal fluids tend to, at least, re-adjust their chemical compositions, or to breakdown forming a series of new hydrothermal phases (e.g. chlorite Chl; sericite Ser). Hydrothermal alteration of Ti-bearing minerals, leaves behind residual rutile (Rut) as one of their sub-products. Note that most hydrothermal rutile occupies textural positions controlled by the location of its Ti-bearing parental phase. Sulfides (Py), sulfates (Anh) and silicates (Qz) commonly fill the veins where fluid circulated.

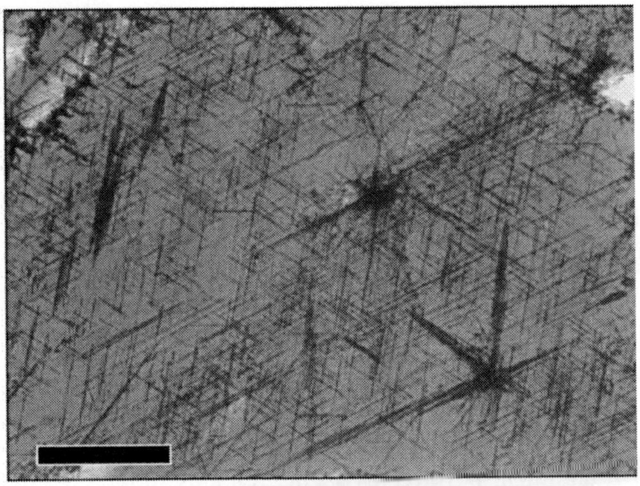

Figure 3. Image under optical microscope of large crystal of biotite displaying "sagenitic" rutile (needle-like) texture. This type of rutile results from titanium exsolution from the parental biotite during high temperature hydrothermal re-equilibration. Transmitted light, open Nichols. Black bar size 10 microns.

Similarly, pre-existing primary magmatic titano-magnetite re-equilibrates to Ti-poor hydrothermal magnetite releasing additional Ti as residual rutile during this hydrothermal alteration stage (Rabbia, et al., 2009).

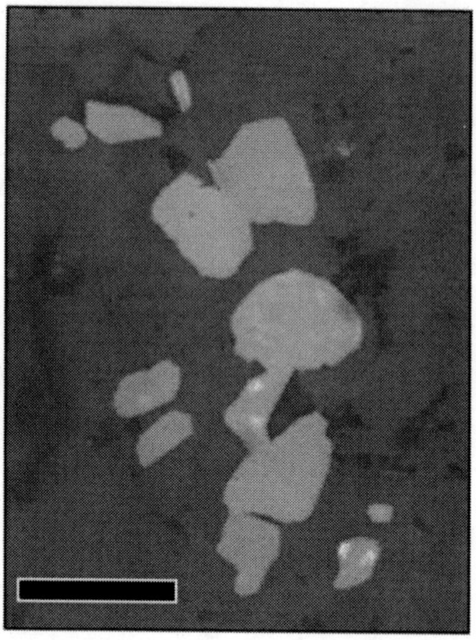

Figure 4. Image under optical microscope of larger and more equant rutile crystals (light gray) formed by dissolution re-precipitation of tiny sagenitc rutile during biotite breakdown to Ti-free chlorite at lower temperature. Transmitted light, open Nichols. Black bar size 40 microns.

During subsequent alteration events under low-temperature (below ~500 ºC), low salinity and more acidic pH (< ~6) conditions (the so called phyllic alteration), biotite becomes unstable and is completely replaced by a rutile-bearing association including sericite ± quartz ± chlorite. This rutile tends to be more prismatic and slightly coarser than sagenitic rutile, but smaller than those formed from sphene (or ilmenite) breakdown (Figure 4).

In summary, superimposed episodes of different hydrothermal alteration processes (mainly potassic and phyllic) acting on contrasting lithologies and different precursor minerals, produce different generations of hydrothermal rutile. It is during these alteration stages that rutile incorporates trace elements from both the Ti-bearing parental phases and the hydrothermal fluids, explaining the variety of chemical compositions observed in natural hydrothermal rutile found in porphyry copper deposits.

The origin of anatase in porphyry systems is comparatively less understood. Anatase occurrence is rare and tends to be restricted to late, low-temperature, carbonate veinlets located on the fringe of porphyry copper deposits, where calcium-rich mineral associations dominate (propylitic alteration zone). However, like rutile, it may also be present as a residual phase after the breakdown of a Ti-bearing mineral (Rabbia et al. 2003).

Although most titanium in porphyry systems remains immobile, a minor proportion of hydrothermal rutile and anatase may be found filling veins, which is indicative of titanium solubilization in aqueous fluids (Rabbia et al. 1998, 2003, Rabbia 2002). The origin of rutile by direct precipitation from fluids contrasts with the most general case of *in situ* nucleation and growth of rutile from a Ti-bearing precursor phase as previously described. One possible explanation for this behaviour of titanium lies in the fact that this nominally insoluble element may form chloride, fluoride, hydroxide, carbonate, phosphate and sulphate complexes in solution, enhancing its mobility (Ciavatta and Pirozzi 1983, Manning and Henderson 1984, Ciavatta et al. 1985, Aja et al. 1995). Although there is increasing supporting evidence of titanium mobilization by aqueous fluids (Gao et al. 2007, Manning et al. 2007, 2008, Antignano and Manning 2008, Schmidt et al. 2009, Luvizotto and Zack 2009, Wu et al. 2009, van Sijl et al. 2010), the precise physicochemical conditions that favour this behaviour of titanium in porphyry systems are not clear.

Physical Characteristics of Natural Hydrothermal Rutile

Under the optical microscope, most hydrothermal rutiles from porphyry Cu-Mo-Au deposits are typically black-brown, deep reddish to brownish yellow in colour, fine-grained, mostly ~10-100 μm (but up to 600 microns), and granular with anhedral to euhedral shapes. Large rutile crystals usually show heterogeneous distribution of impurities expressed as well-developed zoning patterns (e.g.: oscillatory zoning, Figure 5), which is consistent with rutile growth under fluctuating physicochemical conditions characteristic of an open system. Crystallographic control of the incorporation of impurities into the rutile lattice during crystal growth may also cause compositional variations (e.g.: sector zoning, Figure 5 and 6). The combination of more than one controlling factor for trace element incorporation results in more complex patterns of dopant distribution (e.g.: irregular or patchy zoning patterns, Figure 7). Exsolution-like inclusions in these rutiles are not frequent but intergrowth with sulfides or other mineral may be present (Figure 8). Due to the common presence of elements heavier than Ti, such as W, Nb, V and Fe, the calculated density for natural hydrothermal rutile (Figure 9) tends to be slightly higher (mean 4.310 g/cm^3, ranging from 4.286 to 4.545 g/cm^3) than for the impurity-free rutile (4.248 g/cm^3).

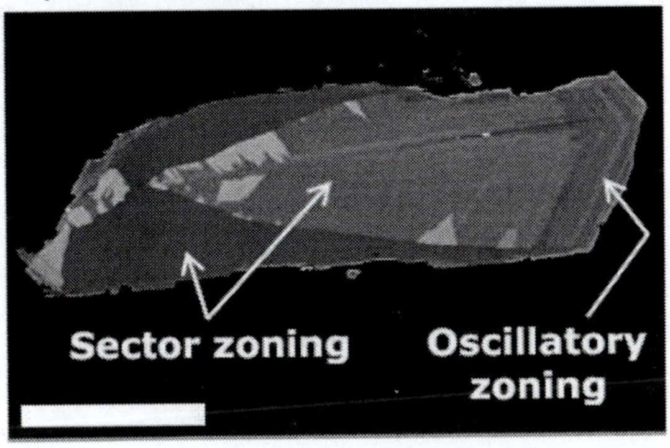

Figure 5. Backscattered electron image of a large rutile crystal (up to 300 microns) showing heterogeneous chemical composition (zonation pattern). Brighter areas represent sectors with higher trace element contents. The preferential incorporation of impurities on certain faces of this crystal, results on a sector zoning pattern of the sand-clock type. Note to the right of the crystal the alternation of impurity-rich and impurity-poor bands defining an oscillatory zoning pattern, possible related to changes in physicochemical conditions during the final stage of rutile growth. White bar size 100 microns.

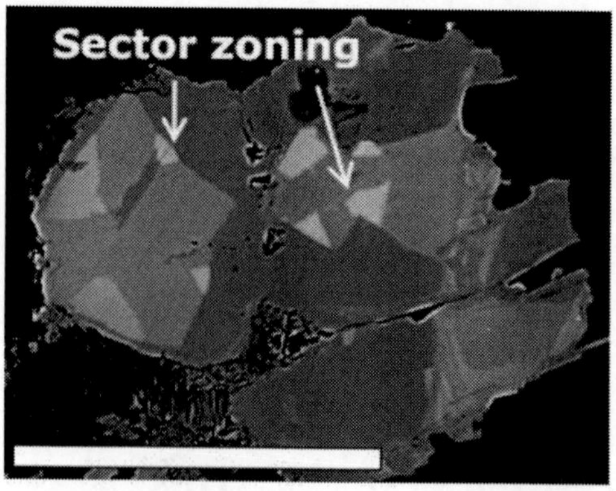

Figure 6. Backscattered electron image of a rutile crystal showing sector zoning pattern in the grains cores, amalgamated by an impurity-poor unzoned overgrowth (dark gray). White bar size 100 microns.

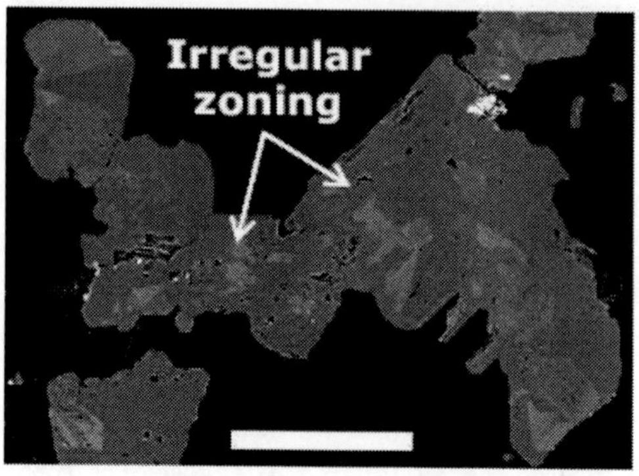

Figure 7. Backscattered electron image of rutile crystals showing irregular or patchy zoning. White bar size 100 microns.

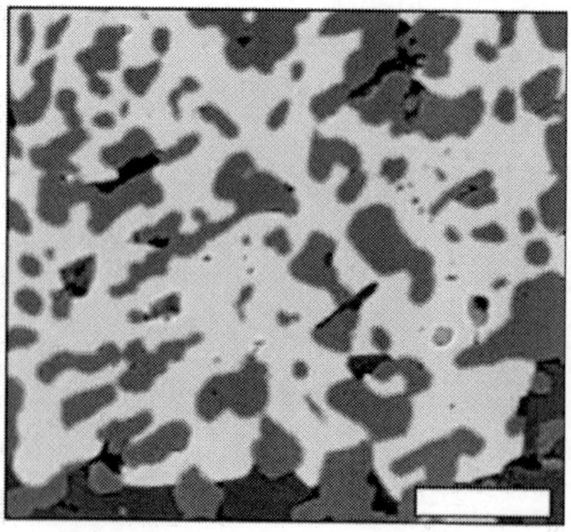

Figure 8. Backscattered electron image of rutile crystal (dark grey) intergrown with chalcopyrite ($CuFeS_2$; light grey). White bar size 25 microns.

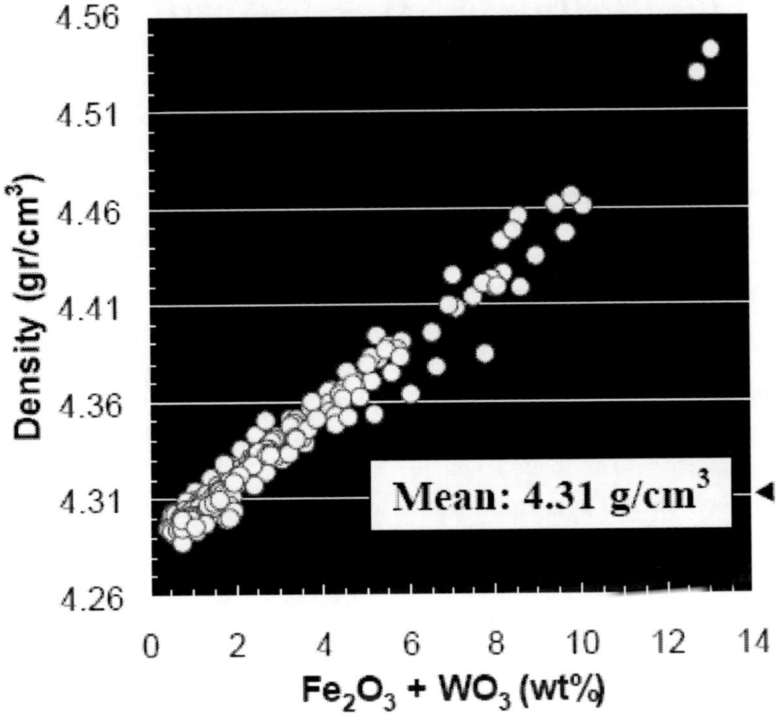

Figure 9. Calculated density versus the sum of WO_3 and Fe_2O_3 wt% for hydrothermal rutiles from Andean porphyry copper deposits. The formula used for density calculation is: (formula wt * Z * 1.6603E^{-24}) / (UC Vol * E^{-24}), where formula wt is the sum of the mass of its constituent atoms calculated after EPM analyses of rutile; Z is the number of formulae in one unit cell, that for rutile are 2; 1.6603E^{-24} is the reciprocal of Avogadro's number; UC Vol is the volume of the unit cell in cubic Armstrong and E^{-24} is to convert the volume to cm^3. Atoms where assumed to be spherical and the net volume change of the unit cell, due to the incorporation of larger and smaller atoms than titanium, was nil. The total electron microprobe analyses of rutile used in this calculation is 329. Data scattered are due to variable amounts of W in these rutiles. All data are after Rabbia (2002).

The heavily doped nature of hydrothermal rutile (mean: ~3 wt% as metal oxides; Rabbia et al 1999; Rabbia 2002), particularly in ions such as Fe^{3+} and W^{6+}, increases the covalent character of their chemical bonds in the crystalline structure, which in turn affects its optical properties. As a consequence, the refractive index of multi-doped rutile may increase in comparison with impurity-free rutile. Although it has not been determined yet, this important optical property may enhance the applicability of natural rutiles in the optoelectrical field technology.

The ability of hydrothermal rutile from porphyry systems to incorporate important quantities of Fe (mean Fe content expressed as Fe_2O_3 of ~1.3 wt%; Rabbia 2002) into its lattice, presupposes a magnetic susceptibility stronger than impurity-free rutile, which is considered as a paramagnetic phase with low magnetic susceptibility. Thus, equipment with permanent magnet technology could be used for low cost-recovery of natural rutile.

Since the rutile trace element composition is relevant for its performance as a catalyst semiconductor, a detailed chemical characterization is presented below.

Chemical Composition of Natural Hydrothermal Rutile

The assessment of rutile composition constitutes an essential step for the evaluation of its potentiality as a catalyst semiconductor. In this section, the focus will be more on mineral chemistry aspects rather than on compositional variations of rutile within and between different ore deposits. Compiled mineral chemical information for this section mainly corresponds to rutiles from Andean porphyry copper deposits (Figure 1).

The variability of dopant content within and between crystals is a characteristic feature of natural hydrothermal rutiles from porphyry systems. For example, W concentrations may range from a few ppm up to more than 6 wt%, although most values are about 0.8 to 0.9 wt% (Rabbia 2002). Taking this aspect into consideration, the impurities in these rutiles are grouped into three categories arbitrarily designed according to their mean contents: > 1000 ppm; 1000 to 100 ppm and < 100 ppm (Table 1).

Table 1. Mean contents of common impurities present in natural hydrothermal rutile from porphyry copper deposits

Mean abundance	Impurity
> 1000 ppm	Fe, W, V, Nb
1000 - 100 ppm	Si, Cr, Zr, Sn, Al, H*
< 100 ppm	Mn, Ta, Sc, Mo, U, Sb, Zn, Th, Cu, Pb, Hf, As, Ni, Co

Data from Gunow 1983, Rabbia et al. 1996b, 1998, 1999, 2009, Rabbia 2002, Scott 2005, Scott and Radford 2007.
*The range content for H^+ is estimated. Elements in the range above 100 ppm were measured by EPMA and those below 100 ppm were quantified by PIXE.

The total content of impurities, expressed as oxides, ranges from 1.0 to 14.2 wt% with a mean value of 3.1 wt% (Rabbia et al. 1999; Rabbia 2002). Iron, tungsten, vanadium and niobium oxides commonly represent about 90 wt% of the total dopants in these rutiles. Besides, the hydrothermal rutile is the main W-bearing phase in altered rocks from porphyry copper deposits (Rabbia 2002).

Maximum contents for some impurities from these natural hydrothermal rutiles are as follow: WO_3: 8,66 wt%; Nb_2O_5: 7.00 wt%; Fe_2O_3: 6.61 wt%; V_2O_3: 3.23 wt%; Sb_2O_5: 2.66 wt%; SnO_2: 1.13 wt%; Cr_2O_3: 0.78 wt%, ZrO_2: 0.68 wt%; Sc_2O_3: 0.56 wt%; Ta_2O_5: 0.28 wt%; MnO: 0.13 wt% (Rabbia 2002, Scott and Radford 2007).

Elemental mapping done in rutile crystals using the Particle Induced X-ray Emission (PIXE) technique revealed that most impurities are differentially partitioned among non-equivalent crystallographic faces, and thereby related to growth sector zoning in the rutile crystal (Rabbia et al. 2009). The preservation of sharp compositional gradients (e.g., sector and oscillatory zoning; Figures 5 and 6) in natural rutiles exposed to multiple pulses of hydrothermal fluids, implies that sub-solidus diffusion of dopants is minor after rutile formation (Rabbia et al. 2009).

In natural rutile, commonly there are several possible substitutions for Ti^{4+} such as hexa-, penta-, tetra-, tri-, di- and monovalent cations (Graham and Morris 1973, Urban et al. 1992, Brenan et al. 1994, Hassan 1994, Fett 1995, Murad et al. 1995, Rabbia et al. 1996b, 1998,

1999, Smith and Perseil 1997, Rice et al. 1998, Zack et al. 2002, Rabbia 2002, Bromiley and Hilairet 2005, Scott 2005, Carruzzo et al. 2006).

Impurities such as Zr^{4+}, Sn^{4+} or Hf^{4+} can enter the rutile structure by direct (or homovalent) substitution of Ti^{4+}:

$$^{VI}M^{4+} \leftrightarrow {}^{VI}Ti^{+4}.$$

A charge balancing substitution mechanism has to be considered for element incorporation with charges other than 4+ into the octahedral site of the rutile structure. This is the case of highly charged (e.g.: W^{6+}, Nb^{5+}, etc.) and low charged (e.g.: Fe^{3+}) ions replacing Ti^{4+} via coupled (or heterovalent) substitutions (Figure 10). The most frequent type of heterovalent substitutions observed in these rutiles includes:

$$2^{VI}M^{3+} + {}^{VI}M^{6+} \leftrightarrow 3^{VI}Ti^{+4},$$
$$^{VI}M^{3+} + {}^{VI}M^{5+} \leftrightarrow 2^{VI}Ti^{+4},$$

where M^{3+} is mainly represented by Fe, but also may include additional quantities of other elements such as V, Cr, Sc, Al, and As; M^{5+} is represented by Nb, Sb and Ta and the most common M^{6+} cations are W and Mo. It is also possible that U could enter as 6+ in these rutiles (Rabbia 2002).

The M^{2+} impurities, such as Mn and Zn, may enter the rutile structure by virtue of the following substitutions:

$$2^{VI}M^{5+} + {}^{VI}M^{2+} \leftrightarrow 3^{VI}Ti^{+4},$$
$$^{VI}M^{6+} + {}^{VI}M^{2+} \leftrightarrow 2^{VI}Ti^{+4},$$

However, larger M^{2+} cations, such as Pb, cannot be incorporated in the octahedral site, but rather are accommodated into the tunnel structure along C-axe of this mineral by:

$$2^{VI}M^{3+} + {}^{tunnel}M^{2+} \leftrightarrow 2^{VI}Ti^{+4} + {}^{tunnel}\text{vacancy}.$$

A notable feature of these natural hydrothermal rutiles is that the total content of M^{3+} ions is always in excess in order to balance the presence of M^{6+}, M^{5+} and M^{2+} ions. Figure 10 shows that the intercept on the Y-axis (M^{3+} ions) $\neq 0$ when the X-axis ($M^{6+} + M^{5+} + M^{2+}$) = 0. This excess in M^{3+} ions may be partly or totally balanced by proton (H^+) accommodation into the tunnel structure along C-axis according to:

$$^{VI}M^{3+} + {}^{tunnel}H^{+} \leftrightarrow {}^{VI}Ti^{+4} + {}^{tunnel}\text{vacancy}.$$

and the concentration of H^+ would be:

$$[H^+] = \Sigma\,[M^{3+}]_{Ti} + 2\Sigma\,[M^{2+}]_{Ti} - \Sigma\,[M^{5+}]_{Ti} \qquad \text{(Vlassopoulos et al. 1993).}$$

The incorporation of hydrogen in the rutile structure was first proposed by Johnson et al (1968) as a mechanism for charge-balancing titanium substitution by low valence cations. The presence of structural water in the nominally anhydrous rutile has been confirmed by Fourier Transform Infrared Spectroscopy (FTIR) technique, for rutiles from different geological environments (Rossmann and Smyth 1990, Hammer and Beran 1991, Vlassopoulos et al. 1993, Rabbia et al. 1996a, Bromiley and Hilairet 2005). Using atomic force microscopy and scanning tunnelling microscopy, Enevoldsen et al. (2009) concluded that hydrogen may exist at a stable interstitial site in rutile as a subsurface OH.

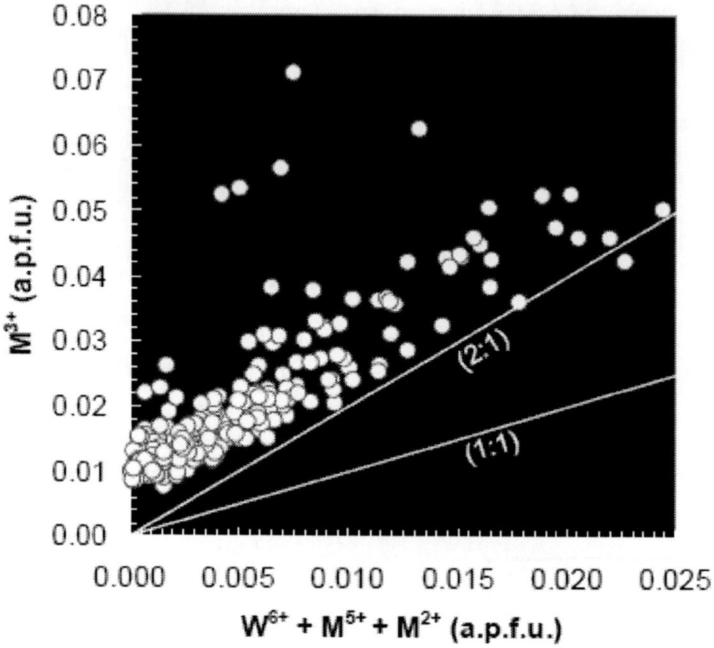

Figure 10. This diagram shows the co-variation of M^{3+} ions versus the sum of M^{6+}, M^{5+} and M^{2+} for natural hydrothermal rutiles from Andean porphyry copper deposits. Data from Rabbia (2002). a.p.f.u.: atoms per formula unit.

All discussed substitutions result in lattice distortions and defects of the rutile crystalline structure. Charge compensation may also involve metal-deficient point defects. Impurities controlled by these different substitutional mechanisms should affect polymorphic transformation kinetics of natural TiO_2 in quite different ways (Banfield et al. 1993).

The temperature of rutile formation exerts a first order control on the amount of impurities, such as Zr^{4+}, that are incorporated into the rutile structure (Zack et al. 2004). The positive correlation of Zr content in rutile and temperature, only slightly affected by pressure (Tomkins et al. 2007), has been calibrated as a geothermometer (Zr-in-rutile; Watson et al 2006). The substitution of Ti by other High Field Strength Elements (e.g.: Mo^{6+}; Rabbia et al. 2009) in the rutile structure may also be temperature dependent.

Additionally, the content of chalcophile elements (e.g.: Cu, Mo, Sb, Zn etc.) present in natural hydrothermal rutiles may vary with the timing of sulfide precipitation due to their affinity with sulfur. For example, in the El Teniente porphyry copper deposit, it was observed that the mean Mo content in high-temperature (~550-700°C) rutiles, formed before molybdenite (MoS_2) precipitation, was 186 ± 20 ppm; in contrast the mean Mo content in low-temperature (below ~550°C) rutiles, formed during and after molybdenite precipitation, was only 5.4 ± 1.1 ppm (Rabbia et al. 2009).

The provenance of trace metals in natural hydrothermal rutiles includes both local and distal sources (Figure 11). Local sources are mostly represented by mineral reactions in the host rock, including the breakdown of pre-existing Ti-bearing mineral phases from which rutile is formed. These mineral reactions are promoted by fluid-rock interaction associated with mineralization processes. The silicate melt in the magmatic reservoir at depth, from where metal-bearing hydrothermal fluids are exsolved, represent the main distal source for rutile impurities.

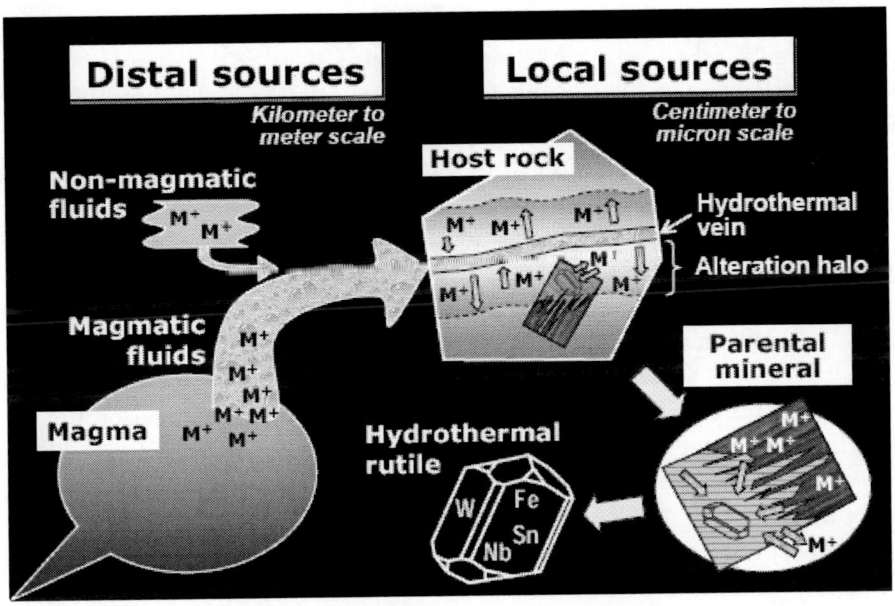

Figure 11. Schematic representation, not to scale, of how Nature "dopes" hydrothermal rutiles. All major sources of impurities for natural hydrothermal rutiles are represented, including both, local and distal ones.

Rutile composition also varies from deposit to deposit (e.g.: Rabbia et al. 1998, 1999, Scott 2005, Scott and Radford 2007). Although the extent of this variation is not fully constrained, there is some consistency in the ranges of trace element abundances for these natural hydrothermal rutiles (Table 1).

General Uses of Natural Rutile

Commercially, natural rutile is mainly consumed by producers of microcrystalline titanium dioxide for the manufacture of white pigment. The global production capacity is estimated to be 5.6 million metric tonnes per year (Gambogi 2011). According to the 2009 Minerals Yearbook of the United States Geological Survey, about 94% of titanium mineral concentrates in USA are used for white pigment production. The leading applications of this pigment are: paint and coatings (58.6%), plastics and rubber (25.5%), and paper (9.4%).

Other uses (6.5%) included catalysts, ceramics, coated fabrics and textiles, floor coverings, printing ink, and roofing granules. The remaining 6% is used in welding rod coatings and for manufacturing carbides, chemicals, and the production of metallic titanium, mainly used in aerospace applications (Gambogi 2011).

Large-scale photocatalytic applications of natural rutile will certainly modify the above statistics, impacting the market of titanium by increasing its global demand.

DOPED TiO_2 AS A CATALYST: NATURE DID IT FIRST

Titanium dioxide is by far the most extensively studied and best known semiconductor catalyst (Thiruvenkatachari et al. 2008 and references there in). It is a key material for a variety of applications, such as catalytic devices, solar cells, and other optoelectronic devices (Taramasso et al. 1983, O'Regan and Gratzel 1991, Hoffmann et al. 1995). Titanium dioxide is widely used due to its unique properties that include: a strong oxidative ability due to a band gap about 3.2eV; long-term stability to photocorrosion; re-usability and non-toxicity. This oxide semiconductor is transparent to visible light and has a high refractive index. Owing to its unique optical properties and chemical stability, TiO_2 may be used in the splitting of water (e.g.: Rao et al. 1980).

However, despite being an attractive, non-selective room-temperature oxidative process, TiO_2 photocatalysis has attracted limited commercial attention. Its principal disadvantages are largely related to: the low reaction rates exhibited by pure TiO_2, light adsorption only in the ultraviolet-light spectrum region and the high electron (e^-) - hole (h^+) recombination rate (Surender et al. 1998).

A large number of studies have been performed to extend TiO_2 photoactive range into the visible-light spectrum region, mostly by doping it with transition metal ions, such as Fe, W, and V (e.g.: Ashokkumar and Maruthamuthu 1991, Pal and Sharon 1998, Serpone et al. 1994, Navio et al. 1999, Klosek and Raftery 2001, Lu et al. 2004, Wang et al. 2004, Yang et al. 2007, Ambrus et al. 2008, Kim et al. 2010). All these research works have systematically shown the benefits of transition-metal ions as single-dopants in TiO_2, enhancing its photocatalytic activity. Even when the synergetic effect of multiple metal dopants on the optical shift of TiO_2 has not been fully evaluated, the co-doped TiO_2 has shown superior photocatalysis owing to the cooperative actions of the two dopants (e.g.: Yang et al. 2002). Despite the action of dopant ions as recombination centres for the photogenerated holes and electrons, as well as traps for the electrons, there is increasing evidence suggesting that multi-doped rutile enhances its photocatalytic efficiency under visible-light (Lu et al. 2004, Chuan et al. 2008, Lu et al. 2010). This is a promising aspect for natural semiconductor minerals.

Increasing effort has been dedicated to the development of more efficient (and less expensive) catalysts, either by synthesizing new compounds or optimizing existing ones. Nature, on the other hand, has already "prepared" a large variety of multi-doped rutile ready to be tested as natural semiconductor photocatalysts. Compared with synthetic catalysts, the complexity of natural ones is always a challenging factor, and perhaps, this complexity could be the source of new knowledge and inspiration for future laboratory research. It wouldn't be the first time we learn from Nature.

From a practical point of view, it is relevant to link the photocatalytic activity of hydrothermal rutile to specific crystallochemical features. These features could, then, be related to conditions of rutile formation and/or to a specific type of ore deposit, providing a valuable guide for prospecting natural rutiles with particular photocatalytic properties and applications. In order to evaluate the catalytic performance of rutile from a given ore deposit, a representative sampling is essential in order to account for the expected compositional variability of this mineral.

APPLICATIONS OF MULTI-DOPED NATURAL RUTILE AS A SEMICONDUCTOR

To date, there are few investigations on photocatalytic properties and applications of natural rutile (e.g.: Lu et al. 2004, 2007, 2010, Chuan et al. 2008). Despite its abundance and potentiality, none of these studies refer to hydrothermal rutile from porphyry copper deposits. In contrast, synthetic anatase has concentrated most of the attention mainly because the poor photocatalytic performance shown by pure rutile in early studies. This situation, however, is progressively changing in view of the favourable effect caused by the presence of dopants on rutile photocatalytic activity.

Most natural rutiles reported in photocatalytic studies, so far, are from a rather uncommon type of mineral deposit from China. These rutiles are V-rich (1.22 wt% of V_2O_5) but also have minor quantities of Fe (0.39 wt% of Fe_2O_3), Zn (0.35 wt% of ZnO) and Cu (0.22 wt% of CuO). These rutiles display photocatalytic activity on degrading halohydrocarbon pollutants (Lu et al. 2004 and Chuan et al. 2008). It has also been demonstrated that after treatment (e.g.: grinding, heating and quenching), these rutiles modify their surface characteristics, increasing their photoactivity by enhancing visible-light absorption. This, in turn, improves the degradation rate of organic pollutants. These results offer a possible cost-effective way for clean-up environmental pollution under visible-light with naturally impure rutiles.

A novel application in microbial fuel cells (MFC) for V-rich natural multi-doped rutile was recently proposed (Lu et al. 2010). This rutile was successfully used as a cathodic catalyst (rutile-coated cathode) for energy production with MFC reactor. The maximum power density achieved by this reactor using rutile as a cathode catalyst was comparable to that obtained with the highly expensive platinum cathode. It offers a cost-effective alternative catalyst to noble metals in MFC applications. The high visible-light photocatalytic activity of this V-rich natural rutile is related to the Ti substitution by transition ions. These impurities in the crystal lattice of rutile generate defect energy levels in the band gap and alter the band structure, enhancing the light absorbance and broadening the spectral response from ultraviolet to visible-light (Lu et al, 2007, 2010).

Given that transition metals as dopants in TiO_2 have shown good visible-light induced reactions, the high content of WO_3, V_2O_3 and Fe_2O_3 (mean: ~2.6 wt%; Rabbia 2002) in hydrothermal rutile from porphyry copper deposits is seen as an encouraging chemical feature. The possible synergetic role of the remainder ions in these natural rutiles (see Table 1), however, is still unclear.

The presence of hydrogen in rutile is particularly important because it may change the material properties by acting as a dopant or by quenching defect states and thereby modify charge carrier lifetimes important for photocatalytic or photoluminescence applications (Enevoldsen et al. 2009). If hydrogen bounded to structural oxygen by a strong O-H bond is a ubiquitous impurity in natural rutiles from porphyry deposits, as inferred from the calculated charge deficiency (excess in M^{3+} ions), then its role on the photocatalytic activity of this mineral needs to be more deeply investigated.

The still poorly explored field of natural multi-doped rutile photocatalysis seems promising considering the preceding theoretical and experimental data and the very wide potential technological applications of this low-cost natural semiconductor.

Recovery of Rutile as a By-Product from Porphyry Copper Deposits

Porphyry copper deposits are on the top of the list of the geologic resource options for rutile. Indeed, the recovery of multi-doped rutile from copper mining industry is very attractive because the rutile present in the tailings is already comminuted down to a granulometric range of 10 to 100 microns, implying that at least a fraction of the rutile grains may be released (mono-mineral grains) from other gangue minerals.

Additionally, it is worth mentioning that just in the Peruvian and Chilean Andes (Figure 1), one of the more productive mining regions in the world, about $1x10^6$ tonnes of rutile is wasted yearly in tailings. To put this quantity into perspective, it is necessary to mention that the current global production of rutile concentrate is on the order of $0.6x10^6$ tonnes per year (Gambogi 2011).

Despite some difficulty to recycle rutile from porphyry copper ore tailings, mainly arising from its fine dissemination size and complex relationship of association with other gangue minerals, it can be efficiently recovered by froth flotation process (Llewellyn and Sullivan 1981, Bulatovic and Wyslouzil 1999, Bulatovic 2007, Ye et al. 2011). Due to the common presence of iron in the highly doped rutiles described in this Chapter, its recovery by magnetic techniques or a combination with other techniques should be explored.

Conclusions

In summary, based on the information discussed in this Chapter, it is clear that natural rutiles, and particularly the metal-multi-doped hydrothermal rutile from porphyry copper deposits, may also have potential applications as a photocatalyst in addition to being an important source of titanium oxide pigments.

Even when the photocatalytic reactors based on naturally existing semiconductors do not necessarily have optimal efficiency, this technology is still attractive due to its cost-benefit ratio, especially when envisaged as large-scale, long-term and low operating-cost solutions for a wide-range of environmental problems. The compelling evidence presented in this Chapter suggests the need for a deeper evaluation of natural rutiles as a photocatalytic semiconductor. This task will require a collaborative effort between specialists from different disciplines in order to cover not only the development of low-cost technological applications but also the exploration and mining of this promising mineral.

Acknowledgments

We specially thank Dr Pablo Guerstein and Dr Leopoldo López-Escobar, for a detailed revision of the English of the original manuscript of this Chapter. We also thank to Ana Dalmasso de Rabbia for logistic support during writing of this manuscript. This work was partially supported by the Chilean Science Fundation (Project FONDECYT #198-0511 and FONDECYT Líneas Complementarias #800-0006).

References

Ambrus, Z.; Balazs, N.; Alapi, T.; Wittmann, G.; Sipos, P.; Dombi, A.; Mogyorosi, K. *Applied Catalysis B: Environmental.* 2008, vol 81, 27-37.

Antignano, A.; Manning, C.E. *Chem. Geol.* 2008, vol 255, 283-293.

Aja, S. U.; Wood, S. A.; Williams Jones A. E. *Appl. Geochem.* 1995, vol 10, 603-620.

Ashokkumar, M.; Maruthamuthu, P. *Int. J. Hydrogen Energ.* 1991, vol 16, 591-595.

Ayers, J.C.; Watson, E. B. *Contrib. Mineral. Petrol.* 1993, vol 114, 321-330.

Banfield, J. F., Bischoff, B. L., Anderson, M.A. *Chem. Geol.* 1993, vol 110, 211-231.

Beane, R. E.; Titley, S. R. *Econ. Geol.* 1981, vol 75, 214-269.

Bohlen, S. R.; Liotta, J. J. *J. Petrol.* 1986, vol 27, 1025-1034.

Brenan, J. M.; Shaw, H. F.; Phinney, D. L.; Ryerson, F. J. *EPSL.* 1994, vol 128, 327-339.

Bromiley, G. D.; Hilairet, N. *Mineral. Mag.* 2005, vol 69, 345-358.

Bulatovic, S.; Wyslouzil, D. M. *Min. Eng.* 1999, vol 12, 1407-1417.

Bulatovic, S. M.; *Handbook of Flotation Reagents, Chemistry, Theory and Practice: Flotation of Sulfide Ores*, Elsevier Science & Technology Books, Berlin, Vol. 1, 2007.

Carruzzo, S.; Clarke, D. B.; Pelrine, K. M.; MacDonald, M. A. *Can. Mineral.* 2006, vol 44, 715-729.

Černý, P.; Paul, B. J.; Hawthorne, F. C.; Chapman, R. *Can. Mineral.* 1981, vol 19, 541-548.

Chuan, X. Y.; Lu, A. H.; Chen, J.; Li, N.; Guo, Y. J. *Mineral. Petrol.* 2008, vol 93, 143-152.

Ciavatta, L.; Ferri, D.; Riccio, G. *Polyhedron.* 1985, vol 4, 15-22.

Ciavatta, L.; Pirozzi, A. *Polyhedron.* 1983, vol 2, 769-774.

Czamanske, G. K.; Force, E. R; Moore, W. J. *Econ. Geol.* 1981, vol 76, 2240-2245.

Enevoldsen, G. H.; Pinto, H. P.; Foster, A. S.; Jensen, M. C. R.; Hofer, W. A.; Hammer, B.; Lauritsen, J. V.; Besenbacher, F. *Phys. Rev. Lett.* 2009, vol 102, 136103-136106.

Fett, A. Elementverteilung zwischen Granat, Klinopyroxen und Rutil in Eklogiten - Experiment und Natur. PhD Thesis, University of Mainz, Germany. 1995, pp 227.

Force, E. R. *Geol. Soc. Am. Spec. Pap.* 1991, n 259, 1-112.

Gambogi, J. *Mineral Commodity Summaries.* U.S. Geological Survey, 2011, 174-175.

Gao, J.; John, T.; Klemd, R.; Xiong, X. M. *Geochim. Cosmochim. Acta.* 2007, vol 71, 4974-4996.

Graham, J.; Morris, R. C. *Mineral. Mag.* 1973, vol 39, 470-473.

Gunow, A. J. Trace element mineralogy in the porphyry molybdenum environment. PhD Thesis, University of Colorado. 1983, pp 304.

Haggerty, S.E. In *Oxide minerals: Petrologic and magnetic significance*; Lindsley, D. H.; Ed.; *Oxide textures - A miniatlas*; Mineralogical Society of America: Washington, D.C., WA, 1991, vol 25, pp 129-137.

Hammer, M. F. V.; Beran A. *Mineral. Petrol.* 1991, vol 45, 1-9.

Hassan, W. F., *J. Southeast Asian Earth Sci.* 1994, vol 10, 11-23.

Hoffmann, M. R.; Martin, S. T.; Choi, W.; Bahneman, D. W. *Chem. Rev.* 1995, vol 95, 69-95.

Hunt, J. P. *Econ. Geol. Monogr.* 1991, n 8, 192-206.

Johnson, O.; Ohlsen, W.; Kingsbury, P. Jr. *Phys. Rev.* 1968, vol 175, 1102-1108.

Kim, H. J.; Bae, S. B.; Bae, D. S. *Adv. Tech. Mat. Mat. Proc.* 2010, vol 12, 9-12.

Klosek, S.; Raftery, D. *J. Phys. Chem.* 2001, vol B 105, 2815-2819.

Laing, M. *A tetrahedron of bonding*; *Educ. Chem.* 1993, vol 30, 160-163

Llewellyn, T. O.; Sullivan, G. V. *US Patent.* 1981, n 4,362,615.

Lu, A.; Guo, Y.; Liu, J.; Liu, F.; Wang, C.; Li, N.; Li, Q. *Chinese Sci. Bull.* 2004, vol 49, 2284-2287.

Lu, A.; Li, Y.; Lv, M.; Wang, C.; Yang, L.; Liu, J.; Wang, Y.; Wong, K.; Wong, P. *Sol. Energy Mater. Sol. Cells*, 2007, vol 91, 1849-1855.

Lu, A.; Li, Y.; Jin, S.; Ding, H.; Zeng, C.; Wang, X.; Wang, C. *Energy Fuels.* 2010, vol 24, 1184-1190.

Luvizotto, G. L.; Zack, T. *Chem. Geol.* 2009, vol 261, 303-317.

Manning, D. A. C.; Henderson, P. *Contrib. Mineral. Petrol.* 1984, vol 86, 286-293.

Manning, C. E.; Antignano, A.; Wilke, M.; Schmidt, C.; Sutton, S.; Newville, M. *Geochim. Cosmochim. Acta* 2007, vol 71, A620.

Manning, C. E.; Wilke, M.; Schmidt, C.; Cauzid, J. *EPSL.* 2008, vol 272, 730-737.

Marchand, R.; Brohan, L.; Tournoux, M. *Mater. Res. Bull.* 1980, vol 15, 1129-1133.

Meinhold, G. *Earth Sci. Rev.* 2010, vol 102, 1-28.

Murad, E.; Cashion, J. D.; Noble, C. J.; Pilbrow, J. R. *Mineral. Mag.* 1995, vol 59, 557-560.

Navio, J. A.; Testa, J. J.; Djedjeian, P.; Padron, J. R.; Rodriguez, D.; Litter, M. I. *Appl. Cat. A. General.* 1999, vol 178, 191-203.

O'Regan, B.; Gratzel, M. *Nature.* 1991, vol 353, 737-740.

Pal, B.; Sharon, M. *J. Chem. Technol. Biotechnol.* 1998, vol 73, 269-273.

Rabbia, O. M.; Hernandez, L. B.; Kojima, S. Hydroxyl contents of W-Cr-Fe-bearing rutile from the Sierras de Córdoba ultrabasic rocks, Pampean Ranges, Argentina. III Reunión de Mineralogía y Metalogénia. La Plata, Argentina, 1996a.

Rabbia, O. M.; Hernandez, L. B.; Wetten, A. Rutilos portadores de Sc, Cr, Fe y W del distrito minero de Tocota, San Juan, Argentina. III Reunión de Mineralogía y Metalogénia. La Plata, Argentina, 1996b.

Rabbia, O. M.; Hernández, L. B.; Olivares, O. Preliminary results on mineral chemistry of rutile-anatasa from Cerro Pelado propspect, El Salvador, Chile. IV Reunión de Mineralogía y Metalogénia, Bahía Blanca, Argentina, 1998.

Rabbia, O. M.; French, D.; Hernández, L. B. *Comunicaciones.* 1999, n 50, 49-53.

Rabbia, OM. Crystallochemistry of rutile and anatase in Andean porphyry copper deposits: evaluation of its use as a monitor of metal activity in crustal hydrothermal fluids. PhD Thesis, Universidad de Chile. 2002, pp 147.

Rabbia, O. M.; Hernández, L. B.; Townley, B.; King, R. W.; Ayers, J. C. Anatase-bearing veins in the El Teniente Cu-Mo porphyry system. Special Symposium on Supergiant Andean Porphyry Copper Deposits. 10° Cong. Geol. Chileno, Concepción, Chile, 2003.

Rabbia, O. M.; Hernández, L. B.; French, D. H.; King, R. W.; Ayers, J. C. *Mineral. Deposita*. 2009, vol 44, 849-866.

Rao, M. V.; Rajeshwar, K.; Pal Verneker, V. R.; Du Bow, J. *J. Phys. Chem*. 1980, vol 84, 1987-1991.

Rice, C.; Darke, K.; Still, J. *Mineral. Mag*. 1998, vol 62, 421-429.

Rossman, G.; Smyth, J. *Am. Mineral*. 1990, vol 75, 775-780.

Rudnick, R. L.; Gao, S. In *The Crust;* Rudnick, R. L.; Ed.; in *Treatise on Geochemistry* Holland, H. D., Turekian, K. K.; Eds.; *The Composition of the Continental Crust*; Elsevier-Pergamon, Oxford, UK, 2003, Vol. 3, pp 1-64.

Schmidt, A.; Weyer, S.; John, T.; Brey, G. P. *Geochim. Cosmochim. Acta*. 2009, vol 73, 455-468.

Scott, K. M. *GEEA*. 2005, vol 5, 247-253.

Scott, K. M.; Radford, N. W. *GEEA*. 2007, vol 7, 353-361.

Serpone N, Lawless D, Disdier J, Herrmann JM. *Langmuir*. 1994, vol 10, 643-52.

Sillitoe, R. H. Econ. Geol. 2010, vol 105, 3-41.

Skewes, A.; Arevalo, A.; Floody, R.; Zuñiga, P.; Stern, C. R. *Soc. Econ. Geol. Spec. Publ*. 2002, vol 9, 299-332.

Smith, D.; Perseil, E. A. *Mineral. Mag*. 1997, vol 61, 655-669.

Surender, G. D., Fotou, G. P., Pratsinis, S. E. *Trends Chem. Eng*. 1998, vol 4, 145-159.

Taramasso, M.; Perego, G.; Notari, B. 1983 US Patent No. 4410501.

Thiruvenkatachari, R.; Vigneswaran, S.; Moon, I. S. *Korean J. Chem. Eng*. 2008, vol 25, 64-72.

Tomkins, H. S.; Powell, R.; Ellis, D. J. *J. Metam. Geol*. 2007, vol 25, 703-713.

Udubasa, G. In *Ore Genesis-The state of the art;* Amstutz, G. C.; *Rutile of post-magmatic mineral formation*; Springer-Verlag: Berlin, 1982, pp 784-793.

Urban, A. J.; Hoskins, B. F.; Grey, I. E. *Can. Mineral*. 1992, vol 30, 319-326.

van Sijl, J.; Allan, N. L.; Davies, G. R.; van Westrenen, W. *Geochim. Cosmochim. Acta*. 2010, vol 74, 2797-2810.

Vlassopoulos, D.; Rossman, G.; Haggerty, S. *A. Mineral*. 1993, vol 78, 1181-1191.

Wang, C. Y.; Bottcher, C.; Bahnemann, D. W.; Dohrmann, J. K. *J. Nanoparticle Res*. 2004, vol 6, 119-122.

Watson, E. B.; Harrison, T. M. *Science*. 2005, vol 308, 841-844.

Watson, E. B.; Wark, D. A.; Thomas, J. B. *Contrib. Mineral. Petrol*. 2006, vol 151, 413-433.

Waychunas, G. A. *Am. Mineral*. 1987, vol 72, 89-101.

Williams, S.; Cesbron, F. *Mineral. Mag*. 1977, vol 41, 288-292.

Wong, K.; Wong, P. *Sol. Energy Mater. Sol. Cells*, 2007, vol 91, 1849-1855.

Wu, Y. B.; Gao, S.; Zhang, H. F.; Yang, S. H.; Liu, X. C.; Jiao, W. F. *Am. Mineral*. 2009, vol 94, 303-312.

Yang, P.; Lu, C.; Hua, N.; Du, Y *Materials Letters*. 2002, vol 57, 794-801.

Yang, X.; Cao, C.; Hohn, K.; Erickson, L.; Maghirang, R.; Hamal, D.; Klabunde, K. *J. Catal*. 2007, vol 252, 296-302.

Ye, X-J.; Xiong, L.; Cui, Z-H.; Hu, C.; Jiang, H-Y. *Nonferrous Met. Sci. Eng*. 2011, vol 2, 55-59.

Zack, T. Trace element mineral analysis in high pressure metamorphic rocks from trescolmen, Central Alps. PhD Thesis, Göttingen University, Germany. 1999, pp 97.

Zack, T.; Moraes, R.; Kronz, A. *Contrib. Mineral. Petrol.* 2004, vol 148, 471-488.

Zack, T.; Kronz, A.; Foley, S.F.; Rivers, T.C. *Chem. Geol.* 2002, vol 184, 97-122.

INDEX

A

B

C

D

E

F

G

H

I

J

K

L

M

N

O

P

Q

R

S

T

U

V

W

X

Y

Z